Springer Texts in Statistics

Series Editors:
G. Casella
S. Fienberg
I. Olkin

Springer Texts in Statistics

For other titles published in this series, go to
www.springer.com/series/417

Simon J. Sheather

A Modern Approach to Regression with R

 Springer

Simon J. Sheather
Department of Statistics
Texas A&M University
College Station, TX, USA

ISBN: 978-1-4419-1872-7   e-ISBN: 978-0-387-09608-7
DOI: 10.1007/978-0-387-09608-7

Dedicated to
My mother, Margaret,
and my wife, Filomena

Preface

This book focuses on tools and techniques for building regression models using real-world data and assessing their validity. A key theme throughout the book is that *it makes sense to base inferences or conclusions only on valid models*.

Plots are shown to be an important tool for both building regression models and assessing their validity. We shall see that deciding what to plot and how each plot should be interpreted will be a major challenge. In order to overcome this challenge we shall need to understand the mathematical properties of the fitted regression models and associated diagnostic procedures. As such this will be an area of focus throughout the book. In particular, we shall carefully study the properties of residuals in order to understand when patterns in residual plots provide direct information about model misspecification and when they do not.

The regression output and plots that appear throughout the book have been generated using R. The output from R that appears in this book has been edited in minor ways. On the book web site you will find the R code used in each example in the text. You will also find SAS-code and Stata-code to produce the equivalent output on the book web site. Primers containing expanded explanation of R, SAS and Stata and their use in this book are also available on the book web site. Purpose-built functions have been written in SAS and Stata to cover some of the regression procedures discussed in this book. Examples include a multivariate version of the Box-Cox transformation method, inverse response plots and marginal model plots.

The book contains a number of new real data sets from applications ranging from rating restaurants, rating wines, predicting newspaper circulation and magazine revenue, comparing the performance of NFL kickers and comparing finalists in the Miss America pageant across states. In addition, a number of real data sets that have appeared in other books are also considered. The practice of considering contemporary real data sets was begun based on questions from students about how regression can be used in real life. One of the aspects of the book that sets it apart from many other regression books is that complete details are provided for each example. This completeness helps students better understand how regression is used in practice to build different models and assess their validity.

Included in the Exercises are two different types of problems involving data. In the first, a situation is described and it is up to the students to develop a valid regression model. In the second type of problem a situation is described and then output from one or models is provided and students are asked to comment and provide conclusions. This has been a conscious choice as I have found that both types of problems enhance student learning.

Chapters 2, 3 and 4 look at the case when there is a single predictor. This again has been a conscious choice as it enables students to look at many aspects of regression in the simplest possible setting. Chapters 5, 6, 7 and 9 focus on regression models with multiple predictors. In Chapter 8 we consider logistic regression. Chapter 9 considers regression models with correlated errors. Finally, Chapter 10 provides an introduction to random effects and mixed models.

Throughout the book specific suggestions are given on how to proceed when performing a regression analysis. Flow charts providing step-by-step instructions are provided first for regression problems involving a single predictor and later for multiple regression problems. The flow charts were first produced in response to requests from students when this material was first taught. They have been used with great success ever since.

Chapter 1 contains a discussion of four real examples. The first example highlights a key message of the book, namely, it is only sensible to base decisions of inferences on a valid regression model. The other three examples provide an indication of the practical problems one can solve using the regression methods discussed in the book.

In Chapter 2 we consider problems involving modeling the relationship between two variables. Throughout this chapter we assume that the model under consideration is a valid model (i.e., correctly specified.)

In Chapter 3 we will see that when we use a regression model we implicitly make a series of assumptions. We then consider a series of tools known as regression diagnostics to check each assumption. Having used these tools to diagnose potential problems with the assumptions, we look at how to first identify and then overcome or deal with problems with assumptions due to nonconstant variance or nonlinearity. A primary aim of Chapter 3 is to understand what actually happens when the standard assumptions associated with a regression model are violated, and what should be done in response to each violation.

In Chapter 3, we show that it is sometimes possible to overcome nonconstant error variance by transforming the response and/or the predictor variables. In Chapter 4 we consider an alternative way of coping with nonconstant error variance, namely weighted least squares.

Chapter 5 considers multiple linear regression problems involving modeling the relationship between a dependent variable and two or more predictor variables. Throughout Chapter 5, we assume that the multiple linear regression model under consideration is a valid model for the data. Chapter 6 considers regression diagnostics to check each of these assumptions associated with having a valid multiple regression model.

In Chapter 7 we consider methods for choosing the "best" model from a class of multiple regression models, using what are called variable selection methods. We discuss the consequences of variable selection on subsequent inferential procedures, (i.e., tests and confidence intervals).

Chapter 8 considers the situation in which the response variable follows a binomial distribution rather than a continuous distribution. We show that an appropriate model in this circumstance is a logistic regression model. We consider both inferential and diagnostic procedures for logistic regression models.

In many situations data are collected over time. It is common for such data sets to exhibit serial correlation, that is, results from the current time period are correlated with results from earlier time periods. Thus, these data sets violate the assumption that the errors are independent, an important assumption necessary for the validity of least squares based regression methods. Chapter 9 considers regression models when the errors are correlated over time. Importantly, we show how to re-specify a regression model with correlated errors as a different but equivalent regression model with uncorrelated errors. We shall discover that this allows us to use the diagnostic methods discussed in earlier chapters on problems with correlated errors.

Chapter 10 contains an introduction to random effects and mixed models. We again stress the use of re-specifying such models to obtain equivalent models with uncorrelated errors.

Finally, the Appendix discusses two nonparametric smoothing techniques, namely, kernel density estimation and nonparametric regression for a single predictor.

The book is aimed at first-year graduate students in statistics. It could also be used for a senior undergraduate class. The text grew out of a set of class notes, used for both a graduate and a senior undergraduate semester-long regression course at Texas A&M University. I am grateful to the students who took these courses. I would like to make special mention of Brad Barney, Dana Bergstresser, Charles Lindsey, Andrew Redd and Elizabeth Young. Charles Lindsey wrote the Stata code that appears in the Stata primer that accompanies the book. Elizabeth Young, along with Brad Barney and Charles Lindsey, wrote the SAS code that appears in the SAS primer that accompanies the book. Brad Barney kindly provided the analyses of the NFL kicker data in Chapter 1. Brad Barney and Andrew Redd contributed some of the R code used in the book.

Readers of this book will find that the work of Cook and Weisberg has had a profound influence on my thinking about regression. In particular, this book contains many references to the books by Cook and Weisberg (1999b) and Weisberg (2005).

The content of the book has also been influenced by a number of people. Robert Kohn and Geoff Eagleson, my colleagues for more than 10 years at the University of New South Wales, taught me a lot about regression but more importantly about the importance of thoroughness when it comes to scholarship. My long-time collaborators on nonparametric statistics, Tom Hettmansperger and Joe McKean have helped me enormously both professionally and personally for more than 20 years. Lively discussions with Mike Speed about valid models and residual plots lead to dramatic changes to the examples and the discussion of this subject in Chapter 6. Mike Longnecker, kindly acted as my teaching mentor when I joined Texas A&M University in 2005. A number of reviewers provided valuable comments and suggestions. I would like to especially acknowledge Larry Wasserman, Bruce Brown and Fred Lombard in this regard. Finally, I am grateful to Jennifer South who painstakingly proofread the whole manuscript.

The web site that accompanies the book contains R, SAS and Stata code and primers, along with all the data sets from the book can be found at www.stat.tamu.edu/~sheather/book. Also available at the book web site are online tutorials on matrices, R and SAS.

College Station, Texas
October 2008

Simon Sheather

Contents

Chapter 1
Introduction

1.1 Building Valid Models

This book focuses on tools and techniques for building valid regression models for real-world data. We shall see that a key step in any regression analysis is assessing the validity of the given model. When weaknesses in the model are identified the next step is to address each of these weaknesses. A key theme throughout the book is that *it makes sense to base inferences or conclusions only on valid models.*

Plots will be an important tool for both building regression models and assessing their validity. We shall see that deciding what to plot and how each plot should be interpreted will be a major challenge. In order to overcome this challenge we shall need to understand the mathematical properties of the fitted regression models and associated diagnostic procedures. As such this will be an area of focus throughout the book.

1.2 Motivating Examples

Throughout the book we shall carefully consider a number of real data sets. The following examples provide examples of four such data sets and thus provide an indication of what is to come.

1.2.1 Assessing the Ability of NFL Kickers

The first example illustrates the importance of only basing inferences or conclusions on a valid model. In other words, any conclusion is only as sound as the model on which it is based.

S.J. Sheather, *A Modern Approach to Regression with R*,
DOI: 10.1007/978-0-387-09608-7_1, © Springer Science + Business Media LLC 2009

In the Keeping Score column by Aaron Schatz in the Sunday November 12, 2006 edition of the *New York Times* entitled "N.F.L. Kickers Are Judged on the Wrong Criteria" the author makes the following claim:

> There is effectively no correlation between a kicker's field goal percentage one season and his field goal percentage the next.

Put briefly, we will show that once the different ability of field goal kickers is taken into account, there is a highly statistically significant **negative correlation** between a kicker's field goal percentage one season and his field goal percentage the next.

In order to examine the claim we consider data on the 19 NFL field goal kickers who made at least ten field goal attempts in each of the 2002, 2003, 2004, 2005 seasons and at the completion of games on Sunday, November 12, in the 2006 season. The data were obtained from the following web site http://www.rototimes. com/nfl/stats (accessed November 13, 2006). The data are available on the book web site, in the file FieldGoals2003to2006.csv.

Figure 1.1 contains a plot of each kicker's field goal percentage in the current year against the corresponding result in the previous year for years 2003, 2004, 2005 and for 2006 till November 12.

It can be shown that the resulting correlation in Figure 1.1 of -0.139 is not statistically significantly different from zero (p-value $= 0.230$). This result is in line with Schatz's claim of "effectively no correlation." However, this approach is **fundamentally flawed** as it fails to take into account the potentially

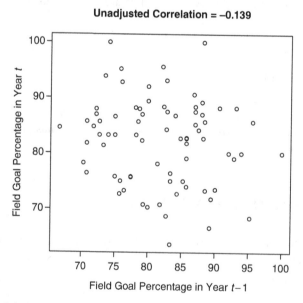

Figure 1.1 A plot of field goal percentages in the current and previous year

different abilities of the 19 kickers. In other words this approach is based on an **invalid model**.

In order to take account of the potentially different abilities of the 19 kickers we used linear regression to analyze the data in Figure 1.1. In particular, a separate regression line can be fit for each of the 19 kickers. There is very strong evidence that the intercepts of the 19 lines differ (*p*-value = 0.006) but little evidence that the slopes of the 19 lines differ (*p*-value = 0.939). (Details on how to perform these calculations will be provided in Chapter 5.) Thus, a valid way of summarizing the data in Figure 1.1 is to allow a different intercept for each kicker, but to force the same slope across all kickers. This slope is estimated to be –0.504. Statistically, it is highly significantly different from zero (*p*-value < 0.001).

Figure 1.2 shows the data in Figure 1.1 with a regression line for each kicker such that each line has the same slope but a different intercept.

There are two notable aspects of the regression lines in Figure 1.2. Firstly, the common slope of each line is negative. This means that if a kicker had a high field goal percentage in the previous year then they are predicted to have a lower field goal percentage in the current year. Let θ_i denote the true average field goal percentage of kicker i, the negative slope means that a field goal percentage one year above θ_i is likely to be followed by a lower field goal percentage, i.e., one that has shrunk back toward θ_i. (We shall discuss the concept of shrinkage in Chapter 10.) Secondly, the difference in the heights of the lines (i.e., in the intercepts) is as much as 20%, indicating a great range in performance across the 19 kickers.

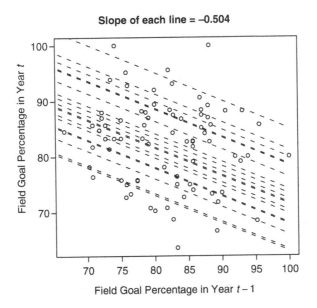

Figure 1.2 Allowing for different abilities across the 19 field goal kickers

1.2.2 Newspaper Circulation

This example illustrates the use of so-called dummy variables along with transformations to overcome skewness.

Imagine that the company that publishes a weekday newspaper in a mid-size American city has asked for your assistance in an investigation into the feasibility of introducing a Sunday edition of the paper. The current circulation of the company's weekday newspaper is 210,000. Interest centers on developing a regression model that enables you to predict the Sunday circulation of a newspaper with a weekday circulation of 210,000.

Actual circulation data from September 30, 2003 are available for 89 US newspapers that publish both weekday and Sunday editions. The first 15 rows of the data are given in Table 1.1. The data are available on the book web site, in the file circulation.txt.

The situation is further complicated by the fact that in some cities there is more than one newspaper. In particular, in some cities there is a tabloid newspaper along with one or more so-called "serious" newspapers as competitors.

The last column in Table 1.1 contains what is commonly referred to as a **dummy variable**. In this case it takes value 1 when the newspaper is a tabloid with a serious competitor in the same city and value 0 otherwise. For example, the *Chicago Sun-Times* is a tabloid while the *Chicago Herald* and the *Chicago Tribune* are serious competitors.

Given in Figure 1.3 is a plot of the Sunday circulation versus weekday circulation with the dummy variable tabloid identified. We see from Figure 1.3 that the data for the four tabloid newspapers are separated from the rest of the data and that the variability in Sunday circulation increases as weekday circulation increases. Given below in Figure 1.4 is a plot of log(Sunday circulation) versus log(weekday circulation). Here, and throughout the book, "log" stands for log to the base e. Taking logs has made the variability much more constant.

We shall return to this example in Chapter 6.

Table 1.1 Partial list of the newspaper circulation data (circulation.txt) (http://www.editorand publisher.com/eandp/yearbook/reports_trends.jsp". Accessed November 8, 2005)

Newspaper	Sunday circulation	Weekday circulation	Tabloid with a serious competitor
Akron Beacon Journal (OH)	185,915	134,401	0
Albuquerque Journal (NM)	154,413	109,693	0
Allentown Morning Call (PA)	165,607	111,594	0
Atlanta Journal-Constitution (GA)	622,065	371,853	0
Austin American-Statesman (TX)	233,767	183,312	0
Baltimore Sun (MD)	465,807	301,186	0
Bergen County Record (NJ)	227,806	179,270	0
Birmingham News (AL)	186,747	148,938	0
Boston Herald (MA)	151,589	241,457	1
Boston Globe (MA)	706,153	450,538	0

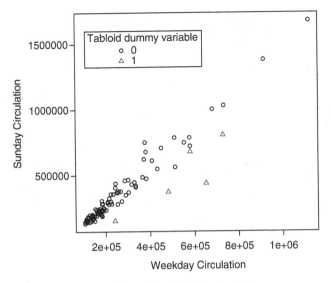

Figure 1.3 A plot of Sunday circulation against Weekday circulation

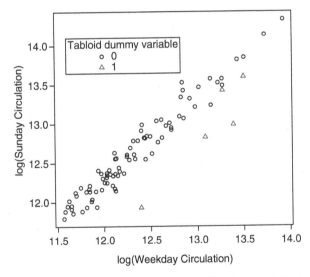

Figure 1.4 A plot of log(Sunday Circulation) against log(Weekday Circulation)

1.2.3   Menu Pricing in a New Italian Restaurant in New York City

This example highlights the use of multiple regression in a practical business setting. It will be discussed in detail in Chapters 5 and 6.

Imagine that you have been asked to join the team supporting a young New York City chef who plans to create a new Italian restaurant in Manhattan. The stated aims of the restaurant are to provide the highest quality Italian food utilizing state-of-the-art décor while setting a new standard for high-quality service in Manhattan. The creation and the initial operation of the restaurant will be the basis of a reality TV show for the US and international markets (including Australia). You have been told that the restaurant is going to be located no further south than the Flatiron District and it will be either east or west of Fifth Avenue.

You have been asked to determine the pricing of the restaurants dinner menu such that it is competitively positioned with other high-end Italian restaurants in the target area. In particular, your role in the team is to analyze the pricing data that have been collected in order to produce a regression model to predict the price of dinner. Actual data from surveys of customers of 168 Italian restaurants in the target area are available. The data are in the form of the average of customer views on

Y = Price = the price (in $US) of dinner (including one drink & a tip)
x_1 = Food = customer rating of the food (out of 30)
x_2 = Décor = customer rating of the decor (out of 30)
x_3 = Service = customer rating of the service (out of 30)
x_4 = East = dummy variable = 1 (0) if the restaurant is east (west) of Fifth Avenue

Figures 1.5 and 1.6 contain plots of the data.

Whilst the situation described above is imaginary, the data are real ratings of New York City diners. The data are given on the book web site in the file nyc.csv. The source of the data is:

Zagat Survey 2001: New York City Restaurants, Zagat, New York.

According to www.zagat.com, Tim and Nina Zagat (two lawyers in New York City) started Zagat restaurant surveys in 1979 by asking 20 of their friends to rate and review restaurants in New York City. The survey was an immediate success and the Zagats have produced a guide to New York City restaurants each year since. In less than 30 years, Zagat Survey has expanded to cover restaurants in more than 85 cities worldwide and other activities including travel, nightlife, shopping, golf, theater, movies and music.

In particular you have been asked to:

• Develop a regression model that ***directly predicts*** the price of dinner (in dollars) using a subset or all of the four potential predictor variables listed above.
• Determine which of the predictor variables Food, Décor and Service has the largest estimated effect on Price? Is this effect also the most statistically significant?
• If the aim is to choose the location of the restaurant so that the price achieved for dinner is maximized, should the new restaurant be on the east or west of Fifth Avenue?
• Does it seem possible to achieve a price premium for "setting a new standard for high-quality service in Manhattan" for Italian restaurants?
• Identify the restaurants in the data set which, given the customer ratings, are (i) unusually highly priced; and (ii) unusually lowly priced.

We shall return to this example in Chapters 5 and 6.

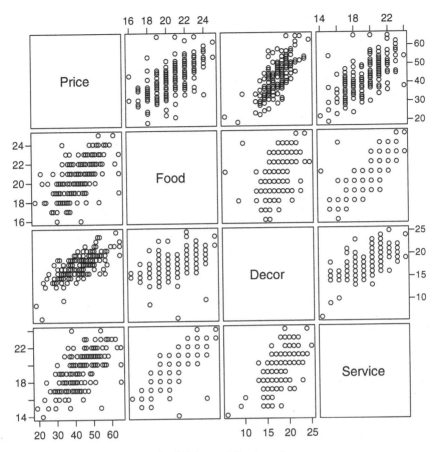

Figure 1.5 Matrix plot of Price, Food, Décor and Service ratings

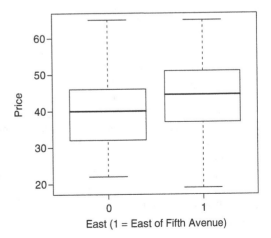

Figure 1.6 Box plots of Price for the
two levels of the dummy variable East

East (1 = East of Fifth Avenue)

1.2.4   Effect of Wine Critics' Ratings on Prices of Bordeaux Wines

In this example we look at the effects two wine critics have on Bordeaux wine prices in the UK. The two critics are Robert Parker from the US and Clive Coates from the UK. Background information on each appears below:

> The most influential critic in the world today happens to be a critic of wine. ... His name is Robert Parker ... and he has no formal training in wine. ... ... many people now believe that Robert Parker is single-handedly changing the history of wine. ... He is a self-employed consumer advocate, a crusader in a peculiarly American tradition. ... Parker samples 10,000 wines a year. ... he writes and publishes an un-illustrated journal called The Wine Advocate, (which) ... accepts no advertising. ... *The Wine Advocate* has 40,000 subscribers (at $50 each) in every US-state and 37 foreign countries. Rarely, Parker has given wine a perfect score of 100 – seventy-six times out of 220,000 wines tasted. ... he remembers every wine he has tasted over the past thirty-two years and, within a few points, every score he has given as well. ... Even his detractors admit that he is phenomenally consistent – that after describing a wine once he will describe it in nearly the same way if he retastes it 'blind' (without reference to the label) .... (Langewiesche 2000)
>
> Clive Coates MW (Master of Wine) is one of the world's leading wine authorities. Coates' lifetime of distinguished activity in the field has been recognised by the French government, which recently awarded him the Chevalier de l'Ordre du Mérite Agricole, and he's also been honoured with a "Rame d'Honneur" by Le Verre et L'Assiette, the Ruffino/Cyril Ray Memorial Prize for his writings on Italian wine, and the title of "Wine Writer of the Year" for 1998/1999 in the Champagne Lanson awards. ...Coates has published *The Vine*, his independent fine wine magazine, since 1985. Prior to his career as an author, Coates spent twenty years as a professional wine merchant. (http://www. clive-coates.com/)
>
> The courtier Eric Samazeuilh puts it plainly: "... Parker is the wine writer who matters. Clive Coates is very serious and well respected, but in terms of commercial impact his influence is zero. It's an amazing phenomenon." ...The pseudo-certainties of the 100-point (Parker) system have immense appeal in markets where a wine culture is either non-existent or very new. The German wine collector Hardy Rodenstock recalls: "I know very rich men in Hong Kong who have caught the wine bug. ...the only thing they buy are wines that Parker scores at ninety five or above ..." .... (Brook 2001)

Parker (2003) and Coates (2004) each contain numerical ratings and reviews of the wines of Bordeaux. In this example we look at the effect of these ratings on the prices (in pounds sterling) on the wholesale brokers' auction market per dozen bottles, duty paid but excluding delivery and VAT in London in September 2003. In particular, we consider the prices for 72 wines from the 2000 vintage in Bordeaux. The prices are taken from Coates (2004, Appendix One). The 2000 vintage has been chosen since it is ranked by both critics as a "great vintage." For example, Parker (2003, pages 30–31) claims that the 2000 vintage "produced many wines of exhilarating quality ... at all levels of the Bordeaux hierarchy. ... The finest 2000s appear to possess a staggering 30–40 years of longevity." In addition, Coates (2004, page 439) describes the 2000 vintage as follows: "Overall it is a splendid vintage."

Data are available on the ratings by Parker and Coates for each of the 72 wines. Robert Parker uses a 100-point rating system with wines given a whole number score between 50 and 100 as follows:

96–100 points	Extraordinary
90–95 points	Outstanding
80–89 points	Above average to very good
70–79 points	Average
50–69 points	Below average to poor

On the other hand, Clive Coates uses a 20-point rating system with wines given a score between 12.5 and 20 that ends in 0 or 0.5 as follows:

20	Excellent. 'Grand vin'	16	Very good
19.5	Very fine indeed	15.5	Good plus
19	Very fine plus	15	Good
18.5	Very fine	14.5	Quite good plus
18	Fine plus	14	Quite good
17.5	Fine	13.5	Not bad plus
17	Very good indeed	13	Not bad
16.5	Very good plus	12.5	Poor

Data are available on the following other potentially important predictor variables:

- P95andAbove is a dummy variable which is 1 if the wine scores 95 or above from Robert Parker (and 0 otherwise). This variable is included as potential predictor in view of the comment by Hardy Rodenstock.
- FirstGrowth is a dummy variable which is 1 if the wine is a First Growth (and 0 otherwise). First Growth is the highest classification given to a wine from Bordeaux. The classification system dates back to at least 1855 and it is based on the "selling price and vineyard condition" (Parker, 2003, page 1148). Thus, first-growth wines are expected to achieve higher prices than other wines.
- CultWine is a dummy variable which is 1 if the wine is a cult wine (and 0 otherwise). Cult wines (such as Le Pin) have limited availability and as such demand way outstrips supply. As such cult wines are among the most expensive wines of Bordeaux.
- Pomerol is a dummy variable which is 1 if the wine is from Pomerol (and 0 otherwise). According to Parker (2003, page 610):

  The smallest of the great red wine districts of Bordeaux, Pomerol produces some of the most expensive, exhilarating, and glamorous wines in the world. ..., wines are in such demand that they must be severely allocated.

- VintageSuperstar is a dummy variable which is 1 if the wine is a vintage superstar (and 0 otherwise). Superstar status is awarded by Robert Parker to a few wines in certain vintages. For example, Robert Parker (2003, page 529) describes the 2000 La Mission Haut-Brion as follows:

  A superstar of the vintage, the 2000 La Mission Haut-Brion is as profound as such recent superstars as 1989, 1982 and 1975. ... The phenomenal aftertaste goes on for more than a minute.

In summary, data are available on the following variables:

Y = Price = the price (in pounds sterling) of 12 bottles of wine
x_1 = ParkerPoints = Robert Parker's rating of the wine (out of 100)
x_2 = CoatesPoints = Clive Coates' rating of the wine (out of 20)
x_3 = P95andAbove = 1 (0) if the Parker score is 95 or above (otherwise)
x_4 = FirstGrowth = 1 (0) if the wine is a First Growth (otherwise)
x_5 = CultWine = 1 (0) if the wine is a cult wine (otherwise)
x_6 = Pomerol = 1 (0) if the wine is from Pomerol (otherwise)
x_7 = VintageSuperstar = 1 (0) if the wine is a superstar (otherwise)

The data are given on the book web site in the file Bordeaux.csv.

Figure 1.7 contains a matrix plot of price, Parker's ratings and Coates' ratings, while Figure 1.8 shows box plots of Price against each of the dummy variables.

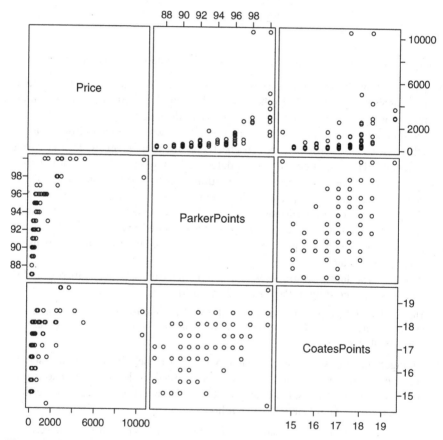

Figure 1.7 Matrix plot of Price, ParkerPoints and CoatesPoints

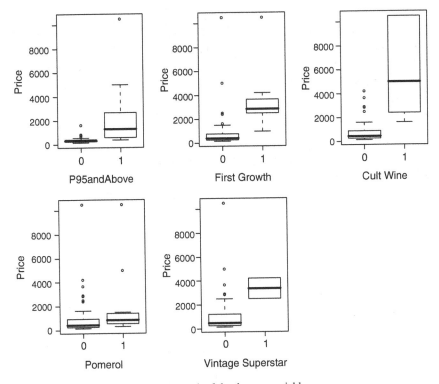

Figure 1.8 Box plots of Price against each of the dummy variables

In particular you have been asked to:

1. Develop a regression model that enables you to estimate the percentage effect on price of a 1% increase in ParkerPoints and a 1% increase in Coates-Points using a subset, or all, of the seven potential predictor variables listed above.

2. Using the regression model developed in part (1), specifically state your estimate of the percentage effect on price of

   (i) A 1% increase in ParkerPoints
   (ii) A 1% increase in CoatesPoints

3. Using the regression model developed in part (1), decide which of the predictor variables ParkerPoints and CoatesPoints has the largest estimated percentage effect on Price. Is this effect also the most statistically significant?

4. Using your regression model developed in part (1), comment on the following claim from Eric Samazeuilh:

   Parker is the wine writer who matters. Clive Coates is very serious and well respected, but in terms of commercial impact his influence is zero.

5. Using your regression model developed in part (1), decide whether there is a statistically significant extra price premium paid for Bordeaux wines from the 2000 vintage with a Parker score of 95 and above.

6. Identify the wines in the data set which, given the values of the predictor variables, are:

  (i) Unusually highly priced
  (ii) Unusually lowly priced

In Chapters 3 and 6, we shall see that a log transformation will enable us to estimate percentage effects. As such, Figure 1.9 contains a matrix plot of log(Price), log(ParkerPoints) and log(CoatesPoints), while Figure 1.10 shows box plots of log(Price) against each of the dummy variables. We shall return to this example in Chapter 6.

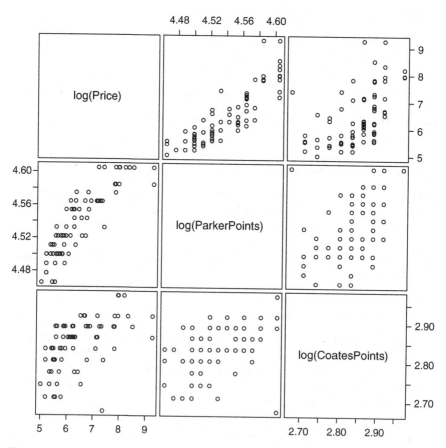

Figure 1.9 Matrix plot of log(Price), log(ParkerPoints) and log(CoatesPoints)

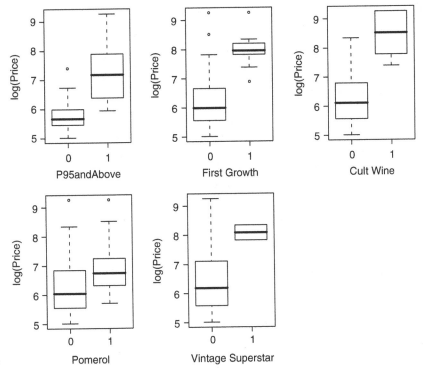

Figure 1.10   Box plots of log(Price) against each of the dummy variables

1.3   Level of Mathematics

Throughout the book we will focus on understanding the properties of a number of regression procedures. An important component of this understanding will come from the mathematical properties of regression procedures.

The following excerpt from Chapter 5 on the properties of least squares estimates demonstrates the level of mathematics associated with this book:

Consider the linear regression model written in matrix form as

$$\mathbf{Y} = \mathbf{X}\beta + \mathbf{e}$$

with $Var(\mathbf{e}) = \sigma^2 \mathbf{I}$, where $\mathbf{I}$ is the $(n \times n)$ identity matrix and the $(n \times 1)$ vectors, $\mathbf{Y}$, β, $\mathbf{e}$ and the $n \times (p + 1)$ matrix, $\mathbf{X}$, are given by

$$\mathbf{Y} = \begin{pmatrix} y_1 \\ y_2 \\ \vdots \\ y_n \end{pmatrix}, \mathbf{X} = \begin{pmatrix} 1 & x_{11} \cdots x_{1p} \\ 1 & x_{21} \cdots x_{2p} \\ \vdots & \vdots \\ 1 & x_{n1} \cdots x_{np} \end{pmatrix}, \beta = \begin{pmatrix} \beta_0 \\ \beta_1 \\ \vdots \\ \beta_p \end{pmatrix}, \mathbf{e} = \begin{pmatrix} e_1 \\ e_2 \\ \vdots \\ e_n \end{pmatrix}$$

The least squares estimates are given by

$$\hat{\beta} = (\mathbf{X'X})^{-1}\mathbf{X'Y}$$

We next derive the conditional mean of the least squares estimates:

$$
\begin{aligned}
\mathrm{E}\left(\hat{\beta} \mid \mathbf{X}\right) &= \mathrm{E}\left((\mathbf{X'X})^{-1}\mathbf{X'Y} \mid \mathbf{X}\right) \\
&= (\mathbf{X'X})^{-1}\mathbf{X'}\mathrm{E}(\mathbf{Y} \mid \mathbf{X}) \\
&= (\mathbf{X'X})^{-1}\mathbf{X'X}\beta \\
&= \beta
\end{aligned}
$$

Chapter 2
Simple Linear Regression

2.1 Introduction and Least Squares Estimates

Regression analysis is a method for investigating the functional relationship among variables. In this chapter we consider problems involving modeling the relationship between two variables. These problems are commonly referred to as simple linear regression or straight-line regression. In later chapters we shall consider problems involving modeling the relationship between three or more variables.

In particular we next consider problems involving modeling the relationship between two variables as a straight line, that is, when Y is modeled as a linear function of X.

Example: A regression model for the timing of production runs
We shall consider the following example taken from Foster, Stine and Waterman (1997, pages 191–199) throughout this chapter. The original data are in the form of the time taken (in minutes) for a production run, Y, and the number of items produced, X, for 20 randomly selected orders as supervised by three managers. At this stage we shall only consider the data for one of the managers (see Table 2.1 and Figure 2.1). We wish to develop an equation to model the relationship between Y, the run time, and X, the run size.

A scatter plot of the data like that given in Figure 2.1 should **ALWAYS** be drawn to obtain an idea of the sort of relationship that exists between two variables (e.g., linear, quadratic, exponential, etc.).

2.1.1 Simple Linear Regression Models

When data are collected in pairs the standard notation used to designate this is:

$$(x_1, y_1), (x_2, y_2), \ldots, (x_n, y_n)$$

where x_1 denotes the first value of the so-called X-variable and y_1 denotes the first value of the so-called Y-variable. The X-variable is called the **explanatory** or **predictor variable**, while the Y-variable is called the **response variable** or the **dependent variable**. The X-variable often has a different status to the Y-variable in that:

S.J. Sheather, *A Modern Approach to Regression with R*,
DOI: 10.1007/978-0-387-09608-7_2, © Springer Science+Business Media LLC 2009

Table 2.1 Production data (production.txt)

Case	Run time	Run size	Case	Run time	Run size
1	195	175	11	220	337
2	215	189	12	168	58
3	243	344	13	207	146
4	162	88	14	225	277
5	185	114	15	169	123
6	231	338	16	215	227
7	234	271	17	147	63
8	166	173	18	230	337
9	253	284	19	208	146
10	196	277	20	172	68

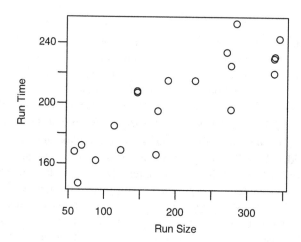

Figure 2.1 A scatter plot of the production data

- It can be thought of as a potential predictor of the Y-variable
- Its value can sometimes be chosen by the person undertaking the study

Simple linear regression is typically used to model the relationship between two variables Y and X so that given a specific value of X, that is, $X = x$, we can predict the value of Y. Mathematically, the regression of a random variable Y on a random variable X is

$$E(Y \mid X = x),$$

the expected value of Y when X takes the specific value x. For example, if $X =$ Day of the week and $Y =$ Sales at a given company, then the regression of Y on X represents the mean (or average) sales on a given day.

The regression of Y on X is linear if

$$E(Y \mid X = x) = \beta_0 + \beta_1 x \qquad (2.1)$$

where the unknown parameters β_0 and β_1 determine the intercept and the slope of a specific straight line, respectively. Suppose that $Y_1, Y_2, \ldots, Y_n$ are independent realizations of the random variable Y that are observed at the values $x_1, x_2, \ldots, x_n$ of a random variable X. If the regression of Y on X is linear, then for $i = 1, 2, \ldots, n$

$$Y_i = E(Y \mid X = x) + e_i = \beta_0 + \beta_1 x + e_i$$

where e_i is the random error in Y_i and is such that $E(e \mid X) = 0$.

The random error term is there since there will almost certainly be some variation in Y due strictly to random phenomenon that cannot be predicted or explained. In other words, all unexplained variation is called **random error**. Thus, the random error term does not depend on x, nor does it contain any information about Y (otherwise it would be a systematic error).

We shall begin by assuming that

$$\mathrm{Var}(Y \mid X = x) = \sigma^2. \qquad (2.2)$$

In Chapter 4 we shall see how this last assumption can be relaxed.

Estimating the population slope and intercept

Suppose for example that X = height and Y = weight of a randomly selected individual from some population, then for a straight line regression model the mean weight of individuals of a given height would be a linear function of that height. In practice, we usually have a sample of data instead of the whole population. The slope β_1 and intercept β_0 are unknown, since these are the values for the whole population. Thus, we wish to use the given data to estimate the slope and the intercept. This can be achieved by finding the equation of the line which "best" fits our data, that is, choose b_0 and b_1 such that $\hat{y}_i = b_0 + b_1 x_i$ is as "close" as possible to y_i. Here the notation $\hat{y}_i$ is used to denote the value of the line of best fit in order to distinguish it from the observed values of y, that is, y_i. We shall refer to $\hat{y}_i$ as the ith **predicted value** or the **fitted value** of y_i.

Residuals

In practice, we wish to minimize the difference between the actual value of y (y_i) and the predicted value of y $(\hat{y}_i)$. This difference is called the residual, $\hat{e}_i$, that is,

$$\hat{e}_i = y_i - \hat{y}_i.$$

Figure 2.2 shows a hypothetical situation based on six data points. Marked on this plot is a **line of best fit**, $\hat{y}_i$ along with the residuals.

Least squares line of best fit

A very popular method of choosing b_0 and b_1 is called the method of least squares. As the name suggests b_0 and b_1 are chosen to minimize the sum of squared residuals (or residual sum of squares [RSS]),

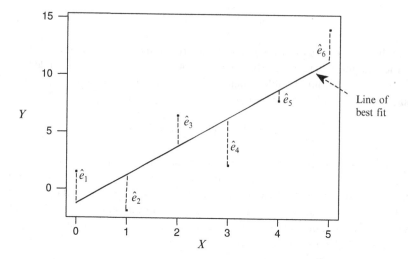

Figure 2.2 A scatter plot of data with a line of best fit and the residuals identified

$$\text{RSS} = \sum_{i=1}^{n} \hat{e}_i^2 = \sum_{i=1}^{n} (y_i - \hat{y}_i)^2 = \sum_{i=1}^{n} (y_i - b_0 - b_1 x_i)^2.$$

For RSS to be a minimum with respect to b_0 and b_1 we require

$$\frac{\partial \text{RSS}}{\partial b_0} = -2 \sum_{i=1}^{n} (y_i - b_0 - b_1 x_i) = 0$$

and

$$\frac{\partial \text{RSS}}{\partial b_1} = -2 \sum_{i=1}^{n} x_i (y_i - b_0 - b_1 x_i) = 0$$

Rearranging terms in these last two equations gives

$$\sum_{i=1}^{n} y_i = b_0 n + b_1 \sum_{i=1}^{n} x_i$$

and

$$\sum_{i=1}^{n} x_i y_i = b_0 \sum_{i=1}^{n} x_i + b_1 \sum_{i=1}^{n} x_i^2.$$

These last two equations are called the **normal equations**. Solving these equations for b_0 and b_1 gives the so-called **least squares estimates** of the intercept

$$\hat{\beta}_0 = \bar{y} - \hat{\beta}_1 \bar{x} \tag{2.3}$$

and the slope

$$\hat{\beta}_1 = \frac{\sum_{i=1}^{n} x_i y_i - n\overline{x}\overline{y}}{\sum_{i=1}^{n} x_i^2 - n\overline{x}^2} = \frac{\sum_{i=1}^{n}(x_i - \overline{x})(y_i - \overline{y})}{\sum_{i=1}^{n}(x_i - \overline{x})^2} = \frac{SXY}{SXX}. \qquad (2.4)$$

Regression Output from R

The least squares estimates for the production data were calculated using R, giving the following results:

```
Coefficients:
              Estimate Std. Error  t value  Pr(>|t|)
(Intercept)  149.74770    8.32815    17.98  6.00e-13 ***
RunSize        0.25924    0.03714     6.98  1.61e-06 ***
---
Signif. codes:0 `***' 0.001 `**' 0.01 `*' 0.05 `.' 0.1 ` ` 1

Residual standard error: 16.25 on 18 degrees of freedom
Multiple R-Squared: 0.7302,       Adjusted R-squared: 0.7152
F-statistic: 48.72 on 1 and 18 DF, p-value: 1.615e-06
```

The least squares line of best fit for the production data

Figure 2.3 shows a scatter plot of the production data with the least squares line of best fit. The equation of the least squares line of best fit is

$$y = 149.7 + 0.26x.$$

Let us look at the results that we have obtained from the line of best fit in Figure 2.3. The intercept in Figure 2.3 is 149.7, which is where the line of best fit crosses the run time axis. The slope of the line in Figure 2.3 is 0.26. Thus, we say that each additional unit to be produced is predicted to add 0.26 minutes to the run time. The intercept in the model has the following interpretation: for any production run, the average set up time is 149.7 minutes.

Estimating the variance of the random error term

Consider the linear regression model with constant variance given by (2.1) and (2.2). In this case,

$$Y_i = \beta_0 + \beta_1 x_i + e_i \quad (i = 1,2,...,n)$$

where the random error e_i has mean 0 and variance σ^2. We wish to estimate $\sigma^2 = \mathrm{Var}(e)$. Notice that

$$e_i = Y_i - (\beta_0 + \beta_1 x_i) = Y_i - \text{unknown regression line at } x_i.$$

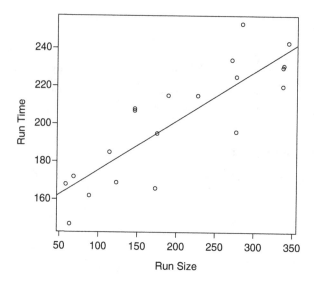

Figure 2.3 A plot of the production data with the least squares line of best fit

Since β_0 and β_1 are unknown all we can do is estimate these errors by replacing β_0 and β_1 by their respective least squares estimates $\hat{\beta}_0$ and $\hat{\beta}_1$ giving the residuals

$$\hat{e}_i = Y_i - (\hat{\beta}_0 + \hat{\beta}_1 x_i) = Y_i - \text{estimated regression line at } x_i.$$

These residuals can be used to estimate σ^2. In fact it can be shown that

$$S^2 = \frac{RSS}{n-2} = \frac{1}{n-2} \sum_{i=1}^{n} \hat{e}_i^2$$

is an unbiased estimate of σ^2.

Two points to note are:

1. $\bar{\hat{e}} = 0$ (since $\sum \hat{e}_i = 0$ as the least squares estimates minimize $RSS = \sum \hat{e}_i^2$)
2. The divisor in S^2 is $n-2$ since we have estimated two parameters, namely β_0 and β_1.

2.2   Inferences About the Slope and the Intercept

In this section, we shall develop methods for finding confidence intervals and for performing hypothesis tests about the slope and the intercept of the regression line.

2.2.1 Assumptions Necessary in Order to Make Inferences About the Regression Model

Throughout this section we shall make the following assumptions:

1. Y is related to x by the simple linear regression model
   $Y_i = \beta_0 + \beta_1 x_i + e_i \ (i = 1,...,n)$, i.e., $E(Y \mid X = x_i) = \beta_0 + \beta_1 x_i$
2. The errors $e_1, e_2,...,e_n$ are independent of each other
3. The errors $e_1, e_2,...,e_n$ have a common variance σ^2
4. The errors are normally distributed with a mean of 0 and variance σ^2, that is, $e \mid X \sim N(0, \sigma^2)$

Methods for checking these four assumptions will be considered in Chapter 3. In addition, since the regression model is conditional on X we can assume that the values of the predictor variable, $x_1, x_2, ..., x_n$ are known fixed constants.

2.2.2 Inferences About the Slope of the Regression Line

Recall from (2.4) that the least squares estimate of β_1 is given by

$$\hat{\beta}_1 = \frac{\sum_{i=1}^{n} x_i y_i - n\overline{xy}}{\sum_{i=1}^{n} x_i^2 - n\overline{x}^2} = \frac{\sum_{i=1}^{n}(x_i - \overline{x})(y_i - \overline{y})}{\sum_{i=1}^{n}(x_i - \overline{x})^2} = \frac{SXY}{SXX}$$

Since, $\sum_{i=1}^{n}(x_i - \overline{x}) = 0$ we find that

$$\sum_{i=1}^{n}(x_i - \overline{x})(y_i - \overline{y}) = \sum_{i=1}^{n}(x_i - \overline{x})y_i - \overline{y}\sum_{i=1}^{n}(x_i - \overline{x}) = \sum_{i=1}^{n}(x_i - \overline{x})y_i$$

Thus, we can rewrite $\hat{\beta}_1$ as

$$\hat{\beta}_1 = \sum_{i=1}^{n} c_i y_i \text{ where } c_i = \frac{x_i - \overline{x}}{SXX} \tag{2.5}$$

We shall see that this version of $\hat{\beta}_1$ will be used whenever we study its theoretical properties.

Under the above assumptions, we shall show in Section 2.7 that

$$E(\hat{\beta}_1 \mid X) = \beta_1 \tag{2.6}$$

$$Var(\hat{\beta}_1 \mid X) = \frac{\sigma^2}{SXX} \tag{2.7}$$

$$\hat{\beta}_1 \mid X \sim N\left(\beta_1, \frac{\sigma^2}{SXX}\right) \tag{2.8}$$

Note that in (2.7) the variance of the least squares slope estimate decreases as SXX increases (i.e., as the variability in the X's increases). This is an important fact to note if the experimenter has control over the choice of the values of the X variable.

Standardizing (2.8) gives

$$Z = \frac{\hat{\beta}_1 - \beta_1}{\sigma / \sqrt{SXX}} \sim N(0,1)$$

If σ were known then we could use a Z to test hypotheses and find confidence intervals for β_1. When σ is unknown (as is usually the case) replacing σ by S, the standard deviation of the residuals results in

$$T = \frac{\hat{\beta}_1 - \beta_1}{S / \sqrt{SXX}} = \frac{\hat{\beta}_1 - \beta_1}{\text{se}(\hat{\beta}_1)}$$

where $\text{se}(\hat{\beta}_1) = S / \sqrt{SXX}$ is the estimated standard error (se) of $\hat{\beta}_1$, which is given directly by R. In the production example the X-variable is *RunSize* and so $\text{se}(\hat{\beta}_1) = 0.03714$.

It can be shown that under the above assumptions that T has a t-distribution with $n - 2$ degrees of freedom, that is

$$T = \frac{\hat{\beta}_1 - \beta_1}{\text{se}(\hat{\beta}_1)} \sim t_{n-2}$$

Notice that the degrees of freedom satisfies the following formula

degrees of freedom = sample size – number of mean parameters estimated.

In this case we are estimating two such parameters, namely, β_0 and β_1.

For **testing the hypothesis** $H_0 : \beta_1 = \beta_1^0$ the test statistic is

$$T = \frac{\hat{\beta}_1 - \beta_1^0}{\text{se}(\hat{\beta}_1)} \sim t_{n-2} \text{ when } H_0 \text{ is true.}$$

R provides the value of T and the p-value associated with testing $H_0 : \beta_1 = 0$ against $H_A : \beta_1 \neq 0$ (i.e., for the choice $\beta_1^0 = 0$). In the production example the X-variable is *RunSize* and $T = 6.98$, which results in a p-value less than 0.0001.

A $100(1-\alpha)\%$ **confidence interval** for β_1, the slope of the regression line, is given by

$$(\hat{\beta}_1 - t(\alpha/2, n-2)\,\mathrm{se}\,(\hat{\beta}_1), \hat{\beta}_1 + t(\alpha/2, n-2)\,\mathrm{se}\,(\hat{\beta}_1))$$

where $t(\alpha/2, n-2)$ is the $100(1-\alpha/2)$th quantile of the t-distribution with $n-2$ degrees of freedom.

In the production example the X-variable is *RunSize* and $\hat{\beta}_1 = 0.25924$, $\mathrm{se}(\hat{\beta}_1) = 0.03714$, $t(0.025, 20-2 = 18) = 2.1009$. Thus a 95% confidence interval for β_1 is given by

$$(0.25924 \pm 2.1009 \times 0.03714) = (0.25924 \pm 0.07803) = (0.181, 0.337)$$

2.2.3   Inferences About the Intercept of the Regression Line

Recall from (2.3) that the least squares estimate of β_0 is given by

$$\hat{\beta}_0 = \bar{y} - \hat{\beta}_1 \bar{x}$$

Under the assumptions given previously we shall show in Section 2.7 that

$$E(\hat{\beta}_0 \mid X) = \beta_0 \qquad (2.9)$$

$$\mathrm{Var}(\hat{\beta}_0 \mid X) = \sigma^2 \left(\frac{1}{n} + \frac{\bar{x}^2}{SXX} \right) \qquad (2.10)$$

$$\hat{\beta}_0 \mid X \sim N\left(\beta_0, \sigma^2 \left(\frac{1}{n} + \frac{\bar{x}^2}{SXX} \right) \right) \qquad (2.11)$$

Standardizing (2.11) gives

$$Z = \frac{\hat{\beta}_0 - \beta_0}{\sigma \sqrt{1/n + \bar{x}^2/SXX}} \sim N(0,1)$$

If σ were known then we could use Z to test hypotheses and find confidence intervals for β_0. When σ is unknown (as is usually the case) replacing σ by S results in

$$T = \frac{\hat{\beta}_0 - \beta_0}{S\sqrt{1/n + \bar{x}^2/SXX}} = \frac{\hat{\beta}_0 - \beta_0}{\mathrm{se}(\hat{\beta}_0)} \sim t_{n-2}$$

where $\mathrm{se}(\hat{\beta}_0) = S\sqrt{1/n + \bar{x}^2/SXX}$ is the estimated standard error of $\hat{\beta}_0$, which is given directly by R. In the production example the intercept is called *Intercept* and so $\mathrm{se}(\hat{\beta}_0) = 8.32815$.

For **testing the hypothesis** $H_0 : \beta_0 = \beta_0^0$ the test statistic is

$$T = \frac{\hat{\beta}_0 - \beta_0^0}{se(\hat{\beta}_0)} \sim t_{n-2} \text{ when } H_0 \text{ is true.}$$

R provides the value of T and the p-value associated with testing $H_0 : \beta_0 = 0$ against $H_A : \beta_0 \neq 0$. In the production example the intercept is called *Intercept* and $T = 17.98$ which results in a p-value < 0.0001.

A $100(1 - \alpha)\%$ **confidence interval** for β_0, the intercept of the regression line, is given by

$$(\hat{\beta}_0 - t(\alpha/2, n-2)\, se(\hat{\beta}_0), \hat{\beta}_0 + t(\alpha/2, n-2) se(\hat{\beta}_0))$$

where $t(\alpha/2, n-2)$ is the $100(1-\alpha/2)$th quantile of the t-distribution with $n-2$ degrees of freedom.

In the production example,

$$\hat{\beta}_0 = 149.7477, se(\hat{\beta}_0) = 8.32815, t(0.025, 20-2 = 18) = 2.1009.$$

Thus a 95% confidence interval for β_0 is given by

$$(149.7477 \pm 2.1009 \times 8.32815) = (149.748 \pm 17.497) = (132.3, 167.2)$$

Regression Output from R: 95% confidence intervals

```
                  2.5%    97.5%
(Intercept)  132.251   167.244
RunSize        0.181     0.337
```

2.3 Confidence Intervals for the Population Regression Line

In this section we consider the problem of finding a confidence interval for the unknown population regression line at a given value of X, which we shall denote by x^*. First, recall from (2.1) that the population regression line at $X = x^*$ is given by

$$E(Y \mid X = x^*) = \beta_0 + \beta_1 x^*$$

An estimator of this unknown quantity is the value of the estimated regression equation at $X = x^*$, namely,

$$\hat{y}^* = \hat{\beta}_0 + \hat{\beta}_1 x^*$$

Under the assumptions stated previously, it can be shown that

$$E(\hat{y}^*) = E(\hat{y} \mid X = x^*) = \beta_0 + \beta_1 x^* \tag{2.12}$$

$$\text{Var}(\hat{y}^*) = \text{Var}(\hat{y} \mid X = x^*) = \sigma^2 \left(\frac{1}{n} + \frac{(x^* - \bar{x})^2}{SXX} \right) \qquad (2.13)$$

$$\hat{y}^* = \hat{y} \mid X = x^* \sim N\left(\beta_0 + \beta_1 x^*, \sigma^2 \left(\frac{1}{n} + \frac{(x^* - \bar{x})^2}{SXX} \right) \right) \qquad (2.14)$$

Standardizing (2.14) gives

$$Z = \frac{\hat{y}^* - (\beta_0 + \beta_1 x^*)}{\sigma \sqrt{(\frac{1}{n} + \frac{(x^* - \bar{x})^2}{SXX})}} \sim N(0,1)$$

Replacing σ by S results in

$$T = \frac{\hat{y}^* - (\beta_0 + \beta_1 x^*)}{S \sqrt{(\frac{1}{n} + \frac{(x^* - \bar{x})^2}{SXX})}} \sim t_{n-2}$$

A $100(1 - \alpha)\%$ **confidence interval** for $E(Y \mid X = x^*) = \beta_0 + \beta_1 x^*$, the population regression line at $X = x^*$, is given by

$$\hat{y}^* \pm t(\alpha/2, n-2) S \sqrt{(\frac{1}{n} + \frac{(x^* - \bar{x})^2}{SXX})}$$

$$= \hat{\beta}_0 + \hat{\beta}_1 x^* \pm t(\alpha/2, n-2) S \sqrt{(\frac{1}{n} + \frac{(x^* - \bar{x})^2}{SXX})}$$

where $t(\alpha/2, n-2)$ is the $100(1-\alpha/2)$th quantile of the t-distribution with $n - 2$ degrees of freedom.

2.4   Prediction Intervals for the Actual Value of Y

In this section we consider the problem of finding a prediction interval for the actual value of Y at x^*, a given value of X.

Important Notes:

1. $E(Y \mid X = x^*)$, the expected value or average value of Y for a given value x^* of X, is what one would expect Y to be in the long run when $X = x^*$. $E(Y \mid X = x^*)$ is therefore a fixed but unknown quantity whereas Y can take a number of values when $X = x^*$.

2. $E(Y \mid X = x^*)$, the value of the regression line at $X = x^*$, is entirely different from Y^*, a single value of Y when $X = x^*$. In particular, Y^* need not lie on the population regression line.

3. A **confidence interval** is always reported for a **parameter** (e.g., $E(Y \mid X = x^*)$ $= \beta_0 + \beta_1 x^*$) and a *prediction interval* is reported for the value of a *random variable* (e.g., Y^*).

We base our prediction of Y when $X = x^*$ (that is of Y^*) on

$$\hat{y}^* = \hat{\beta}_0 + \hat{\beta}_1 x^*$$

The error in our prediction is

$$Y^* - \hat{y}^* = \beta_0 + \beta_1 x^* + e^* - \hat{y}^* = E(Y \mid X = x^*) - \hat{y}^* + e^*$$

that is, the deviation between $E(Y \mid X = x^*)$ and $\hat{y}^*$ plus the random fluctuation e^* (which represents the deviation of Y^* from $E(Y \mid X = x^*)$). Thus the variability in the error for predicting a single value of Y will exceed the variability for estimating the expected value of Y (because of the random error e^*).

It can be shown that under the previously stated assumptions that

$$E(Y^* - \hat{y}^*) = E(Y - \hat{y} \mid X = x^*) = 0 \tag{2.15}$$

$$\mathrm{Var}(Y^* - \hat{y}^*) = \mathrm{Var}(Y - \hat{y} \mid X = x^*) = \sigma^2 \left[1 + \frac{1}{n} + \frac{(x^* - \overline{x})^2}{SXX} \right] \tag{2.16}$$

$$Y^* - \hat{y}^* \sim N\left(0, \sigma^2 \left[1 + \frac{1}{n} + \frac{(x^* - \overline{x})^2}{SXX} \right] \right) \tag{2.17}$$

Standardizing (2.17) and replacing σ by S gives

$$T = \frac{Y^* - \hat{y}^*}{S\sqrt{(1 + \dfrac{1}{n} + \dfrac{(x^* - \overline{x})^2}{SXX})}} \sim t_{n-2}$$

A $100(1-\alpha)\%$ **prediction interval** for Y^*, the value of Y at $X = x^*$, is given by

$$\hat{y}^* \pm t(\alpha/2, n-2) S \sqrt{(1 + \frac{1}{n} + \frac{(x^* - \overline{x})^2}{SXX})}$$

$$= \hat{\beta}_0 + \hat{\beta}_1 x^* \pm t(\alpha/2, n-2) S \sqrt{(1 + \frac{1}{n} + \frac{(x^* - \overline{x})^2}{SXX})}$$

where $t(\alpha/2, n-2)$ is the $100(1-\alpha/2)$th quantile of the t-distribution with $n - 2$ degrees of freedom.

Regression Output from R

Ninety-five percent confidence intervals for the population regression line (i.e., the average *RunTime*) at *RunSize* = 50, 100, 150, 200, 250, 300, 350 are:

```
        fit        lwr        upr
1  162.7099   148.6204   176.7994
2  175.6720   164.6568   186.6872
3  188.6342   179.9969   197.2714
4  201.5963   193.9600   209.2326
5  214.5585   206.0455   223.0714
6  227.5206   216.7006   238.3407
7  240.4828   226.6220   254.3435
```

Ninety-five percent prediction intervals for the actual value of Y (i.e., the actual *RunTime*) at at *RunSize* = 50, 100, 150, 200, 250, 300, 350 are:

```
        fit        lwr        upr
1  162.7099   125.7720   199.6478
2  175.6720   139.7940   211.5500
3  188.6342   153.4135   223.8548
4  201.5963   166.6076   236.5850
5  214.5585   179.3681   249.7489
6  227.5206   191.7021   263.3392
7  240.4828   203.6315   277.3340
```

Notice that each prediction interval is considerably wider than the corresponding confidence interval, as is expected.

2.5 Analysis of Variance

There is a linear association between Y and x if

$$Y = \beta_0 + \beta_1 x + e$$

and $\beta_1 \neq 0$. If we knew that $\beta_1 \neq 0$ then we would predict Y by

$$\hat{y} = \hat{\beta}_0 + \hat{\beta}_1 x$$

On the other hand, if we knew that $\beta_1 = 0$ then we predict Y by

$$\hat{y} = \bar{y}$$

To test whether there is a linear association between Y and X we have to test

$$H_0 : \beta_1 = 0 \text{ against } H_A : \beta_1 \neq 0.$$

We can perform this test using the following t-statistic

$$T = \frac{\hat{\beta}_1 - 0}{\text{se}(\hat{\beta}_1)} \sim t_{n-2} \text{ when } H_0 \text{ is true.}$$

We next look at a different test statistic which can be used when there is more than one predictor variable, that is, in multiple regression. First, we introduce some terminology.

Define the total corrected sum of squares of the Y's by

$$\text{SST} = SYY = \sum_i^n (y_i - \bar{y})^2$$

Recall that the residual sum of squares is given by

$$\text{RSS} = \sum_i^n (y_i - \hat{y}_i)^2$$

Define the regression sum of squares (i.e., sum of squares explained by the regression model) by

$$\text{SSreg} = \sum_i^n (\hat{y}_i - \bar{y})^2$$

It is clear that SSreg is close to zero if for each i, $\hat{y}_i$ is close to $\bar{y}$ while SSreg is large if $\hat{y}_i$ differs from $\bar{y}$ for most values of x.

We next look at the hypothetical situation in Figure 2.4 with just a single data point (x_i, y_i) shown along with the least squares regression line and the mean of y based on all n data points. It is apparent from Figure 2.4 that $y_i - \bar{y} = (y_i - \hat{y}_i) + (\hat{y}_i - \bar{y})$.
Further, it can be shown that

SST	= SSreg	+ RSS
Total sample =	Variability explained by +	Unexplained (or error)
variability	the model	variability

See exercise 6 in Section 2.7 for details.
If

$$Y = \beta_0 + \beta_1 x + e \text{ and } \beta_1 \neq 0$$

then RSS should be "small" and SSreg should be "close" to SST. But how small is "small" and how close is "close"?

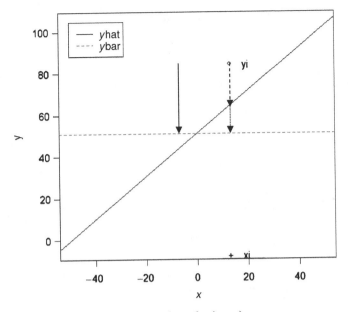

Figure 2.4 Graphical depiction that $y_i - \bar{y} = (y_i - \hat{y}_i) + (\hat{y}_i - \bar{y})$

To test

$$H_0 : \beta_1 = 0 \text{ against } H_A : \beta_1 \neq 0$$

we can use the test statistic

$$F = \frac{\text{SSreg}/1}{\text{RSS}/(n-2)}$$

since RSS has $(n-2)$ degrees of freedom and SSreg has 1 degree of freedom.

Under the assumption that $e_1, e_2, ..., e_n$ are independent and normally distributed with mean 0 and variance σ^2, it can be shown that F has an F distribution with 1 and $n-2$ degrees of freedom when H_0 is true, that is,

$$F = \frac{\text{SSreg}/1}{\text{RSS}/(n-2)} \sim F_{1,n-2} \text{ when } H_0 \text{ is true}$$

Form of test: reject H_0 at level α if $F > F_{\alpha,1,n-2}$ (which can be obtained from table of the F distribution). However, all statistical packages report the corresponding p-value.

The usual way of setting out this test is to use an Analysis of variance table

Source of variation	Degrees of freedom (df)	Sum of squares (SS)	Mean square (MS)	F
Regression	1	SSreg	SSreg/1	$F = \dfrac{\text{SSreg}/1}{\text{RSS}/(n-2)}$
Residual	$n-2$	RSS	RSS/$(n-2)$	
Total	$n-1$	SST		

Notes:

1. It can be shown that in the case of simple linear regression $T = \dfrac{\hat{\beta}_1 - 0}{se(\hat{\beta}_1)} \sim t_{n-2}$
   and $F = \dfrac{\text{SSreg}/1}{\text{RSS}/(n-2)} \sim F_{1,n-2}$ are related via $F = T^2$

2. R^2, the coefficient of determination of the regression line, is defined as the proportion of the total sample variability in the Y's explained by the regression model, that is,

$$R^2 = \frac{\text{SSreg}}{\text{SST}} = 1 - \frac{\text{RSS}}{\text{SST}}$$

The reason this quantity is called R^2 is that it is equal to the square of the correlation between Y and X. It is arguably one of the most commonly misused statistics.

Regression Output from R

```
Analysis of Variance Table
Response: RunTime
           Df   Sum Sq   Mean Sq   F value    Pr(>F)
RunSize     1  12868.4   12868.4    48.717  1.615e-06   ***
Residuals  18   4754.6     264.1
---
Signif. codes:  0 `***' 0.001 `**' 0.01 `*' 0.05 `.' 0.1 ` ` 1
```

Notice that the observed F-value of 48.717 is just the square of the observed t-value 6.98 which can be found between Figures 2.2 and 2.3. We shall see in Chapter 5 that Analysis of Variance overcomes the problems associated with multiple t-tests which occur when there are many predictor variables.

2.6 Dummy Variable Regression

So far we have only considered situations in which the predictor or X-variable is quantitative (i.e., takes numerical values). We next consider so-called **dummy variable regression**, which is used in its simplest form when a predictor is categorical with two values (e.g., gender) rather than quantitative. The resulting regression models allow us to test for the difference between the means of two groups. We shall see in a later topic that the concept of a dummy variable can be extended to include problems involving more than two groups.

Using dummy variable regression to compare new and old methods

We shall consider the following example throughout this section. It is taken from Foster, Stine and Waterman (1997, pages 142–148). In this example, we consider a large food processing center that needs to be able to switch from one type of package to another quickly to react to changes in order patterns. Consultants have developed a new method for changing the production line and used it to produce a sample of 48 change-over times (in minutes). Also available is an independent sample of 72 change-over times (in minutes) for the existing method. These two sets of times can be found on book web site in the file called changeover_times.txt. The first three and the last three rows of the data from this file are reproduced below in Table 2.2. Plots of the data appear in Figure 2.5.

We wish to develop an equation to model the relationship between Y, the change-over time and X, the dummy variable corresponding to New and hence test whether the mean change-over time is reduced using the new method.

We consider the simple linear regression model

$$Y = \beta_0 + \beta_1 x + e$$

where Y = change-over time and x is the dummy variable (i.e., $x = 1$ if the time corresponds to the new change-over method and 0 if it corresponds to the existing method).

Regression Output from R

```
Coefficients:
              Estimate  Std. Error   t value    Pr(>|t|)
(Intercept)   17.8611      0.8905    20.058      <2e-16    ***
New           -3.1736      1.4080    -2.254      0.0260     *
---
Signif. codes: 0 `***' 0.001 `**' 0.01 `*' 0.05 `.' 0.1 ` ` 1
Residual standard error: 7.556 on 118 degrees of freedom
Multiple R-Squared: 0.04128, Adjusted R-squared: 0.03315
F-statistic: 5.081 on 1 and 118 DF, p-value: 0.02604
```

We can test whether there is significant reduction in the change-over time for the new method by testing the significance of the dummy variable, that is, we wish to test whether the coefficient of x is zero or less than zero, that is:

$$H_0 : \beta_1 = 0 \text{ against } H_A : \beta_1 < 0$$

We use the one-sided "<" alternative since we are interested in whether the new method has lead to a reduction in mean change-over time. The test statistic is

$$T = \frac{\hat{\beta}_1 - 0}{se(\hat{\beta}_1)} \sim t_{n-2} \text{ when } H_0 \text{ is true.}$$

Table 2.2 Change-over time data (changeover_times.txt)

Method	Y, Change-over time	X, New
Existing	19	0
Existing	24	0
Existing	39	0
.	.	.
New	14	1
New	40	1
New	35	1

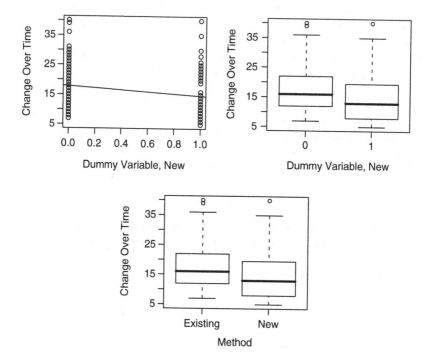

Figure 2.5 A scatter plot and box plots of the change-over time data

In this case,

$$T = -2.254.$$

(This result can be found in the output in the column headed '*t* value'). The associated *p*-value is given by

$$p-value = P(T < -2.254 \text{ when } H_0 \text{ is true}) = \frac{0.026}{2} = 0.013$$

as the two-sided *p-value* $= P(T \neq -2.254$ when H_0 is true$) = 0.026$.

This means that there is significant evidence of a reduction in the mean change-over time for the new method.

Next consider the group consisting of those times associated with the **new change-over method**. For this group, the dummy variable, x is equal to 1. Thus, we can estimate the mean change-over time for the new method as:

$$17.8611 + (-3.1736) \times 1 = 14.6875 = 14.7 \text{ minutes}$$

Next consider the group consisting of those times associated with the **existing change-over method**. For this group, the dummy variable, x is equal to 0. Thus, we can estimate the mean change-over time for the new method as:

$$17.8611 + (-3.1736) \times 0 = 17.8611 = 17.9 \text{ minutes}$$

The new change-over method produces a reduction in the mean change-over time of 3.2 min from 17.9 to 14.7 minutes (Notice that the reduction in the mean change-over time for the new method is just the coefficient of the dummy variable.) This reduction is **statistically significant**.

A 95% confidence interval for the reduction in mean change-over time due to the new method is given by

$$(\hat{\beta}_1 - t(\alpha/2, n-2)\text{se}(\hat{\beta}_1), \hat{\beta}_1 + t(\alpha/2, n-2)\text{se}(\hat{\beta}_1))$$

where $t(\alpha/2, n-2)$ is the $100(1-\alpha/2)$th quantile of the t-distribution with $n-2$ degrees of freedom. In this example the X-variable is the dummy variable *New* and $\hat{\beta}_1 = -3.1736, \text{se}(\hat{\beta}_1) = 1.4080, t(0.025, 120 - 2 = 118) = 1.9803$. Thus a 95% confidence interval for β_1 (in minutes) is given by

$$(-3.1736 \pm 1.9803 \times 1.4080) = (-3.1736 \pm 2.7883) = (-5.96, -0.39).$$

Finally, the company should adopt the new method if a reduction of time of this size is of **practical significance**.

2.7   Derivations of Results

In this section, we shall derive some results given earlier about the least squares estimates of the slope and the intercept as well as results about confidence intervals and prediction intervals.

Throughout this section we shall make the following assumptions:

1. Y is related to x by the simple linear regression model
   $$Y_i = \beta_0 + \beta_1 x_i + e_i \ (i = 1, ..., n), i.e., \text{E}(Y \mid X = x_i) = \beta_0 + \beta_1 x_i$$
2. The errors $e_1, e_2, ..., e_n$ are independent of each other
3. The errors $e_1, e_2, ..., e_n$ have a common variance σ^2
4. The errors are normally distributed with a mean of 0 and variance σ^2 (especially when the sample size is small), that is, $e \mid X \sim N(0, \sigma^2)$

In addition, since the regression model is conditional on X we can assume that the values of the predictor variable, $x_1, x_2, \ldots, x_n$ are known fixed constants.

2.7.1 Inferences about the Slope of the Regression Line

Recall from (2.5) that the least squares estimate of β_1 is given by

$$\hat{\beta}_1 = \sum_{i=1}^{n} c_i y_i \text{ where } c_i = \frac{x_i - \bar{x}}{SXX}.$$

Under the above assumptions we shall derive (2.6), (2.7) and (2.8).

To derive (2.6) let's consider

$$E(\hat{\beta}_1 \mid X) = E\left[\sum_{i=1}^{n} c_i y_i \mid X = x_i \right]$$

$$= \sum_{i=1}^{n} c_i E\left[y_i \mid X = x_i \right]$$

$$= \sum_{i=1}^{n} c_i \left(\beta_0 + \beta_1 x_i \right)$$

$$= \beta_0 \sum_{i=1}^{n} c_i + \beta_1 \sum_{i=1}^{n} c_i x_i$$

$$= \beta_0 \sum_{i=1}^{n} \left\{ \frac{x_i - \bar{x}}{SXX} \right\} + \beta_1 \sum_{i=1}^{n} \left\{ \frac{x_i - \bar{x}}{SXX} \right\} x_i$$

$$= \beta_1$$

since $\displaystyle\sum_{i=1}^{n} (x_i - \bar{x}) = 0$ and $\displaystyle\sum_{i=1}^{n} (x_i - \bar{x}) x_i = \sum_{i=1}^{n} x_i^2 - n\bar{x}^2 = SXX.$

To derive (2.7) let's consider

$$Var(\hat{\beta}_1 \mid X) = Var\left[\sum_{i=1}^{n} c_i y_i \mid X = x_i \right]$$

$$= \sum_{i=1}^{n} c_i^2 Var(y_i \mid X = x_i)$$

$$= \sigma^2 \sum_{i=1}^{n} c_i^2$$

$$= \sigma^2 \sum_{i=1}^{n} \left\{ \frac{x_i - \bar{x}}{SXX} \right\}^2$$

$$= \frac{\sigma^2}{SXX}$$

Finally we derive (2.8). Under assumption (4), the errors $e_i|X$ are normally distributed. Since $y_i = \beta_0 + \beta_1 x_i + e_i$ $(i = 1, 2, ..., n)$, $Y_i|X$ is normally distributed. Since $\hat{\beta}_1 \mid X$ is a linear combination of the y_i's, $\hat{\beta}_1 \mid X$ is normally distributed.

2.7.2 Inferences about the Intercept of the Regression Line

Recall from (2.3) that the least squares estimate of β_0 is given by

$$\hat{\beta}_0 = \bar{y} - \hat{\beta}_1 \bar{x}.$$

Under the assumptions given previously we shall derive (2.9), (2.10) and (2.11). To derive (2.9) we shall use the fact that

$$\mathrm{E}(\hat{\beta}_0 \mid X) = \mathrm{E}(\bar{y} \mid X) - \mathrm{E}(\hat{\beta}_1 \mid X)\bar{x}$$

The first piece of the last equation is

$$\mathrm{E}(\bar{y} \mid X) = \frac{1}{n}\sum_{i=1}^{n}\mathrm{E}(y_i \mid X = x_i)$$

$$= \frac{1}{n}\sum_{i=1}^{n}\mathrm{E}(\beta_0 + \beta_1 x_i + e_i)$$

$$= \beta_0 + \beta_1\frac{1}{n}\sum_{i=1}^{n}x_i$$

$$= \beta_0 + \beta_1\bar{x}$$

The second piece of that equation is

$$\mathrm{E}(\hat{\beta}_1 \mid X)\bar{x} = \beta_1\bar{x}.$$

Thus,

$$\mathrm{E}(\hat{\beta}_0 \mid X) = \mathrm{E}(\bar{y} \mid X) - \mathrm{E}(\hat{\beta}_1 \mid X)\bar{x} = \beta_0 + \beta_1\bar{x} - \beta_1\bar{x} = \beta_0$$

To derive (2.10) let's consider

$$\mathrm{Var}(\hat{\beta}_0 \mid X) = \mathrm{Var}(\bar{y} - \hat{\beta}_1\bar{x} \mid X)$$

$$= \mathrm{Var}(\bar{y} \mid X) + \bar{x}^2\mathrm{Var}(\hat{\beta}_1 \mid X) - 2\bar{x}\mathrm{Cov}(\bar{y}, \hat{\beta}_1 \mid X)$$

The first term is given by

$$\mathrm{Var}(\bar{y} \mid X) = \mathrm{Var}(\frac{1}{n}\sum_{i=1}^{n}y_i \mid X = x_i) = \frac{n\sigma^2}{n^2} = \frac{\sigma^2}{n}.$$

From (2.7),

$$\mathrm{Var}(\hat{\beta}_1 \mid X) = \frac{\sigma^2}{SXX}$$

Finally,

$$\mathrm{Cov}(\bar{y}, \hat{\beta}_1 \mid X) = \mathrm{Cov}\left(\frac{1}{n}\sum_{i=1}^{n} y_i, \sum_{i=1}^{n} c_i y_i\right) = \frac{1}{n}\sum_{i=1}^{n} c_i \mathrm{Cov}(y_i, y_i) = \frac{\sigma^2}{n}\sum_{i=1}^{n} c_i = 0$$

So,

$$\mathrm{Var}(\hat{\beta}_0 \mid X) = \sigma^2\left(\frac{1}{n} + \frac{\bar{x}^2}{SXX}\right)$$

Result (2.11) follows from the fact that under assumption (4), $Y_i \mid X$ (and hence $\bar{y}$) are normally distributed as is $\hat{\beta}_1 \mid X$.

2.7.3 Confidence Intervals for the Population Regression Line

Recall that the population regression line at $X = x^*$ is given by

$$\mathrm{E}(Y \mid X = x^*) = \beta_0 + \beta_1 x^*$$

An estimator the population regression line at $X = x^*$ (i.e., $\mathrm{E}(Y \mid X = x^*) = \beta_0 + \beta_1 x^*$) is the value of the estimated regression equation at $X = x^*$, namely,

$$\hat{y}^* = \hat{\beta}_0 + \hat{\beta}_1 x^*$$

Under the assumptions stated previously, we shall derive (2.12), (2.13) and (2.14). First, notice that (2.12) follows from the following earlier established results $\mathrm{E}(\hat{\beta}_0 \mid X = x^*) = \beta_0$ and $\mathrm{E}(\hat{\beta}_1 \mid X = x^*) = \beta_1$.
  Next, consider (2.13)

$$\mathrm{Var}(\hat{y} \mid X = x^*)$$
$$= \mathrm{Var}(\hat{\beta}_0 + \hat{\beta}_1 x \mid X = x^*)$$
$$= \mathrm{Var}(\hat{\beta}_0 \mid X = x^*) + x^{*2}\,\mathrm{Var}(\hat{\beta}_1 \mid X = x^*) + 2x^*\,\mathrm{Cov}(\hat{\beta}_0, \hat{\beta}_1 \mid X = x^*)$$

Now,

$$\mathrm{Cov}(\hat{\beta}_0, \hat{\beta}_1 \mid X = x^*) = \mathrm{Cov}(\bar{y} - \hat{\beta}_1\bar{x}, \hat{\beta}_1 \mid X = x^*)$$
$$= \mathrm{Cov}(\bar{y}, \hat{\beta}_1 \mid X = x^*) - \bar{x}\,\mathrm{Cov}(\hat{\beta}_1, \hat{\beta}_1)$$
$$= 0 - \bar{x}\,\mathrm{Var}(\hat{\beta}_1)$$
$$= \frac{-\bar{x}\sigma^2}{SXX}$$

So that,

$$\mathrm{Var}(\hat{y} \mid X = x^*) = \sigma^2 \left(\frac{1}{n} + \frac{\overline{x}^2}{SXX} \right) + x^{*2} \frac{\sigma^2}{SXX} - \frac{2x^* \overline{x} \sigma^2}{SXX}$$

$$= \sigma^2 \left(\frac{1}{n} + \frac{(x^* - \overline{x})^2}{SXX} \right)$$

Result (2.14) follows from the fact that under assumption (4), $\hat{\beta}_0 \mid X$ is normally distributed as is $\hat{\beta}_1 \mid X$.

2.7.4   Prediction Intervals for the Actual Value of Y

We base our prediction of Y when $X = x^*$ (that is of Y^*) on

$$\hat{y}^* = \hat{\beta}_0 + \hat{\beta}_1 x^*$$

The error in our prediction is

$$Y^* - \hat{y}^* = \beta_0 + \beta_1 x^* + e^* - \hat{y}^* = \mathrm{E}(Y \mid X = x^*) - \hat{y}^* + e^*$$

that is, the deviation between $\mathrm{E}(Y \mid X = x^*)$ and $\hat{y}^*$ plus the random fluctuation e^* (which represents the deviation of Y^* from $\mathrm{E}(Y \mid X = x^*)$).

Under the assumptions stated previously, we shall derive (2.15), (2.16) and (2.17). First, we consider (2.15)

$$\mathrm{E}(Y^* - \hat{y}^*) = \mathrm{E}(Y - \hat{y} \mid X = x^*)$$

$$= \mathrm{E}(Y \mid X = x^*) - \mathrm{E}(\hat{\beta}_0 + \hat{\beta}_1 x \mid X = x^*)$$

$$= 0$$

In considering (2.16), notice that $\hat{y}$ is independent of Y^*, a future value of Y. Thus,

$$\mathrm{Var}(Y^* - \hat{y}^*) = \mathrm{Var}(Y - \hat{y} \mid X = x^*)$$

$$= \mathrm{Var}(Y \mid X = x^*) + \mathrm{Var}(\hat{y} \mid X = x^*) - 2\mathrm{Cov}(Y, \hat{y} \mid X = x^*)$$

$$= \sigma^2 + \sigma^2 \left[\frac{1}{n} + \frac{(x^* - \overline{x})^2}{SXX} \right] - 0$$

$$= \sigma^2 \left[1 + \frac{1}{n} + \frac{(x^* - \overline{x})^2}{SXX} \right]$$

Finally, (2.17) follows since both $\hat{y}$ and Y^* are normally distributed.

2.8   Exercises

1. The web site www.playbill.com provides weekly reports on the box office ticket sales for plays on Broadway in New York. We shall consider the data for the week October 11–17, 2004 (referred to below as the current week). The data are in the form of the gross box office results for the current week and the gross box office results for the previous week (i.e., October 3–10, 2004). The data, plotted in Figure 2.6, are available on the book web site in the file playbill.csv.

Fit the following model to the data: $Y = \beta_0 + \beta_1 x + e$ where Y is the gross box office results for the current week (in \$) and x is the gross box office results for the previous week (in \$). Complete the following tasks:

(a) Find a 95% confidence interval for the slope of the regression model, β_1. Is 1 a plausible value for β_1? Give a reason to support your answer.
(b) Test the null hypothesis $H_0 : \beta_0 = 10000$ against a two-sided alternative. Interpret your result.
(c) Use the fitted regression model to estimate the gross box office results for the current week (in \$) for a production with \$400,000 in gross box office the previous week. Find a 95% prediction interval for the gross box office

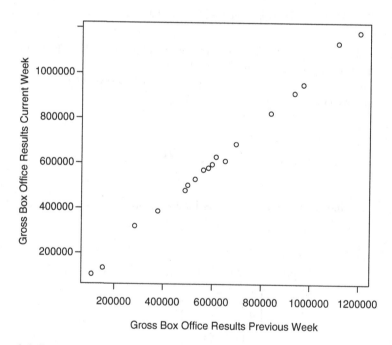

Figure 2.6 Scatter plot of gross box office results from Broadway

results for the current week (in $) for a production with $400,000 in gross box office the previous week. Is $450,000 a feasible value for the gross box office results in the current week, for a production with $400,000 in gross box office the previous week? Give a reason to support your answer.

(d) Some promoters of Broadway plays use the prediction rule that next week's gross box office results will be equal to this week's gross box office results. Comment on the appropriateness of this rule.

2. A story by James R. Hagerty entitled *With Buyers Sidelined, Home Prices Slide* published in the Thursday October 25, 2007 edition of the *Wall Street Journal* contained data on so-called fundamental housing indicators in major real estate markets across the US. The author argues that... *prices are generally falling and overdue loan payments are piling up*. Thus, we shall consider data presented in the article on

Y = Percentage change in average price from July 2006 to July 2007 (based on the S&P/Case-Shiller national housing index); and

x = Percentage of mortgage loans 30 days or more overdue in latest quarter (based on data from Equifax and Moody's).

The data are available on the book web site in the file indicators.txt. Fit the following model to the data: $Y = \beta_0 + \beta_1 x + e$. Complete the following tasks:

(a) Find a 95% confidence interval for the slope of the regression model, β_1. On the basis of this confidence interval decide whether there is evidence of a significant negative linear association.

(b) Use the fitted regression model to estimate $E(Y|X=4)$. Find a 95% confidence interval for $E(Y|X=4)$. Is 0% a feasible value for $E(Y|X=4)$? Give a reason to support your answer.

3. The manager of the purchasing department of a large company would like to develop a regression model to predict the average amount of time it takes to process a given number of invoices. Over a 30-day period, data are collected on the number of invoices processed and the total time taken (in hours). The data are available on the book web site in the file invoices.txt. The following model was fit to the data: $Y = \beta_0 + \beta_1 x + e$ where Y is the processing time and x is the number of invoices. A plot of the data and the fitted model can be found in Figure 2.7. Utilizing the output from the fit of this model provided below, complete the following tasks.

(a) Find a 95% confidence interval for the start-up time, i.e., β_0.

(b) Suppose that a best practice benchmark for the average processing time for an additional invoice is 0.01 hours (or 0.6 minutes). Test the null hypothesis $H_0 : \beta_1 = 0.01$ against a two-sided alternative. Interpret your result.

(c) Find a point estimate and a 95% prediction interval for the time taken to process 130 invoices.

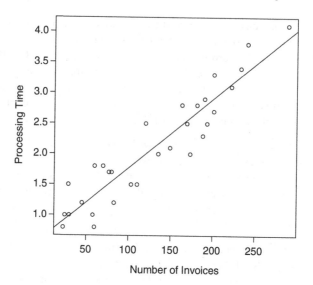

Figure 2.7 Scatter plot of the invoice data

Regression output from R for the invoice data

```
Call:
lm(formula = Time ~ Invoices)

Coefficients:
              Estimate   Std. Error   t value    Pr(>|t|)
(Intercept) 0.6417099    0.1222707     5.248     1.41e-05   ***
Invoices    0.0112916    0.0008184    13.797     5.17e-14   ***
---

Residual standard error: 0.3298 on 28 degrees of freedom
Multiple R-Squared: 0.8718, Adjusted R-squared: 0.8672
F-statistic: 190.4 on 1 and 28 DF, p-value: 5.175e-14

mean(Time)
2.1
median(Time)
2
mean(Invoices)
130.0
median(Invoices)
127.5
```

4. Straight-line regression through the origin:
   In this question we shall make the following assumptions:

   (1) Y is related to x by the simple linear regression model $Y_i = \beta x_i + e_i$ $(i = 1, 2, ..., n)$,
       i.e., $E(Y \mid X = x_i) = \beta x_i$

(2) The errors $e_1, e_2,..., e_n$ are independent of each other

(3) The errors $e_1, e_2,..., e_n$ have a common variance σ^2

(4) The errors are normally distributed with a mean of 0 and variance σ^2 (especially when the sample size is small), i.e., $e \mid X \sim N(0, \sigma^2)$

In addition, since the regression model is conditional on X we can assume that the values of the predictor variable, $x_1, x_2, ..., x_n$ are known fixed constants.

(a) Show that the least squares estimate of β is given by

$$\hat{\beta} = \frac{\sum_{i=1}^{n} x_i y_i}{\sum_{i=1}^{n} x_i^2}$$

(b) Under the above assumptions show that

(i) $E(\hat{\beta} \mid X) = \beta$

(ii) $Var(\hat{\beta} \mid X) = \dfrac{\sigma^2}{\sum_{i=1}^{n} x_i^2}$

(iii) $\hat{\beta} \mid X \sim N(\beta, \dfrac{\sigma^2}{\sum_{i=1}^{n} x_i^2})$

5. Two alternative straight line regression models have been proposed for Y. In the first model, Y is a linear function of x_1, while in the second model Y is a linear function of x_2. The plot in the first column of Figure 2.8 is that of Y against x_1, while the plot in the second column below is that of Y against x_2. These plots also show the least squares regression lines. In the following statements RSS stands for residual sum of squares, while SSreg stands for regression sum of squares. Which one of the following statements is true?

(a) RSS for model 1 is greater than RSS for model 2, while SSreg for model 1 is greater than SSreg for model 2.

(b) RSS for model 1 is less than RSS for model 2, while SSreg for model 1 is less than SSreg for model 2.

(c) RSS for model 1 is greater than RSS for model 2, while SSreg for model 1 is less than SSreg for model 2.

(d) RSS for model 1 is less than RSS for model 2, while SSreg for model 1 is greater than SSreg for model 2.

Give a detailed reason to support your choice.

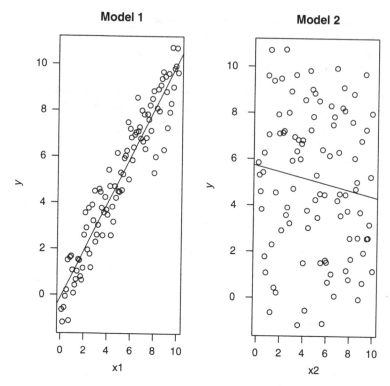

Figure 2.8 Scatter plots and least squares lines

6. In this problem we will show that SST=SSreg+RSS . To do this we will show

   that $\sum_{i=1}^{n}(y_i - \hat{y}_i)(\hat{y}_i - \bar{y}) = 0$.

   (a) Show that $(y_i - \hat{y}_i) = (y_i - \bar{y}) - \hat{\beta}_1(x_i - \bar{x})$.

   (b) Show that $(\hat{y}_i - \bar{y}) = \hat{\beta}_1(x_i - \bar{x})$.

   (c) Utilizing the fact that $\hat{\beta}_1 = \dfrac{SXY}{SXX}$, show that $\sum_{i=1}^{n}(y_i - \hat{y}_i)(\hat{y}_i - \bar{y}) = 0$.

7. A statistics professor has been involved in a collaborative research project with two entomologists. The statistics part of the project involves fitting regression models to large data sets. Together they have written and submitted a manuscript to an entomology journal. The manuscript contains a number of scatter plots with each showing an estimated regression line (based on a valid model) and

associated individual 95% confidence intervals for the regression function at each x value, as well as the observed data. A referee has asked the following question:

I don't understand how 95% of the observations fall outside the 95% CI as depicted in the figures.

Briefly explain how it is entirely possible that 95% of the observations fall outside the 95% CI as depicted in the figures.

Chapter 3
Diagnostics and Transformations for Simple Linear Regression

In Chapter 2 we studied the simple linear regression model. Throughout Chapter 2, we assumed that the simple linear regression model was a valid model for the data, that is, the conditional mean of Y given X is a linear function of X and the conditional variance of Y given X is constant. In other words,

$$E(Y \mid X = x) = \beta_0 + \beta_1 x \text{ and } \mathrm{Var}(Y \mid X = x) = \sigma^2.$$

In Section 3.1, we start by examining the important issue of deciding whether the model under consideration is indeed valid. In Section 3.2, we will see that when we use a regression model we implicitly make a series of assumptions. We then consider a series of tools known as regression diagnostics to check each assumption. Having used these tools to diagnose potential problems with the assumptions, we look at how to first identify and then overcome or deal with a common problem, namely, nonconstant error variance.

The section on transformations shows how transformations can be used in some situations to overcome problems with assumptions due to nonconstant variance or nonlinearity, as well as enabling us to fit models for specific purposes, such as to estimate percentage effects.

A primary aim of this chapter is to understand what actually happens when the standard assumptions associated with a regression model are violated, and what should be done in response to each violation.

3.1 Valid and Invalid Regression Models: Anscombe's Four Data Sets

Throughout this section we shall consider four data sets constructed by Anscombe (1973). This example illustrates dramatically the point that looking only at the numerical regression output may lead to very misleading conclusions about the data, and lead to adopting the wrong model. The data are given in the table below (Table 3.1) and are plotted in Figure 3.1. Notice that the Y-values differ in each of the four data sets, while the X-values are the same for data sets 1, 2 and 3.

S.J. Sheather, *A Modern Approach to Regression with R*,
DOI: 10.1007/978-0-387-09608-7_3, © Springer Science+Business Media LLC 2009

Table 3.1 Anscombe's four data sets

Case	x1	x2	x3	x4	y1	y2	y3	y4
1	10	10	10	8	8.04	9.14	7.46	6.58
2	8	8	8	8	6.95	8.14	6.77	5.76
3	13	13	13	8	7.58	8.74	12.74	7.71
4	9	9	9	8	8.81	8.77	7.11	8.84
5	11	11	11	8	8.33	9.26	7.81	8.47
6	14	14	14	8	9.96	8.1	8.84	7.04
7	6	6	6	8	7.24	6.13	6.08	5.25
8	4	4	4	19	4.26	3.1	5.39	12.5
9	12	12	12	8	10.84	9.13	8.15	5.56
10	7	7	7	8	4.82	7.26	6.42	7.91
11	5	5	5	8	5.68	4.74	5.73	6.89

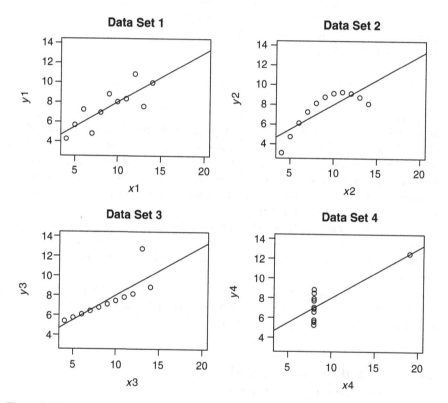

Figure 3.1 Plots of Anscombe's four data sets

When a regression model is fitted to data sets 1, 2, 3 and 4, in each case the fitted regression model is

$$\hat{y} = 3.0 + 0.5x.$$

The regression output for data sets 1 to 4 is given below. The regression output for the four constructed data sets is identical (to two decimal places) in every respect.

Looking at Figure 3.1 it is obvious that a straight-line regression model is appropriate only for Data Set 1, since it is the only data set for which $E(Y \mid X = x) = \beta_0 + \beta_1 x$ and $\text{Var}(Y \mid X = x) = \sigma^2$ seem reasonable assumptions. On the other hand, the data in Data Set 2 seem to have a curved rather than a straight-line relationship. The third data set has an extreme outlier that should be investigated. For the fourth data set, the slope of the regression line is solely determined by a single point, namely, the point with the largest x-value.

Regression output from R

```
Coefficients:
            Estimate  Std. Error  t value  Pr(>|t|)
(Intercept)   3.0001     1.1247    2.667    0.02573   *
x1            0.5001     0.1179    4.241    0.00217   **
---
Signif. codes: 0 '***' 0.001 '**' 0.01 '*' 0.05 '.' 0.1 ' ' 1

Residual standard error: 1.237 on 9 degrees of freedom
Multiple R-Squared: 0.6665,  Adjusted R-squared: 0.6295
F-statistic: 17.99 on 1 and 9 DF,  p-value: 0.002170

Coefficients:
            Estimate  Std. Error  t value  Pr(>|t|)
(Intercept)   3.001      1.125     2.667    0.02576   *
x2            0.500      0.118     4.239    0.00218   **
---
Signif. codes: 0 '***' 0.001 '**' 0.01 '*' 0.05 '.' 0.1 ' ' 1

Residual standard error: 1.237 on 9 degrees of freedom
Multiple R-Squared: 0.6662,  Adjusted R-squared: 0.6292
F-statistic: 17.97 on 1 and 9 DF,  p-value: 0.002179

Coefficients:
            Estimate  Std. Error  t value  Pr(>|t|)
(Intercept)   3.0025     1.1245    2.670    0.02562   *
x3            0.4997     0.1179    4.239    0.00218   **
---
Signif. codes: 0 '***' 0.001 '**' 0.01 '*' 0.05 '.' 0.1 ' ' 1

Residual standard error: 1.236 on 9 degrees of freedom
Multiple R-Squared: 0.6663,  Adjusted R-squared: 0.6292
F-statistic: 17.97 on 1 and 9 DF,  p-value: 0.002176

Coefficients:
            Estimate  Std. Error  t value  Pr(>|t|)
(Intercept)   3.0017     1.1239    2.671    0.02559   *
x4            0.4999     0.1178    4.243    0.00216   **
---
Signif. codes: 0 '***' 0.001 '**' 0.01 '*' 0.05 '.' 0.1 ' ' 1

Residual standard error: 1.236 on 9 degrees of freedom
Multiple R-Squared: 0.6667,  Adjusted R-squared: 0.6297
F-statistic: 18 on 1 and 9 DF,  p-value: 0.002165
```

This example demonstrates that the numerical regression output should always be supplemented by an analysis to ensure that an appropriate model has been fitted to the data. In this case it is sufficient to look at the scatter plots in Figure 3.1 to determine whether an appropriate model has been fit. However, when we consider situations in which there is more than one predictor variable, we shall need some additional tools in order to check the appropriateness of the fitted model.

3.1.1 Residuals

One tool we will use to validate a regression model is one or more plots of residuals (or standardized residuals, which will be defined later in this chapter). These plots will enable us to assess visually whether an appropriate model has been fit to the data no matter how many predictor variables are used.

Figure 3.2 provides plots of the residuals against X for each of Anscombe's four data sets. There is no discernible pattern in the plot of the residuals from data set 1 against $x1$. We shall see next that this indicates that an appropriate model has been

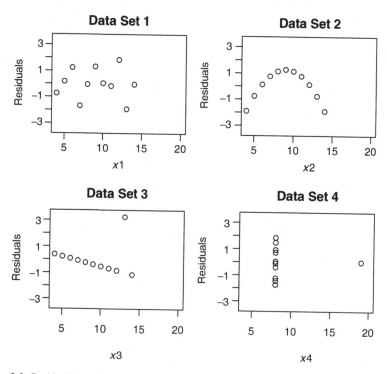

Figure 3.2 Residual plots for Anscombe's data sets

fit to the data. We shall see that a plot of residuals against X that produces a random pattern indicates an appropriate model has been fit to the data. Additionally, we shall see that a plot of residuals against X that produces a discernible pattern indicates an incorrect model has been fit to the data.

Recall that a valid simple linear regression model is one for which $E(Y \mid X = x) = \beta_0 + \beta_1 x$ and $\mathrm{Var}(Y \mid X = x) = \sigma^2$.

3.1.2   Using Plots of Residuals to Determine Whether the Proposed Regression Model Is a Valid Model

One way of checking whether a valid simple linear regression model has been fit is to plot residuals versus x and look for patterns. If no pattern is found then this indicates that the model provides an adequate summary of the data, i.e., is a valid model. If a pattern is found then the shape of the pattern provides information on the function of x that is missing from the model.

For example, suppose that the true model is a straight line

$$Y_i = E(Y_1 \mid X_1 = x_i) + e_i = \beta_0 + \beta_1 x_i + e_i$$

where $e_i =$ random fluctuation (or error) in Y_i and is such that $E(e_i) = 0$ and that we fit a straight line $\hat{y}_i = \hat{\beta}_0 + \hat{\beta}_1 x_i$.

Then, assuming that the least squares estimates $\hat{\beta}_0$ and $\hat{\beta}_1$ are close to the unknown population parameters β_0 and β_1, we find that

$$\hat{e}_i = y_i - \hat{y}_i = (\beta_0 - \hat{\beta}_0) + (\beta_1 - \hat{\beta}_1)x_i + e_i \approx e_i,$$

that is, the residuals should resemble random errors. If the residuals vary with x then this indicates that an incorrect model has been fit. For example, suppose that the true model is a quadratic

$$y_i = \beta_0 + \beta_1 x_i + \beta_2 x_i^2 + e_i$$

and that we fit a straight line

$$\hat{y}_i = \hat{\beta}_0 + \hat{\beta}_1 x_i$$

Then, somewhat simplistically assuming that the least squares estimates $\hat{\beta}_0$ and $\hat{\beta}_1$ are close to the unknown population parameters β_0 and β_1, we find that

$$\hat{e}_i = y_i - \hat{y}_i = (\beta_0 - \hat{\beta}_0) + (\beta_1 - \hat{\beta}_1)x_i + \beta_2 x_i^2 + e_i \approx \beta_2 x_i^2 + e_i,$$

that is, the residuals show a pattern which resembles a quadratic function of x. In Chapter 6 we will study the properties of least squares residuals more carefully.

3.1.3   Example of a Quadratic Model

Suppose that Y is a quadratic function of X without any random error. Then, the residuals from the straight-line fit of Y and X will have a quadratic pattern. Hence, we can conclude that there is need for a quadratic term to be added to the original straight-line regression model. Anscombe's data set 2 is an example of such a situation. Figure 3.3 contains scatter plots of the data and the residuals from a straight-line model for data set 2. As expected, a clear quadratic pattern is evident in the residuals in Figure 3.3.

3.2   Regression Diagnostics: Tools for Checking the Validity of a Model

We next look at tools (called regression diagnostics) which are used to check the validity of all aspects of regression models. When fitting a regression model we will discover that it is important to:

1. Determine whether the proposed regression model is a valid model (i.e., determine whether it provides an adequate fit to the data). The main tools we will

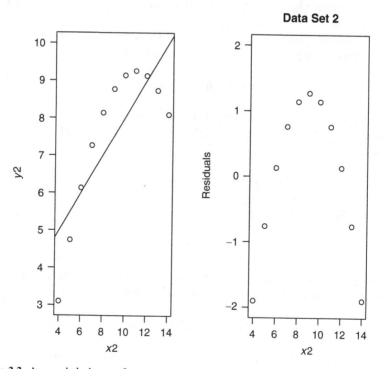

Figure 3.3 Anscombe's data set 2

use to validate regression assumptions are plots of **standardized residuals**.[1] The plots enable us to assess visually whether the assumptions are being violated and point to what should be done to overcome these violations.

2. Determine which (if any) of the data points have x-values that have an unusually large effect on the estimated regression model (such points are called **leverage points**).

3. Determine which (if any) of the data points are **outliers**, that is, points which do not follow the pattern set by the bulk of the data, when one takes into account the given model.

4. If leverage points exist, determine whether each is a **bad leverage point**. If a bad leverage point exists we shall assess its influence on the fitted model.

5. Examine whether the assumption of constant variance of the errors is reasonable. If not, we shall look at how to overcome this problem.

6. If the data are collected over time, examine whether the data are correlated over time.

7. If the sample size is small or prediction intervals are of interest, examine whether the assumption that the errors are normally distributed is reasonable.

We begin by looking at the second item of the above list, leverage points, as these will be needed in the explanation of standardized residuals.

3.2.1 Leverage Points

Data points which exercise considerable influence on the fitted model are called *leverage points*. To make things as simple as possible, we shall begin somewhat unrealistically, by describing leverage points as either "good" or "bad."

McCulloch's example of a "good" and a "bad" leverage point

Robert McCulloch from the University of Chicago has produced a web-based applet[2] to illustrate leverage points. The applet randomly generates 20 points from a known straight-line regression model. It produces a plot like that shown in Figure 3.4. One of the 20 points has an x-value which makes it distant from the other points on the x-axis. We shall see that this point, which is marked on the plot, is a **good leverage point**. The applet marks on the plot the true population regression line (namely, $\beta_0 + \beta_1 x$) and the least squares regression line (namely, $\hat{y} = \hat{\beta}_0 + \hat{\beta}_1 x$).

Next we use the applet to drag one of the points away from the true population regression line. In particular, we focus on the point with the largest x-value. Dragging this point vertically down (so that its x-value stays the same) produces the results shown in Figure 3.5. Notice how in the least squares regression has changed

---

[1] Standardized residuals will be defined later in this section.
[2] http://faculty.chicagogsb.edu/robert.mcculloch/research/teachingApplets/Leverage/index.html (Accessed 11/25/2007)

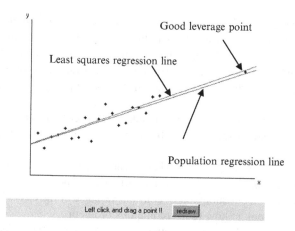

Figure 3.4 A plot showing a good leverage point

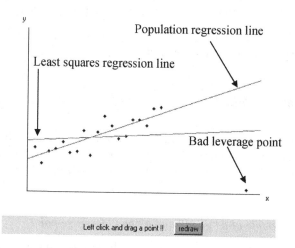

Figure 3.5 A plot showing a bad leverage point

dramatically in response to changing the Y-value of just a single point. The least squares regression line has been levered down by single point. Hence we call this point a **leverage point**. It is a **bad leverage point** since its Y-value does not follow the pattern set by the other 19 points.

In summary, a **leverage point** is a point whose x-value is distant from the other x-values. A point is a **bad leverage point** if its Y-value does not follow the pattern set by the other data points. In other words, **a bad leverage point is a leverage point which is also an outlier**.

Returning to Figure 3.4, the point marked on the plot is said to be a **good leverage point** since its Y-value closely follows the upward trend pattern set by the other 19 points. In other words, **a good leverage point is a leverage point which is NOT also an outlier**.

Next we investigate what happens when we change the Y-value of a point in Figure 3.4 which has a central x-value. We use the applet to drag one of these points away from the true population regression line. In particular, we focus on the point with the 11th largest x-value. Dragging this point vertically up (so that its x-value stays the same) produces the results shown in Figure 3.6. Notice how in the least squares regression has changed relatively little in response to changing the Y-value of centrally located x. This point is said to be an **outlier that is not a leverage point**.

Huber's example of a "good" and a "bad" leverage point

This example is adapted from Huber (1981, pp. 153–155). The data in this example were constructed to further illustrate so-called "good" and "bad" leverage points. The data given in Table 3.2 can be found on the book web site in the file huber.txt.

Notice that the values of x in Table 3.2 are the same for both data sets. Notice that the values of Y are the same for both data sets except when $x = 10$. We shall see that $x = 10$ is a *leverage point* in both data sets in the sense that *this value of x is a long way away from the other values of x* and *the value of Y at this point has a very large effect on the least squares regression line*. The data in Table 3.2 are

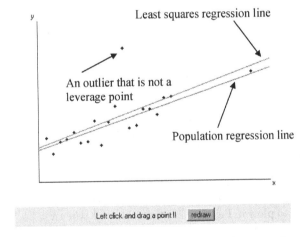

Figure 3.6 A plot of Y against x showing an outlier that is not a leverage point

Table 3.2 Huber's so-called bad and good leverage point data sets

x	YBad	x	YGood
−4	2.48	−4	2.48
−3	0.73	−3	0.73
−2	−0.04	−2	−0.04
−1	−1.44	−1	−1.44
0	−1.32	0	−1.32
10	0.00	10	−11.40

plotted below in Figure 3.7. Regression output from R for the straight-line fits to the two data sets is given below.

Regression output from R

```
Call:
lm(formula = YBad ~ x)

Residuals:
          1          2          3          4          5          6
     2.0858     0.4173    -0.2713    -1.5898    -1.3883     0.7463

Coefficients:
              Estimate    Std. Error    t value    Pr(>|t|)
(Intercept)    0.06833       0.63279      0.108       0.919
x             -0.08146       0.13595     -0.599       0.581

Residual standard error: 1.55 on 4 degrees of freedom
Multiple R-Squared: 0.08237,  Adjusted R-squared: -0.147
F-statistic: 0.3591 on 1 and 4 DF,  p-value: 0.5813

Call:
lm(formula = YGood ~ x)

Residuals:
          1          2          3          4          5          6
    0.47813   -0.31349   -0.12510   -0.56672    0.51167    0.01551

Coefficients:
              Estimate    Std. Error    t value    Pr(>|t|)
(Intercept)   -1.83167       0.19640     -9.326    0.000736   ***
x             -0.95838       0.04219    -22.714    2.23e-05   ***

Residual standard error: 0.4811 on 4 degrees of freedom
Multiple R-Squared: 0.9923,  Adjusted R-squared: 0.9904
F-statistic: 515.9 on 1 and 4 DF,  p-value: 2.225e-05
```

It is clear from Figure 3.7 that $x = 10$ is very distant from the rest of the x's, which range in value from -4 to 0. Next, recall that the only difference between the data in the two plots in Figure 3.7 is the value of Y when $x = 10$. When $x = 10$, YGood $= -11.40$, and YBad $= 0.00$. Comparing the plots in Figure 3.7 allows us to ascertain the effects of changing a single Y value when $x = 10$. This change in Y has produced dramatic changes in the equation of the least squares line. For example looking at the regression output from R above, we see that the slope of the regression for YGood is -0.958 while the slope of the regression line for YBad is -0.081. In addition, this change in a single Y value has had a dramatic effect on the value of R^2 (0.992 versus 0.082).

Our aim is to arrive at a numerical rule that will identify x_i as a leverage point (i.e., a point of high leverage). This rule will be based on:

- The distance x_i is away from the bulk of the x's
- The extent to which the fitted regression line is attracted by the given point

Figure 3.7 Plots of YGood and YBad against x with the fitted regression lines

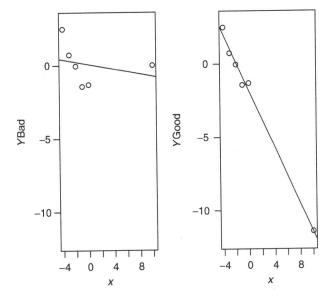

The second bullet point above deals with the extent to which $\hat{y}_i$ (the predicted value of Y at $x = x_i$) depends on y_i (the actual value of Y at $x = x_i$). Recall from (2.3) and (2.5) that

$$\hat{y}_i = \hat{\beta}_0 + \hat{\beta}_1 x_i$$

where $\hat{\beta}_0 = \bar{y} - \hat{\beta}_1 \bar{x}$ and $\hat{\beta}_1 = \sum_{j=1}^{n} c_j y_j$ where $c_j = \dfrac{x_j - \bar{x}}{SXX}$. So that,

$$
\begin{aligned}
\hat{y}_i &= \bar{y} - \hat{\beta}_1 \bar{x} + \hat{\beta}_1 x_i \\
&= \bar{y} + \hat{\beta}_1 (x_i - \bar{x}) \\
&= \frac{1}{n} \sum_{j=1}^{n} y_j + \sum_{j=1}^{n} \frac{(x_j - \bar{x})}{SXX} y_j (x_i - \bar{x}) \\
&= \sum_{j=1}^{n} \left[\frac{1}{n} + \frac{(x_i - \bar{x})(x_j - \bar{x})}{SXX} \right] y_j \\
&= \sum_{j=1}^{n} h_{ij} y_j
\end{aligned}
$$

where

$$h_{ij} = \left[\frac{1}{n} + \frac{(x_i - \bar{x})(x_j - \bar{x})}{SXX} \right]$$

Notice that

$$\sum_{j=1}^{n} h_{ij} = \sum_{j=1}^{n}\left[\frac{1}{n} + \frac{(x_i - \bar{x})(x_j - \bar{x})}{SXX}\right] = \frac{n}{n} + \frac{(x_i - \bar{x})}{SXX}\sum_{j=1}^{n}\left[x_j - \bar{x}\right] = 1$$

since $\sum_{j=1}^{n}\left[x_j - \bar{x}\right] = 0$.

We can express the predicted value, $\hat{y}_i$ as

$$\hat{y}_i = h_{ii} y_i + \sum_{j \neq i} h_{ij} y_j \qquad (3.1)$$

where

$$h_{ii} = \frac{1}{n} + \frac{(x_i - \bar{x})^2}{\sum_{j=1}^{n}(x_j - \bar{x})^2}.$$

The term h_{ii} is commonly called the **leverage** of the ith data point. Consider, for a moment, this formula for leverage (h_{ii}). The top line of the second term in the formula namely, $(x_i - \bar{x})^2$, measures the distance x_i is away from the bulk of the x's, via the squared distance x_i is away from the mean of the x's. Secondly, notice that h_{ii} shows how y_i affects $\hat{y}_i$. For example, if $h_{ii} \cong 1$ then the other h_{ij} terms are close to zero (since $\sum_{j=1}^{n} h_{ij} = 1$), and so

$$\hat{y}_i = 1 \times y_i + \text{other terms} \cong y_i.$$

In this situation, the predicted value, $\hat{y}_i$, will be close to the actual value, y_i, no matter what values of the rest of the data take. Notice also that h_{ii} depends only on the x's. Thus a point of high leverage (or a leverage point) can be found by looking at just the values of the x's and not at the values of the y's.

It can be shown in a straightforward way that for simple linear regression

$$\text{average}(h_{ii}) = \frac{2}{n} \quad (i = 1, 2, \dots, n).$$

Rule for identifying leverage points

A popular rule, which we shall adopt, is to classify x_i as a point of high leverage (i.e., a leverage point) in a simple linear regression model if

$$h_{ii} > 2 \times \text{average}(h_{ii}) = 2 \times \frac{2}{n} = \frac{4}{n}.$$

Huber's example of a 'good' and a 'bad' leverage point

Table 3.3 gives the leverage values for Huber's two data sets. Note that the leverage values are the same for both data sets (i.e., for YGood and YBad) since the *x*-values are the same for both data sets.

Notice that $h_{66} = 0.9359 > 2 \times \text{average}(h_{ii}) = \frac{4}{n} = \frac{4}{6} = 0.67$. Thus, the last point $x_6 = 10$, is a point of high leverage (or a leverage point), while the other points have leverage values much below the cutoff of 0.67.

Recall that a point is a **bad leverage point** if its *Y*-value does not follow the pattern set by the other data points. In other words, **a bad leverage point is a leverage point which is also an outlier**. We shall see in the next section that we can detect whether a leverage point is "bad" based on the value of its standardized residual.

Strategies for dealing with "bad" leverage points

1. Remove invalid data points

Question the validity of the data points corresponding to bad leverage points, that is: **Are these data points unusual or different in some way from the rest of the data?** If so, consider removing these points and refitting the model without them. For example, later in this chapter we will model the price of Treasury bonds. We will discover three leverage points. These points correspond to so-called "flower" bonds, which have definite tax advantages compared to the other bonds. Thus, a reasonable strategy is to remove these cases from the data and refit the model without them.

2. Fit a different regression model

Question the validity of the regression model that has been fitted, that is: **Has an incorrect model been fitted to the data?** If so, consider trying a different model by including extra predictor variables (e.g., polynomial terms) or by transforming *Y* and/or *x* (which is considered later in this chapter). For example, in the case of Huber's bad leverage point, a quadratic model fits all the data very well. See Figure 3.8 and the regression output from R for details.

Table 3.3 Leverage values for Huber's two data sets

i	x_i	Leverage, h_{ii}
1	−4	0.2897
2	−3	0.2359
3	−2	0.1974
4	−1	0.1744
5	0	0.1667
6	10	0.9359

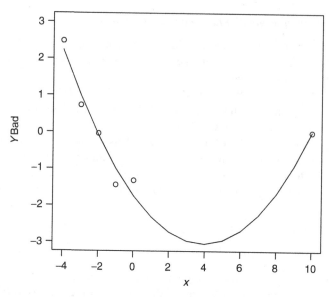

Figure 3.8 Plot of YBad versus x with a quadratic model fit added

Regression output from R

```
Call:
lm(formula = YBad ~ x + I(x^2))

Coefficients:
             Estimate  Std. Error   t value    Pr(>|t|)
(Intercept)  -1.74057     0.29702    -5.860     0.00991 **
x            -0.65945     0.08627    -7.644     0.00465 **
I(x^2)        0.08349     0.01133     7.369     0.00517 **

Residual  standard  error:  0.4096  on  3  degrees  of  freedom
Multiple R-Squared: 0.952, Adjusted R-squared: 0.9199
F-statistic: 29.72 on 2 and 3 DF, p-value: 0.01053
```

"Good" leverage points

Thus, far we have somewhat simplistically classified leverage points as either "bad" or "good". In practice, there is a large gray area between leverage points which do not follow the pattern suggested by the rest of the data (i.e., "bad" leverage points) and leverage points which closely follow the pattern suggested by the rest of the data (i.e., "good" leverage points). Also, while "good" leverage points do not have an adverse effect on the estimated regression coefficients, they do decrease their estimated standard errors as well as increase the value of R^2. Hence, it is important to check extreme leverage points for validity, even when they are so-called "good."

3.2.2   *Standardized Residuals*

Thus far we have discussed the use of residuals to detect any problems with the proposed model. However, as we shall next show, there is a complication that we need to consider, namely, that residuals do not have the same variance. In fact, we shall show below that the ith least squares residual has variance given by

$$\operatorname{Var}(\hat{e}_i) = \sigma^2 \left[1 - h_{ii}\right]$$

where

$$h_{ij} = \frac{1}{n} + \frac{(x_i - \overline{x})(x_j - \overline{x})}{\displaystyle\sum_{j=1}^{n}(x_j - \overline{x})^2} = \frac{1}{n} + \frac{(x_i - \overline{x})(x_j - \overline{x})}{SXX}.$$

Thus, if $h_{ii} \cong 1$ (i.e., h is very close to 1) so that the ith point is a leverage point, then the corresponding residual, $\hat{e}_i$, has small variance (since $1 - h_{ii} \cong 0$). This seems reasonable when one considers that if $h_{ii} \cong 1$ then $\hat{y}_i \cong y_i$ so that $\hat{e}_i$ will always be small (and so it does not vary much).

We shall also show that $\operatorname{Var}(\hat{y}_i) = \sigma^2 h_{ii}$. This again seems reasonable when we consider the fact that when $h_{ii} \cong 1$ then $\hat{y}_i \cong y$. In this case, $\operatorname{Var}(\hat{y}_i) = \sigma^2 h_{ii} \cong \sigma^2 = \operatorname{Var}(y_i)$.

The problem of the residuals having different variances can be overcome by standardizing each residual by dividing it by an estimate of its standard deviation. Thus, the ith *standardized residual*, r_i is given by

$$r_i = \frac{\hat{e}_i}{s\sqrt{1 - h_{ii}}}$$

where $s = \sqrt{\dfrac{1}{n-2}\displaystyle\sum_{j=1}^{n}\hat{e}_j^2}$ is the estimate of σ obtained from the model.

When points of **high leverage** exist, instead of looking at residual plots, it is generally more informative to look at plots of **standardized residuals** since plots of the residuals will have nonconstant variance even if the errors have constant variance. (When points of high leverage do not exist, there is generally little difference in the patterns seen in plots of residuals when compared with those in plots of standardized residuals.) The other advantage of **standardized residuals** is that they immediately tell us how many estimated **standard deviations** any point is away from the fitted regression model. For example, suppose that the 6th point has a standardized residual of 4.3, then this means that the 6th point is an estimated 4.3 standard deviations away from the fitted regression line. If the errors are normally distributed, then observing a point 4.3 standard deviations away from the fitted regression line is highly unusual. Such a point would commonly be referred to as an outlier and as such it should be investigated. We shall follow the common practice of labelling points as **outliers** in small- to moderate-size data sets if the standardized residual for the point falls outside the interval from **–2 to 2**. In very large data sets, we shall change this rule to **–4 to 4**. (Otherwise, many points will be flagged as potential outliers.) *Identification and examination of any outliers is a key part of regression analysis.*

In summary, an **outlier** is a point whose standardized residual falls outside the interval from **–2 to 2**. Recall that a bad leverage point is a leverage point which is also an outlier. Thus, a **bad leverage point** is a leverage point whose standardized residual falls outside the interval from **–2 to 2**. On the other hand, a **good leverage point** is a leverage point whose standardized residual falls inside the interval from **–2 to 2**.

There is a small amount of correlation present in standardized residuals, even if the errors are independent. In fact it can be shown that

$$\text{Cov}(\hat{e}_i, \hat{e}_j) = -h_{ij}\sigma^2 \quad (i \neq j)$$

$$\text{Corr}(\hat{e}_i, \hat{e}_j) = \frac{-h_{ij}}{\sqrt{(1-h_{ii})(1-h_{jj})}} \quad (i \neq j)$$

However, the size of the **correlations inherent in the least squares residuals** are generally so small in situations in which correlated errors is an issue (e.g., data collected over time) that they can be effectively ignored in practice.

Derivation of the variance of the ith residual and fitted value

Recall from (3.1) that,

$$\hat{y}_i = h_{ii}y_i + \sum_{j \neq i} h_{ij}y_j$$

where

$$h_{ij} = \frac{1}{n} + \frac{(x_i - \bar{x})(x_j - \bar{x})}{\sum_{j=1}^{n}(x_j - \bar{x})^2} = \frac{1}{n} + \frac{(x_i - \bar{x})(x_j - \bar{x})}{SXX}.$$

Thus,

$$\hat{e}_i = y_i - \hat{y}_i = y_i - h_{ii}y_i - \sum_{j \ne i} h_{ij}y_j = (1 - h_{ii})y_i - \sum_{j \ne i} h_{ij}y_j$$

So that

$$\text{Var}(\hat{e}_i) = \text{Var}\left((1 - h_{ii})y_i - \sum_{j \ne i} h_{ij}y_j\right)$$

$$= (1 - h_{ii})^2\sigma^2 + \sum_{j \ne i} h_{ij}^2\sigma^2$$

$$= \sigma^2\left[1 - 2h_{ii} + h_{ii}^2 + \sum_{j \ne i} h_{ij}^2\right]$$

$$= \sigma^2\left[1 - 2h_{ii} + \sum_j h_{ij}^2\right]$$

Next, notice that

$$\sum_{j=1}^n h_{ij}^2 = \sum_{j=1}^n\left[\frac{1}{n} + \frac{(x_i - \bar{x})(x_j - \bar{x})}{SXX}\right]^2$$

$$= \frac{1}{n} + 2\sum_{j=1}^n \frac{1}{n} \times \frac{(x_i - \bar{x})(x_j - \bar{x})}{SXX} + \sum_{j=1}^n \frac{(x_i - \bar{x})^2(x_j - \bar{x})^2}{SXX^2}$$

$$= \frac{1}{n} + 0 + \frac{(x_i - \bar{x})^2}{SXX}$$

$$= h_{ii}$$

So that,

$$\text{Var}(\hat{e}_i) = \sigma^2\left[1 - 2h_{ii} + h_{ii}\right] = \sigma^2\left[1 - h_{ii}\right]$$

Next,

$$\text{Var}(\hat{y}_i) = \text{Var}\left(\sum_{j=1}^n h_{ij}y_j\right) = \sum_{j \ne i} h_{ij}^2\text{Var}(y_j) = \sigma^2\sum_j h_{ij}^2 = \sigma^2 h_{ii}$$

Example: US Treasury bond prices

The next example illustrates that a relatively small number of outlying points can have a relatively large effect on the fitted model. We shall look at effect of removing these outliers and refitting the model, producing dramatically different point estimates and confidence intervals. The example is from Siegel (1997, pp. 384–385). The data were originally published in the November 9, 1988 edition of *The Wall Street Journal* (p. C19). According to Siegel:

US Treasury bonds are among the least risky investments, in terms of the likelihood of your receiving the promised payments. In addition to the primary market auctions by the Treasury, there is an active secondary market in which all outstanding issues can be traded. You would expect to see an increasing relationship between the coupon of the bond, which indicates the size of its periodic payment (twice a year), and the current selling price. The ... data set of coupons and bid prices [are] for US Treasury bonds maturing between 1994 and 1998... The bid prices are listed per 'face value' of $100 to be paid at maturity. Half of the coupon rate is paid every six months. For example, the first one listed pays $3.50 (half of the 7% coupon rate) every six months until maturity, at which time it pays an additional $100.

The data are given in Table 3.4 and are plotted in Figure 3.9. They can be found on the book web site in the file bonds.txt. We wish to model the relationship

Table 3.4 Regression diagnostics for the model in Figure 3.9

Case	Coupon rate	Bid price	Leverage	Residuals	Std. Residuals
1	7.000	92.94	0.049	−3.309	−0.812
2	9.000	101.44	0.029	−0.941	−0.229
3	7.000	92.66	0.049	−3.589	−0.881
4	4.125	94.50	0.153	7.066	1.838
5	13.125	118.94	0.124	3.911	1.001
6	8.000	96.75	0.033	−2.565	−0.625
7	8.750	100.88	0.029	−0.735	−0.179
8	12.625	117.25	0.103	3.754	0.949
9	9.500	103.34	0.030	−0.575	−0.140
10	10.125	106.25	0.036	0.419	0.102
11	11.625	113.19	0.068	2.760	0.685
12	8.625	99.44	0.029	−1.792	−0.435
13	3.000	94.50	0.218	10.515	2.848
14	10.500	108.31	0.042	1.329	0.325
15	11.250	111.69	0.058	2.410	0.595
16	8.375	98.09	0.030	−2.375	−0.578
17	10.375	107.91	0.040	1.313	0.321
18	11.250	111.97	0.058	2.690	0.664
19	12.625	119.06	0.103	5.564	1.407
20	8.875	100.38	0.029	−1.618	−0.393
21	10.500	108.50	0.042	1.519	0.372
22	8.625	99.25	0.029	−1.982	−0.482
23	9.500	103.63	0.030	−0.285	−0.069
24	11.500	114.03	0.064	3.983	0.986
25	8.875	100.38	0.029	−1.618	−0.393
26	7.375	92.06	0.041	−5.339	−1.306
27	7.250	90.88	0.044	−6.136	−1.503
28	8.625	98.41	0.029	−2.822	−0.686
29	8.500	97.75	0.030	−3.098	−0.753
30	8.875	99.88	0.029	−2.118	−0.515
31	8.125	95.16	0.032	−4.539	−1.105
32	9.000	100.66	0.029	−1.721	−0.418
33	9.250	102.31	0.029	−0.838	−0.204
34	7.000	88.00	0.049	−8.249	−2.025
35	3.500	94.53	0.187	9.012	2.394

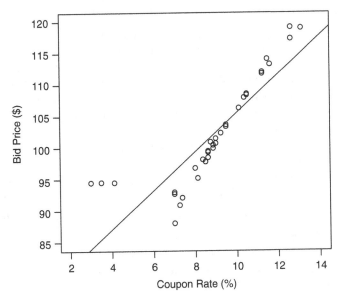

Figure 3.9 A plot of the bonds data with the least squares line included

between bid price and coupon payment. We begin by considering the simple regression model

$$Y = \beta_0 + \beta_1 x + e$$

where Y = bid price and x = coupon rate. Regression output from R is given below.

Regression output from R

```
Call:
lm(formula = BidPrice ~ CouponRate)
Coefficients:
              Estimate    Std. Error    t value    Pr(>|t|)
(Intercept)   74.7866       2.8267       26.458     <2e-16   ***
CouponRate     3.0661       0.3068        9.994    1.64e-11  ***
---
Residual standard error: 4.175 on 33 degrees of freedom
Multiple R-Squared: 0.7516, Adjusted R-squared: 0.7441
F-statistic: 99.87 on 1 and 33 DF, p-value: 1.645e-11

              2.5 %  97.5 %
(Intercept)  69.036  80.537
CouponRate    2.442   3.690
```

Note that a 95% confidence interval for the slope, β_1, is given by (2.44, 3.69). Thus, for every 1 unit increase in x (i.e., every 1% increase in Coupon Rate) then we can be 95% confident that the mean of Y (i.e., mean Bid Price) will increase by between 2.44 and 3.69 units (i.e., $2.44 and $3.69).

Looking at the fitted regression line with the data in Figure 3.9 we see that the fitted model does not describe the data well. The three points on the left of the scatter plot in Figure 3.9 obviously stand out. The least squares line is dragged away from the bulk of the points towards these three points. The data would be better described by a line that has a steeper slope and a lower intercept than the least squares line. In other words, the three points detected earlier distort the least squares line. To better understand the effects of these three points we next examine standardized residuals and leverage values.

Table 3.4 also lists the residuals, the standardized residuals and the leverage values for each of the 35 observations.

Recall that the rule for simple linear regression for classifying a point as a leverage point is $h_{ii} > \frac{4}{n}$. For the bonds data, cases 4, 5, 13 and 35 have leverage values greater than 0.11 $\left(\frac{4}{n} = \frac{4}{35} = 0.11\right)$ and thus can be classified as **leverage points**. Cases 4, 13 and 35 correspond to the three left-most points in Figure 3.9, while case 5 corresponds to the right-most point in this figure.

Recall that we classify points as **outliers** if their standardized residuals have absolute value greater than 2. Cases 13, 34 and 35 have standardized residuals with absolute value greater than 2, while case 4 has a standardized residual equal to 1.8. We next decide whether any of the leverage points are outliers, that is, whether any so-called bad leverage points exist. Cases 13 and 35 (and to a lesser extent case 4) are points of high leverage that are also outliers, i.e., **bad leverage points**.

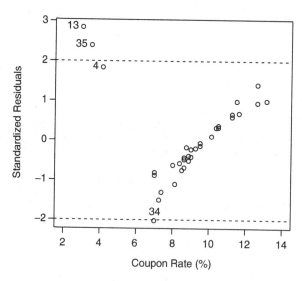

Figure 3.10 Plot of standardized residuals with some case numbers displayed

Next we look at a plot of **standardized residuals** against Coupon Rate, x, in order to assess the overall adequacy of the fitted model. Figure 3.10 provides this plot. There is a clear non-random pattern evident in this plot. The three points marked in the top left hand corner of Figure 3.10 (i.e., cases 4, 13 and 35) stand out from the other points, which seem to follow a linear pattern. These three points are not well-fitted by the model, and should be investigated to see if there was any reason why they do not follow the overall pattern set by the rest of the data.

In this example, further investigation uncovered the fact that cases 4, 13 and 35 correspond to "flower" bonds, which have definite tax advantages compared to the other bonds. Given this information, it is clear that there will be different relationship between coupon rate and bid price for "flower" bonds. It is evident from Figure 3.9 that given the low coupon rate the bid price is higher for "flower" bonds than regular bonds. Thus, a reasonable strategy is to remove the cases corresponding to "flower" bonds from the data and only consider regular bonds. In a later chapter we shall see that an alternative way to cope with points such as "flower" bonds is to add one or more dummy variables to the regression model.

Figure 3.11 shows a scatter plot of the data after the three so-called "flower bonds" have been removed. Marked on Figure 3.11 is the least squares regression line for the data without the "flower bonds." For comparison purposes the horizontal and vertical axes in Figure 3.11 are the same as those in Figure 3.9.

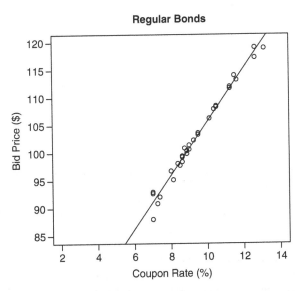

Figure 3.11 A plot of the bonds data with the "flower" bonds removed

Regression output from R

```
lm(formula = BidPrice ~ CouponRate,
subset = (1:35)[-c(4, 13, 35)])

Coefficients:
             Estimate  Std. Error  t value  Pr(>|t|)
(Intercept)  57.2932      1.0358     55.31  <2e-16 ***
CouponRate    4.8338      0.1082     44.67  <2e-16 ***
---
Signif. codes:0 '***'0.001 '**'0.01 '*' 0.05'.'0.1 ' '1

Residual standard error: 1.024 on 30 degrees of freedom
Multiple R-Squared: 0.9852, Adjusted R-squared: 0.9847
F-statistic: 1996 on 1 and 30 DF, p-value: < 2.2e-16
```

Based on all the data, we found previously that the regression coefficient of CouponRate (i.e., the slope of the fitted line in Figure 3.11) is 3.07 while a 95% confidence interval for this slope is (2.44, 3.69). When we remove the three "flower bonds" (i.e., cases 4, 13 and 35), the regression coefficient of CouponRate is 4.83, as given in the table above. You will notice that 4.83 is **NOT** even in the confidence interval based on all 35 cases. We see that a naïve analysis of this data, without removing the three "flower bonds," produces misleading results.

Thus, this example has illustrated **the importance of only basing estimates and confidence intervals on a valid model**.

To get the true picture of the relationship between bid price and coupon rate we include only bonds that have the same tax status. Based on observation of Figure 3.11, the regression model describes regular bonds relatively well. A one unit increase in the coupon rate increases the bond price by an estimated amount of $4.83. However, in Figure 3.12 there is evidence of nonconstant error variance, which is the issue we take up in the next section. It turns out that nonconstant error variance is to be expected in this situation since the maturity date of the bonds varies over a 4-year time period.

3.2.3 Recommendations for Handling Outliers and Leverage Points

We conclude this section with some general advice about how to cope with outliers and leverage points.

- Points should not be routinely deleted from an analysis just because they do not fit the model. Outliers and bad leverage points are signals, flagging potential problems with the model.
- Outliers often point out an important feature of the problem not considered before. They may point to **an alternative model** in which the points are not an outlier. In this case it is then worth considering fitting an alternative model. We shall see

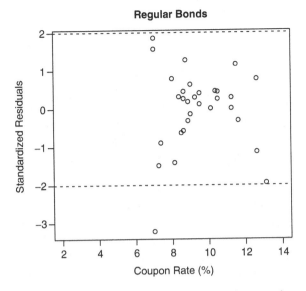

Figure 3.12 Plot of standardized residuals with the "flower" bonds removed

in Chapter 6 that including one or more dummy variables in the regression model is one way of coping with outliers that point to an important feature.

3.2.4   Assessing the Influence of Certain Cases

One or more cases can strongly control or influence the least squares fit of a regression model. For example, in the previous example on US Treasury bond prices, three cases dramatically influenced the least squares regression model. In this section we look at summary statistics that measure the influence of a single case on the least squares fit of a regression model.

We shall use the notation where subscript (i) means that the ith case has been deleted from the fit. In other words, the fit is then based on the other $n-1$ cases $(1, 2, ..., i\text{-}1, i+1, ..., n)$. Thus, $\hat{y}_{j(i)}$ denotes the jth fitted value based on the fit obtained when the ith case has been deleted from the fit.

Cook (1977) proposed a widely used measure of the influence of individual cases which in the case of simple linear regression is given by:

$$D_i = \frac{\sum_{j=1}^{n}(\hat{y}_{j(i)} - \hat{y}_j)^2}{2S^2}$$

where S^2 is the variance of the residuals. It can be shown that

$$D_i = \frac{r_i^2}{2}\frac{h_{ii}}{1-h_{ii}}$$

where r_i is the ith standardized residual and h_{ii} is the ith leverage value, both of which were defined earlier. Thus, Cook's distance can be obtained by multiplying two quantities, namely, the square of the ith standardized residual divided by two and a monotonic function that increases as the ith leverage value increases. The first quantity measures the extent to which the ith case is outlying while the second quantity measures the leverage of the ith case. Thus, a large value of D_i may be due to a large value of r_i, a large value of h_{ii} or both.

According to Weisberg (2005, p. 200), "… if the largest value of D_i is substantially less than one, deletion of a case will not change the estimate … by much".

Fox (2002, p. 198) is among many authors who recommend $\frac{4}{n-2}$ as "a rough cutoff for noteworthy values of D_i" for simple linear regression. In practice, it is important to look for gaps in the values of Cook's distance and not just whether values exceed the suggested cut-off.

Example: US Treasury bond prices (cont.)

Figure 3.13 contains a plot of Cook's distance against Coupon Rate, x. Marked on the plot as a horizontal dashed line is the cutoff value of $4/(35-2) = 0.121$. Cases 4, 35 and 13 exceed this value and as such are worthy of investigation. Note that the Cook's distance for case 13 exceeds 1, which means it deserves special attention.

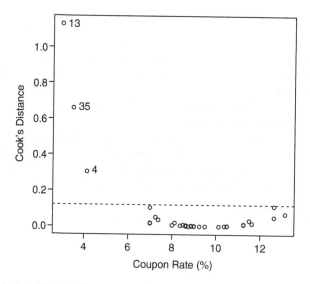

Figure 3.13 A plot of Cook's distance against Coupon Rate

3.2.5 Normality of the Errors

The assumption of normal errors is needed in **small samples** for the validity of t-distribution based hypothesis tests and confidence intervals and for **all sample sizes** for prediction intervals. This assumption is generally checked by looking at the distribution of the residuals or standardized residuals.

Recall that the ith least squares residual is given by $\hat{e}_i = y_i - \hat{y}_i$. We will show below that

$$\hat{e}_i = e_i - \sum_{j=1}^{n} h_{ij} e_j.$$

Derivation:

$$\hat{e}_i = y_i - \hat{y}_i$$

$$= y_i - h_{ii} y_i - \sum_{j \neq i} h_{ij} y_j$$

$$= y_i - \sum_{j=1}^{n} h_{ij} y_j$$

$$= \beta_0 + \beta_1 x_i + e_i - \sum_{j=1}^{n} h_{ij} \left(\beta_0 + \beta_1 x_j + e_j \right)$$

$$= \beta_0 + \beta_1 x_i + e_i - \beta_0 - \beta_1 x_i - \sum_{j=1}^{n} h_{ij} e_j$$

$$= e_i - \sum_{j=1}^{n} h_{ij} e_j$$

since

$$\sum_{j=1}^{n} h_{ij} = 1$$

and

$$\sum_{j=1}^{n} x_j h_{ij} = \sum_{j=1}^{n} \left[\frac{x_j}{n} + \frac{(x_i - \bar{x})(x_j - \bar{x}) x_j}{SXX} \right] = \bar{x} + \frac{(x_i - \bar{x}) SXX}{SXX} = x_i .$$

Thus, the ith least squares residual is equal to e_i minus a weighted sum of all of the e's. *In small to moderate samples*, the second term in the last equation can dominate the first and *the residuals can look like they come from a normal distribution even if the errors do not.* As n increases, the second term in the last equation has a much smaller variance than that of the first term and as such the first term dominates the last equation. This implies that for large samples the residuals can be used to assess normality of the errors.

In spite of what we have just discovered, a common way to assess normality of the errors is to look at what is commonly referred to as a **normal probability plot** or a **normal Q–Q plot** of the standardized residuals. A normal probability plot of the standardized residuals is obtained by plotting the ordered standardized residuals on the vertical axis against the expected order statistics from a standard normal distribution on the horizontal axes. If the resulting plot produces points "close" to a straight line then the data are said to be consistent with that from a normal distribution. On the other hand, departures from linearity provide evidence of non-normality.

Example: Timing of production runs (cont.)

Recall the example from Chapter 2 on the timing of production runs for which we fit a straight-line regression model to run time from run size. Figure 3.14 provides diagnostic plots produced by R when the command plot(m1) is used, where m1 is the result of the "lm" command. The top right plot in Figure 3.14 is a normal Q–Q plot. The bottom right plot of standardized residuals against leverage enables one to readily identify any 'bad' leverage points. We shall see shortly that the bottom left-hand plot provides diagnostic information about whether the variance of the error term appears to be constant.

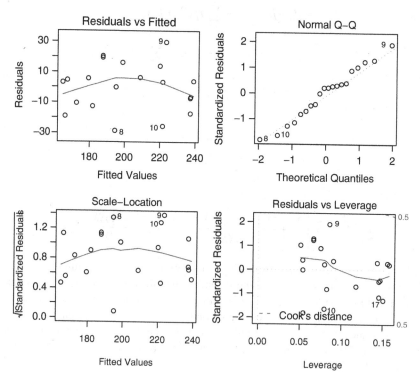

Figure 3.14 A normal Q–Q plot and other plots

3.2.6   Constant Variance

A crucial assumption in any regression analysis is that the errors have constant variance. In this section we examine methods for checking whether the assumption of
constant variance of the errors is reasonable. When the variance is found to be
nonconstant, we can consider two methods for overcoming this, namely, transformations and weighted least squares. *Ignoring nonconstant variance when it exists
invalidates all inferential tools (i.e., p-values, confidence intervals, prediction
intervals etc.).*

Example: Developing a bid on contract cleaning

This example is adapted from Foster, Stine, and Waterman (1997, p. 9). According to
the authors:

> A building maintenance company is planning to submit a bid on a contract to clean corpo-
> rate offices scattered throughout an office complex. The costs incurred by the maintenance
> company are proportional to the number of cleaning crews needed for this task. Recent data
> are available for the number of rooms that were cleaned by varying numbers of crews. For
> a sample of 53 days, records were kept of the number of crews used and the number of
> rooms that were cleaned by those crews.

The data are given in Table 3.5 and are plotted in Figure 3.15. They can be found
on the book web site in the file cleaning.txt.

We want to develop a regression equation to model the relationship between the
number of rooms cleaned and the number of crews, and predict the number of
rooms that can be cleaned by 4 crews and by 16 crews.

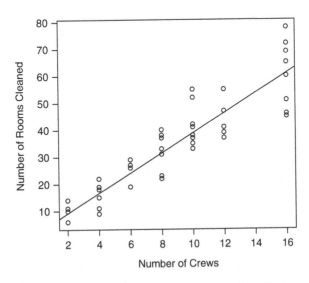

Figure 3.15 Plot of the room cleaning data with the least squares line added

Table 3.5   Data on the cleaning of rooms

Case	Number of crews	Rooms cleaned	Case	Number of crews	Rooms cleaned
1	16	51	28	4	18
2	10	37	29	16	72
3	12	37	30	8	22
4	16	46	31	10	55
5	16	45	32	16	65
6	4	11	33	6	26
7	2	6	34	10	52
8	4	19	35	12	55
9	6	29	36	8	33
10	2	14	37	10	38
11	12	47	38	8	23
12	8	37	39	8	38
13	16	60	40	2	10
14	2	6	41	16	65
15	2	11	42	8	31
16	2	10	43	8	33
17	6	19	44	12	47
18	10	33	45	10	42
19	16	46	46	16	78
20	16	69	47	2	6
21	10	41	48	2	6
22	6	19	49	8	40
23	2	6	50	12	39
24	6	27	51	4	9
25	10	35	52	4	22
26	12	55	53	12	41
27	4	15			

We begin by considering the regression model

$$Y = \beta_0 + \beta_1 x + e$$

where Y = number of rooms cleaned and x = number of cleaning crews. Regression output from R is given below including 95% prediction intervals when $x = 4$ and $x = 16$, respectively.

Regression output from R

```
Call:
lm(formula = Rooms ~ Crews)
Coefficients:
              Estimate    Std. Error    t value    Pr(>|t|)
(Intercept)    1.7847       2.0965        0.851      0.399
Crews          3.7009       0.2118       17.472     <2e-16  ***
---
```

```
Residual standard error: 7.336 on 51 degrees of freedom
Multiple R-Squared: 0.8569, Adjusted R-squared: 0.854
F-statistic: 305.3 on 1 and 51 DF, p-value: < 2.2e-16

      fit              lwr              upr
1  16.58827        1.589410        31.58713
2  60.99899       45.810253        76.18773
```

Figure 3.16 contains a plot of standardized residuals against the number of cleaning crews. It is evident from Figure 3.16 that **the variability in the standardized residuals tends to increase** with the number of crews. Thus, the assumption that the variance of the errors is constant appears to be violated in this case. If, as in Figure 3.16, the distribution of standardized residuals appears to be funnel shaped, there is evidence that the error variance is not constant.

A number of authors (e.g., Cook and Weisberg, 1999b, p. 350) recommend that an effective plot to diagnose nonconstant error variance is a plot of

$|Residuals|^{0.5}$ against x

or

$|Standardized\ Residuals|^{0.5}$ against x
The power 0.5 is used to reduce skewness in the absolute values.

Figure 3.17 contains a plot of the square root of the absolute value of the standardized residuals against x. The least squares regression line has been added to the

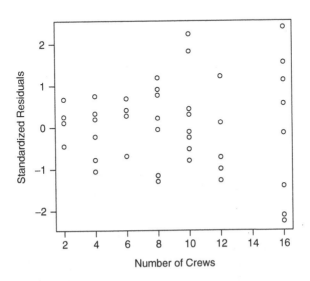

Figure 3.16 Plot of standardized residuals against x, number of cleaning crews

plot. There is clear evidence of an increasing trend in Figure 3.17, which implies that there is evidence that the variance of the errors increases with x.

Figure 3.18 contains four diagnostic plots produced by R. The bottom left-hand plot is a plot of the square root of the absolute value of the standardized residuals against fitted values. The line added to this plot is that obtained using a nonparametric smoothing method. Since the fitted values are given by $\hat{y} = \hat{\beta}_0 + \hat{\beta}_1 x$ the shape of this plot and that in Figure 3.17 are identical.

In this example, the x-variable (number of crews) is discrete with values 2, 4, 6, 8, 10, 12 and16. Notice also that there are multiple measurements of the Y-variable (number of rooms cleaned) at each value of x. In this special case, it is possible to directly calculate the standard deviation of Y at each discrete value of x. Table 3.6 gives the value of each of these standard deviations along with the number of points each is based on. It is evident from Table 3.6 that the standard deviation of the number of rooms cleaned increases as x, the number of crews increases.

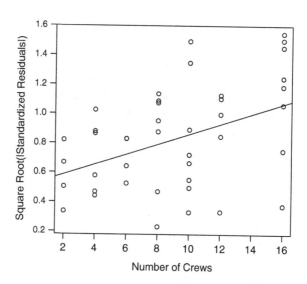

Figure 3.17 A diagnostic plot aimed at detecting nonconstant error variance

Table 3.6 The standard deviation of Y for each value of x

Crews	N	StDev(Rooms cleaned)
2	9	3.00
4	6	4.97
6	5	4.69
8	8	6.64
10	8	7.93
12	7	7.29
16	10	12.00

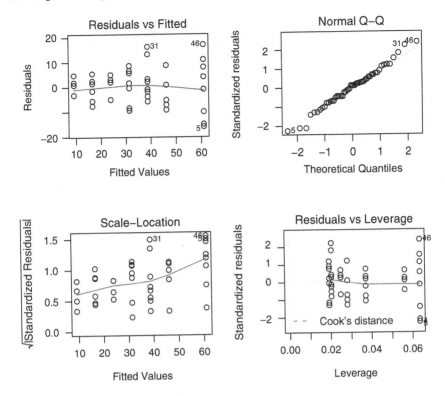

Figure 3.18 Diagnostic plots from R

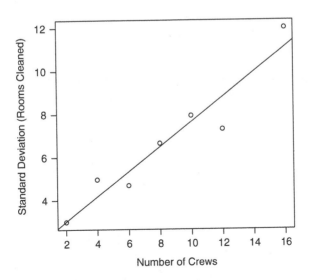

Figure 3.19 Plot of the standard deviation of Y against x

Figure 3.19 shows a plot of the standard deviation of the number of rooms cleaned against the corresponding value of x, number of crews. There is striking evidence from Figure 3.19 that the standard deviation of Y increases as x increases.

When the assumption that the variance of the errors is constant does not hold, we can use weighted least squares to account for the changing variance (with weights equal to 1/variance) or try to transform the data so that the nonconstant variance problem disappears. We begin by considering the approach based on transformations. We shall look at weighted least squares in Chapter 4.

3.3   Transformations

In this section we shall see how transformations can be used to

- Overcome problems due to nonconstant variance
- Estimate percentage effects
- Overcome problems due to nonlinearity

3.3.1   Using Transformations to Stabilize Variance

When nonconstant variance exists, it is often possible to transform one or both of the regression variables to produce a model in which the error variance is constant.

Example: Developing a bid on contract cleaning (cont.)

You will recall that in this example we decided that the assumption that the variance of the errors is constant does not hold. In particular, there is clear evidence that the variance of the errors increases as the value of the predictor variable increases. This is not surprising since the Y-variable in this case (the number of rooms cleaned) is in the form of counts.

Count data are often modelled using the Poisson distribution. Suppose that Y follows a Poisson distribution with mean λ, then it is well-known that the variance of Y is also equal to λ. In this case, the appropriate transformation of Y for stabilizing variance is square root.

Justification:

Consider the following Taylor series expansion

$$f(Y) = f\left(\mathrm{E}(Y)\right) + f'\left(\mathrm{E}(Y)\right)\left(Y - \mathrm{E}(Y)\right) + \dots$$

The well-known delta rule for calculating first-order variance terms is obtained by considering just the terms given in this last expansion. In particular, taking the variance of each side of this equation gives

$$\mathrm{Var}\left(f(Y)\right) \approx \left[f'\left(\mathrm{E}(Y)\right)\right]^2 \mathrm{Var}(Y).$$

Suppose that $f(Y) = Y^{0.5}$ and $\text{Var}(Y) = \lambda = E(Y)$ then

$$\text{Var}\left(Y^{0.5}\right) \approx \left[0.5\left(E(Y)\right)^{-0.5}\right]^2 \text{Var}\left(Y\right) = \left[0.5\lambda^{-0.5}\right]^2 \lambda = \text{constant}\cdot$$

In this case, since the **data** on each axis are **in the form of counts**, we shall try the **square root transformation** of both the predictor variable and the response variable. *When both Y and X are measured in the same units then it is often natural to consider the same transformation for both X and Y.*

Recall that we want to develop a regression equation to model the relationship between the number of rooms cleaned and the number of crews, and predict the number of rooms that can be cleaned by 4 crews and by 16 crews. Given below is the R output from fitting the model

$$Y = \beta_0 + \beta_1 x + e$$

Y = the **square root** of the number of rooms cleaned and
x = the **square root** of the number of cleaning crews

```
Call:
lm(formula = sqrtrooms ~ sqrtcrews)
Coefficients:
                Estimate    Std. Error    t value    Pr(>|t|)
(Intercept)     0.2001      0.2758        0.726      0.471
sqrtcrews       1.9016      0.0936        20.316     <2e-16 ***
---
Residual standard error: 0.594 on 51 degrees of freedom
Multiple R-Squared: 0.89, Adjusted R-squared: 0.8879
F-statistic: 412.7 on 1 and 51 DF, p-value: < 2.2e-16

         fit           lwr           upr
1  4.003286      2.789926      5.216646
2  7.806449      6.582320      9.030578
```

Figure 3.20 contains a scatter plot of the square root of the number of rooms cleaned against the square root of the number of crews and a plot of standardized residuals against the square root of the number of cleaning crews. In Figure 3.20 the variability in the standardized residuals remains relatively constant as the square root of the number of crews increase. The standardized residuals do not have a funnel shape, as they did for the untransformed data.

Figure 3.21 gives the diagnostics plots produced by R, associated with the "lm" command. The bottom left-hand plot further demonstrates the benefit of the square root transformation in terms of stabilizing the error term. Thus, taking the square root of both the x and the y variables has stabilized the variance of the random errors and hence produced a valid model.

Finally, given that we have a valid regression model we are now in the position to be able to predict the number of rooms that can be cleaned by 4 crews and 16 crews (as we set out to do). These results based on the regression model for the

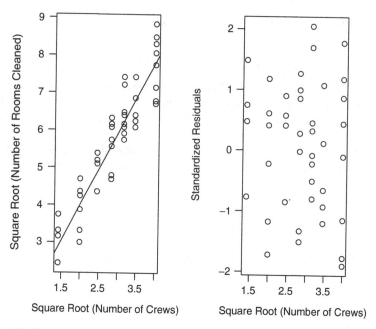

Figure 3.20 Plots of the transformed data and the resulting standardized residuals

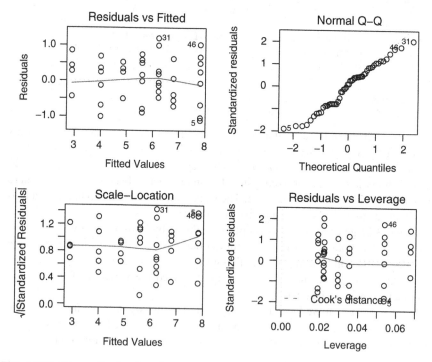

Figure 3.21 Diagnostic plots from R

Table 3.7 Predictions and 95% prediction intervals for the number of rooms

x, Crews	Prediction	Lower limit	Upper limit
4 (transformed data)	$16 = (4.003^2)$	$8 = (2.790^2)$	$27 = (5.217^2)$
4 (raw data)	17	2	32
16 (transformed data)	$61 = (7.806^2)$	$43 = (6.582^2)$	$82 = (9.031^2)$
16 (raw data)	61	46	76

square root transformed data are given in Table 3.7. For comparison purposes the results for the raw data are also presented. Notice that the prediction interval based on the transformed data is narrower than that based on the untransformed data when the number of crews is 4 and wider when the number of crews is 16. This is to be expected in this situation since on the original scale the data have variance which increases as the x-variable increases meaning that realistic prediction intervals will get wider as the x-variable increases. *In summary, ignoring nonconstant variance in the raw data from this example led to invalid prediction intervals.*

3.3.2   Using Logarithms to Estimate Percentage Effects

In this section we illustrate how logarithms can be used to estimate percentage effects. In particular, we shall consider the regression model

$$\log(Y) = \beta_0 + \beta_1 \log(x) + e$$

where here, and throughout the book, log refers to log to the base e or natural logarithms. In this situation the slope

$$\beta_1 = \frac{\Delta \log(Y)}{\Delta \log(x)}$$

$$= \frac{\log(Y_2) - \log(Y_1)}{\log(x_2) - \log(x_1)}$$

$$= \frac{\log(Y_2/Y_1)}{\log(x_2/x_1)}$$

$$\cong \frac{Y_2/Y_1 - 1}{x_2/x_1 - 1} \text{ (using } \log(1+z) \cong z \text{ and assuming } \beta_1 \text{ is small)}$$

$$= \frac{100(Y_2/Y_1 - 1)}{100(x_2/x_1 - 1)}$$

$$= \frac{\%\Delta Y}{\%\Delta x}.$$

That is, the slope approximately equals the ratio of the percentage changes in Y and x. So that, for small β_1

$$\%\Delta Y \approx \beta_1 \times \%\Delta x.$$

Thus for every 1% increase in x, the model predicts a $\beta_1\%$ increase in Y (provided β_1 is small).

An example using logs to estimate the price elasticity of a product

We want to understand the effect of price on sales and in particular to develop a technique to estimate the percentage effect on sales of a 1% increase in price. This effect is commonly referred to as price elasticity.

This example from Carlson (1997) is based on a case involving real supermarket sales data from Nielson SCANTRACK. According to Carlson (1997, p. 37):

Susan Thurston, a product manager for Consolidated Foods Inc., is responsible for developing and evaluating promotional campaigns to increase sales for her canned food product. She has just been appointed product manager for Brand 1, which competes in retail food markets with three other major brands (Brands 2, 3 & 4). Brands 1 and 2 are the dominant brands in the sense that they have a much larger share of the market than Brands 3 and 4. The product is well established in a stable market and is generally viewed by consumers as a food commodity. Successful product performance requires strategies that encourage customers to move to Susan's brand from the various competing brands. She can encourage this movement by various kinds of reduced price specials, in-store displays, and newspaper advertising. Susan's competitors have at their disposal the same strategic tools for increasing their sales.

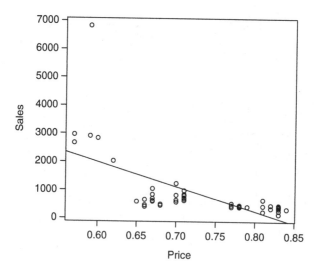

Figure 3.22 A scatter plot of sales against price

We shall start by examining the subset of the available information. The data we first consider are weekly sales (in thousands of units) of Brand 1 at a major US supermarket chain over a year along with the price each week. The data are plotted in Figure 3.22 and can be found on the book web site in the file con-food1.txt.

Notice that the distribution of each variable in Figure 3.22 appears to be skewed. A large outlier is clearly apparent in Figure 3.22. In addition, it is clear that a straight line does not adequately model the relationship between price and sales.

When studying the relationship between price and quantity in economics, it is common practice to take the logarithms of both price and quantity since interest lie in predicting the effect of a 1% increase in price on quantity sold. In this case the model fitted is

$$\log(Q) = \beta_0 + \beta_1 \log(P) + e$$

where P = price and Q = quantity (i.e., sales in thousands of units). The log-transformed data are plotted in Figure 3.23.

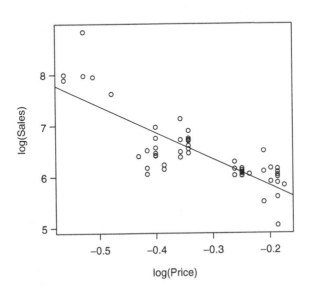

Figure 3.23   A scatter plot of log(sales) against log(price)

The regression output from R is given below:

Regression output from R

```
Call:
lm(formula = log(Sales) ~ log(Price))
Coefficients:
              Estimate  Std. Error  t value   Pr(>|t|)
(Intercept)     4.8029      0.1744    27.53   <2e-16    ***
log(Price)     -5.1477      0.5098   -10.10   1.16e-13  ***
---
Residual standard error: 0.4013 on 50 degrees of freedom
Multiple R-Squared: 0.671, Adjusted R-squared: 0.6644
F-statistic: 102 on 1 and 50 DF, p-value: 1.159e-13
```

The slope $\hat{\beta}_1$ in the fitted model

$$\log(Q) = \hat{\beta}_0 + \hat{\beta}_1 \log(P)$$

approximately equals the ratio of the percentage changes in Q & P. In this equation, the slope $\hat{\beta}_1$ is an estimate of the price elasticity of demand (i.e., the percentage change in quantity demanded in response to the percentage change in price). In this case, $\hat{\beta}_1 = -5.1$ and so we estimate that for every 1% increase in price there will be approximately a 5.1% reduction in demand. Since the magnitude of $\hat{\beta}$ is greater than 1, the quantity demanded is said to be "elastic," meaning that a price change will cause an even larger change in quantity demanded. Now, revenue is price times quantity. Price has risen, but proportionately, quantity has fallen more. Thus revenue has fallen. Increasing the price of the food product will result in a revenue decrease, since the revenue lost from the resulting decrease in quantity sold is more than the revenue gained from the price increase.

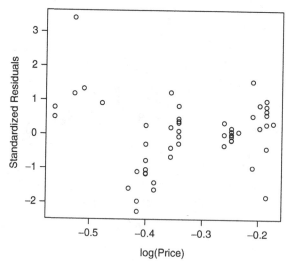

Figure 3.24 A plot of standardized residuals against log(price)

As is our practice, we look at a plot of standardized residuals against the predictor variable given below in Figure 3.24. There is a nonrandom pattern (somewhat similar to a roller coaster) evident in the plot of standardized residuals in Figure 3.24. Thus, we should not be satisfied with the current fitted model. When one considers that other variables (such as advertising) affect sales, it is clear that the current model can be improved by the inclusion of these variables. (In Chapter 9, we will look at models involving more than a single predictor variable to see if we can arrive at a more satisfactory model.)

3.3.3 Using Transformations to Overcome Problems due to Nonlinearity

In this section we consider the following two general methods for transforming the response variable Y and/or the predictor variable X to overcome problems due to nonlinearity:

- Inverse response plots
- Box-Cox procedure

There are three situations we need to consider (a) only the response variable needs to be transformed; (b) only the predictor variable needs to be transformed; and (c) both the response and predictor variables need to be transformed. We begin by looking at the first situation.

Transforming only the response variable Y using inverse regression

Suppose that the true regression model between Y and X is given by

$$Y = g(\beta_0 + \beta_1 x + e)$$

where g is a function which is generally unknown. The previous model can be turned into a simple linear regression model by transforming Y by g^{-1}, the inverse of g, since,

$$g^{-1}(Y) = \beta_0 + \beta_1 x + e.$$

For example suppose that

$$Y = (\beta_0 + \beta_1 x + e)^3$$

then,

$$g(Y) = Y^3 \text{ and so } g^{-1}(Y) = Y^{\frac{1}{3}}.$$

Next, suppose that

$$Y = \exp(\beta_0 + \beta_1 x + e)$$

then,

$$g(Y) = \exp(Y) \text{ and so } g^{-1}(Y) = \log(Y)$$

We next look at methods for estimating g^{-1}.

Generated Example

The 250 data points in this example were generated from the model

$$Y = (\beta_0 + \beta_1 x + e)^3$$

with x and e independently normally distributed. (This situation is very similar to that in Cook and Weisberg (1999b, p. 318).) Our aim is to estimate the function $g^{-1}(Y) = Y^{\frac{1}{3}}$ that transforms Y so that the resulting model is a simple linear regression model. The generated data we shall consider can be found on the course web site in the file responsetransformation.txt. A plot of Y and x is given in Figure 3.25. We begin by fitting the straight line regression model

$$Y = \beta_0 + \beta_1 x + e \tag{3.2}$$

Figure 3.26 contains a scatter plot of the standardized residuals against x and a plot of the square root of the absolute value of the standardized residuals against x. Both plots in Figure 3.26 produce striking patterns. The left-most plot shows that the standardized residuals are strongly related to x in a nonlinear manner. This indicates that model (3.2) is not an appropriate model for the data. The right-most plot shows that the variance of the standardized residuals is not constant, instead, it changes dramatically with x. This indicates that the variance of the errors in model

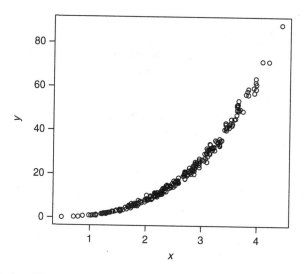

Figure 3.25 A plot of Y vs x for the generated data (responsetransformation.txt)

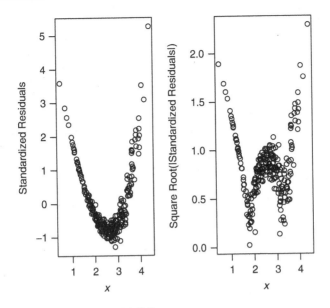

Figure 3.26 Diagnostic plots for model (3.2)

(3.2) is not constant. In view of both of these findings, it is natural to consider transforming Y or x or both.

We begin by looking at the shape of the distributions of Y and x. Figure 3.27 contains box plots, normal Q–Q plots and Gaussian kernel density estimates (based on the Sheather-Jones bandwidth, Sheather and Jones, 1991). A kernel density estimate can be thought as a smoothed histogram. For an introduction to and overview of kernel density estimation see Appendix A.1

It is evident from Figure 3.27 that the distribution of Y is skewed. On the other hand the distribution of x is consistent with a normal distribution. We shall consider transforming only Y since transforming a symmetrically distributed variable will produce a variable with a skewed distribution.

Inverse response plots

Suppose that the true regression model between Y and X is given by

$$Y = g(\beta_0 + \beta_1 x + e)$$

where g is an unknown function. Recall that the previous model can be turned into a simple linear regression model by transforming Y by g^{-1}, the inverse of g, since,

$$g^{-1}(Y) = \beta_0 + \beta_1 x + e.$$

Thus, if we knew β_0 and β_1 we could discover the shape of g^{-1} by plotting Y on the horizontal axis and $\beta_0 + \beta_1 x$ on the vertical axis.

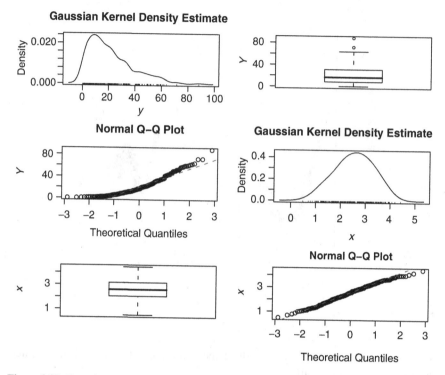

Figure 3.27 Box plots, normal Q–Q plots and kernel density estimates of Y and x

Based on the results of Li and Duan (1989), Cook and Weisberg (1994) showed that if x has an elliptically symmetric distribution then g^{-1} can be estimated from the scatter plot of Y (on the horizontal axis) and the fitted values from model (3.2), i.e., $\hat{y} = \hat{\beta}_0 + \hat{\beta}_1 x$ (on the vertical axis). Such a plot is commonly referred to as an **inverse response plot** (since the usual axis for Y is the vertical axis). A truly remarkable aspect of the inverse response plot is that the fitted values used are from a regression model, which is itself only valid when g is the identity function.

It can be shown that the assumption that the univariate variable x has a normal distribution is much stronger than the assumption that the distribution of x is elliptically symmetric.

Generated example (cont.)

Figure 3.28 contains an inverse response plot for the data in the generated example. (Since x is normally distributed in this example, we should be able to estimate g^{-1} from the inverse response plot.) Marked on the plot are three so-called power curves

$$\hat{y} = y^\lambda \text{ for } \lambda = 0, 0.33, 1$$

where $\hat{y}$ are the fitted values from model (3.2) and, as we shall see below, $\lambda = 0$ corresponds to natural logarithms.

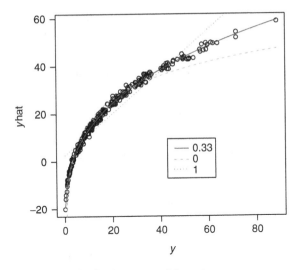

Figure 3.28 Inverse response plot for the generated data set

It is evident that among the three curves in Figure 3.28, the power curve

$$\hat{y} = \left(\hat{\beta}_0 + \hat{\beta}_1 x\right) = y^{0.33}$$

provides the closest fit to the data. This is to be expected since the data were generated from the model

$$Y = g(\beta_0 + \beta_1 x + e) = (\beta_0 + \beta_1 x + e)^3$$

so that

$$g^{-1}(Y) = Y^{\frac{1}{3}} = \beta_0 + \beta_1 x + e.$$

Choosing a power transformation

To estimate g^{-1} in general we consider the following family of *scaled power transformations*, defined for strictly positive Y by

$$\Psi_S(Y, \lambda) = \begin{cases} (Y^\lambda - 1)/\lambda & \text{if } \lambda \neq 0 \\ \log(Y) & \text{if } \lambda = 0 \end{cases}$$

Scaled power transformations have the following three properties:

1. $\Psi_S(Y, \lambda)$ is a continuous function of λ
2. The logarithmic transformation is a member of this family, since

$$\lim_{\lambda \to 0} \Psi_S(Y, \lambda) = \lim_{\lambda \to 0} \frac{\left(Y^\lambda - 1\right)}{\lambda} = \log(Y)$$

3. Scaled transformations preserve the direction of the association between Y and X in the sense that if Y and X are positively (negatively) related than $\Psi_s(Y, \lambda)$ and X are positively (negatively) related for all values of λ

Thus, to estimate g^{-1}, we consider fitting models of the form

$$E(\hat{y} \mid Y = y) = \alpha_0 + \alpha_1 \Psi_s(y, \lambda) \qquad (3.3)$$

For a given value of λ, model (3.3) is just a simple linear regression model with the predictor variable given by $\Psi_s(y, \lambda)$ and the response variable given by $\hat{y}$. Weisberg (2005, p. 151) recommends fitting (3.3) by least squares for a range of values of λ and choosing the estimated optimal value of λ (which we will denote by $\hat{\lambda}$) as the value of λ that minimizes the residual sum of squares (which we will denote by RSS(λ)). Weisberg (2005, p. 151) claims that "selecting λ to minimize RSS(λ) from

$$\lambda \in \left\{ -1, -\tfrac{1}{2}, -\tfrac{1}{3}, 0, \tfrac{1}{3}, \tfrac{1}{2}, 1 \right\}$$

is usually adequate". We shall see that it is sometimes beneficial to also add the values $-\tfrac{1}{4}$ and $\tfrac{1}{4}$ to the set of λ values under consideration.

Generated example (cont.)

Figure 3.29 provides a plot of RSS(λ) for model (3.3) against λ across the set of λ values given above for the data in the generated example.

It is evident from Figure 3.29 that the value of λ that minimizes RSS(λ) in this case falls somewhere between 0 and 0.5. Choosing a smaller range of λ values, one

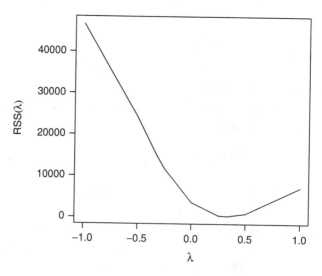

Figure 3.29 A plot of RSS(λ) against λ for the generated data set

can find that the optimal value of λ is given by $\hat{\lambda} = 0.332$. Thus, in this case, the estimated power transformation

$$g^{-1}(Y) = Y^{0.332}$$

is very close to the value used to generate the data

$$g^{-1}(Y) = Y^{\frac{1}{3}}.$$

Alternatively, one can estimate $\alpha_0, \alpha_1, \lambda$ in (3.3) simultaneously using nonlinear least squares and hence obtain the optimal value of λ. The function `inverse.response.plot` in the alr3 R-library of Weisberg (2005) performs this task. Figure 3.28, which provides an example of its use, shows three fits of model (3.3) corresponding to $\lambda = 0, 0.33, 1$ with $\lambda = 0.33$ providing the optimal fit.

Later in this section, we shall apply the inverse response plot technique on a real data set. In the meantime, we next consider an alternative method for transforming Y, namely, the Box-Cox procedure.

Transforming only the response variable Y using the Box-Cox method

In one of the most highly cited papers in statistics[3], Box and Cox (1964) provide a general method for transforming a strictly positive response variable Y. We shall see that this method and its natural extensions can also be used to transform one or more strictly positive predictor variables. The Box-Cox procedure aims to find a transformation that makes the transformed variable close to normally distributed. Before we look into the details of this procedure, we shall look at the properties of least squares and likelihood methods when both X and Y are normally distributed.

Simple Linear Regression when both X and Y are normal

Suppose that y_i and x_i are the observed values of normal random variables with means μ_Y and μ_X, respectively, variances σ_Y^2 and σ_X^2, respectively, and correlation, ρ_{XY}. Then it can be shown (e.g., Casella and Berger, 2002, p. 550) that

$$y_i \mid x_i \sim N\left(\mu_Y - \rho_{XY} \frac{\sigma_Y}{\sigma_X} \mu_X + \rho_{XY} \frac{\sigma_Y}{\sigma_X} x_i, \sigma_Y^2 \left(1 - \rho_{XY}^2\right) \right).$$

This can be rewritten as $y_i \mid x_i \sim N\left(\beta_0 + \beta_1 x_i, \sigma^2\right)$ where

$$\beta_0 = \mu_Y - \beta_1 \mu_X, \beta_1 = \rho_{XY} \frac{\sigma_Y}{\sigma_X}, \sigma^2 = \mathrm{Var}(Y \mid X) = \sigma_Y^2 \left(1 - \rho_{XY}^2\right).$$

In other words, if y_i and x_i are the observed values of normal random variables then the regression of Y on X is

---

[3] According to the Web of Science, the Box and Cox (1964) paper has been cited more than 3000 times as of January 25, 2007.

$$E(Y \mid X = x_i) = \beta_0 + \beta_1 x_i,$$

a linear function of x_i. The Box-Cox transformation method, and its natural extensions, tries to find a transformation that makes Y and X as close to normal distributed as possible, since then the regression of Y on X should be close to linear. The
Box-Cox method is based on likelihood approach and so we briefly review these
principles next.

Maximum Likelihood applied to Simple Linear Regression when both X and Y are normal

As above, suppose that y_i and x_i are the observed values of normal random varia-
bles. Then, $y_i \mid x_i \sim N\left(\beta_0 + \beta_1 x_i, \sigma^2\right)$. Thus, the conditional density of y_i given x_i is
given by

$$f(y_i \mid x_i) = \frac{1}{\sigma\sqrt{2\pi}} \exp\left(-\frac{(y_i - \beta_0 - \beta_1 x_i)^2}{2\sigma^2}\right).$$

Assuming the n observations are independent then given Y the likelihood function
is the function of the unknown parameters β_0, β_1, σ^2 given by

$$L(\beta_0, \beta_1, \sigma^2 \mid Y) = \prod_{i=1}^{n} f(y_i \mid x_i) = \prod_{i=1}^{n} \frac{1}{\sigma\sqrt{2\pi}} \exp\left(-\frac{(y_i - \beta_0 - \beta_1 x_i)^2}{2\sigma^2}\right)$$

$$= \left(\frac{1}{\sigma\sqrt{2\pi}}\right)^n \exp\left(-\frac{1}{2\sigma^2}\sum_{i=1}^{n}(y_i - \beta_0 - \beta_1 x_i)^2\right)$$

The log-likelihood function is given by

$$\log\left(L(\beta_0, \beta_1, \sigma^2 \mid Y)\right)$$

$$= -\frac{n}{2}\log(2\pi) - \frac{n}{2}\log(\sigma^2) - \frac{1}{2\sigma^2}\sum_{i=1}^{n}(y_i - \beta_0 - \beta_1 x_i)^2 \qquad (3.4)$$

We see that β_0 and β_1 are the only unknowns in the third term of this last equation.
Thus, the maximum likelihood estimates of β_0 and β_1 can be obtained by minimizing
this third term, that is, by minimizing the residual sum of squares. Thus, the maximum likelihood estimates of β_0 and β_1 are the same as the least squares estimates.
With β_0 and β_1 equal to their least squares estimates, equation (3.4) becomes

$$\log\left(L(\hat{\beta}_0, \hat{\beta}_1, \sigma^2 \mid Y)\right) = -\frac{n}{2}\log(2\pi) - \frac{n}{2}\log(\sigma^2) - \frac{1}{2\sigma^2}\mathrm{RSS}$$

where in this context RSS is the residual sum of squares corresponding to the least
squares estimates, i.e.,

$$\text{RSS} = \sum_{i=1}^{n} \left(y_i - \hat{\beta}_0 - \hat{\beta}_1 x_i \right)^2$$

Differentiating the last log likelihood equation with respect to σ^2 and setting the result to zero gives the maximum likelihood estimate of σ^2 as

$$\hat{\sigma}^2_{\text{MLE}} = \frac{1}{n} \text{RSS}.$$

Notice that this estimate differs slightly from our usual estimate of σ^2, namely,

$$S^2 = \frac{1}{(n-2)} \text{RSS}.$$

The Box-Cox method for transforming only the response variable

Box and Cox (1964) considered the modified family of power transformations

$$\Psi_M(Y, \lambda) = \Psi_S(Y, \lambda) \times \text{gm}(Y)^{1-\lambda} = \begin{cases} \text{gm}(Y)^{1-\lambda}(Y^\lambda - 1)/\lambda & \text{if } \lambda \neq 0 \\ \text{gm}(Y)\log(Y) & \text{if } \lambda = 0 \end{cases}$$

where $\text{gm}(Y) = \prod_{i=1}^{n} Y_i^{1/n} = \exp\left(\frac{1}{n} \sum_{i=1}^{n} \log(Y_i)\right)$ is the geometric mean of Y. The

Box-Cox method is based on the notion that for some value of λ the transformed version of Y, namely, $\Psi_M(Y, \lambda)$ is normally distributed. As pointed out by Weisberg (2005, p. 153), multiplication of $\Psi_S(Y, \lambda)$ by $\text{gm}(Y)^{1-\lambda}$ ensures that the units of $\Psi_M(Y, \lambda)$ are the same for all values of λ. The log-likelihood function for $\beta_0, \beta_1, \sigma^2, \lambda \mid \Psi_M(Y, \lambda)$ is given by (3.4) with y_i replaced by $\Psi_M(Y, \lambda)$, i. e., by

$$\log\left(L(\beta_0, \beta_1, \sigma^2, \lambda \mid \Psi_M(Y, \lambda))\right)$$

$$= -\frac{n}{2}\log(2\pi) - \frac{n}{2}\log(\sigma^2) - \frac{1}{2\sigma^2} \sum_{i=1}^{n} \left(\Psi_M(y_i, \lambda) - \beta_0 - \beta_1 x_i\right)^2$$

since it can be shown that the Jacobian of the transformation $\Psi_M(Y, \lambda)$ equals 1 for each value of λ. For a given value of λ this last equation is the same as (3.4) with y_i replaced by $\Psi_M(y_i, \lambda)$. Thus, for a fixed value of λ, the maximum likelihood estimates of β_0, β_1 and σ^2 are the same as above with y_i replaced by $\Psi_M(y_i, \lambda)$. With β_0, β_1 and σ^2 replaced by these estimates the log-likelihood is given by

$$\log\left(L(\hat{\beta}_0, \hat{\beta}_1, \hat{\sigma}^2, \lambda \mid \psi_M(Y, \lambda))\right)$$

$$= -\frac{n}{2}\log(2\pi) - \frac{n}{2}\log(\text{RSS}(\lambda)/n) - \frac{1}{2\,\text{RSS}(\lambda)/n}\text{RSS}(\lambda)$$

$$= -\frac{n}{2}\log(2\pi) - \frac{n}{2}\log(\text{RSS}(\lambda)/n) - \frac{n}{2}$$

where RSS (λ) is the residual sum of squares with y_i replaced by $\Psi_M(y_i, \lambda)$, i.e.,

$$RSS(\lambda) = \sum_{i=1}^{n} \left(\Psi_M(y_i, \lambda) - \hat{\beta}_0 - \hat{\beta}_1 x_i \right)^2$$

Since only the second term in the last equation involves the data, maximizing $\log\left(L(\hat{\beta}_0, \hat{\beta}_1, \hat{\sigma}^2, \lambda \,|\, \Psi_M(Y, \lambda))\right)$ with respect to λ is equivalent to minimizing RSS (λ) with respect to λ. Likelihood methods can be used to find a confidence interval for λ.

Generated example (cont.)

Figure 3.30 provides plots of the log-likelihood against λ for the data in the generated example. The value of λ that maximizes the log-likelihood and 95% confidence limits for λ are marked on each plot. The values on the horizontal axis of the right hand plot are restricted so that it is easier to read of the value of λ that maximizes the log-likelihood, namely, 0.333.

In summary, we have found that

- Inverse response/fitted value plot estimated λ to be $\hat{\lambda} = 0.332$
- Box-Cox procedure estimated λ to be $\hat{\lambda} = 0.333$

Given below is the output from R from fitting a linear model to ty where $ty = Y^{\frac{1}{3}}$.

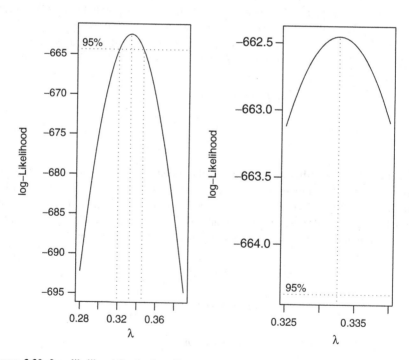

Figure 3.30 Log-likelihood for the Box-Cox transformation method

Regression output from R

```
Call:
lm(formula = ty ~ x)

Coefficients:
             Estimate   Std. Error   t value   Pr(>|t|)
(Intercept)  0.008947   0.011152      0.802     0.423
x            0.996451   0.004186    238.058    <2e-16 ***
---
Residual standard error: 0.05168 on 248 degrees of freedom
Multiple R-Squared: 0.9956, Adjusted R-squared:  0.9956
F-statistic: 5.667e+04 on 1 and 248 DF, p-value: < 2.2e-16
```

Finally, Figure 3.31 contains a box plot, a normal Q–Q plot and a Gaussian kernel density estimate of the transformed version of Y, namely, $Y^{1/3}$. It is evident that the transformed data are consistent with that from a normal distribution. Also given below is a scatter plot of $Y^{1/3}$ against x, which shows striking evidence of a linear relationship.

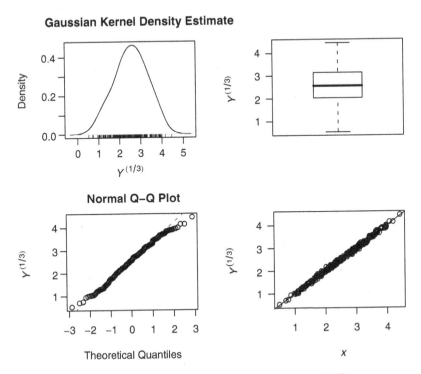

Figure 3.31 Box plots, normal Q–Q plots and kernel density estimates of $Y^{1/3}$

Transforming only the predictor variable X

We again consider the following family of *scaled power transformations*, this time defined for strictly positive X by

$$\Psi_S(X, \lambda) = \begin{cases} (X^\lambda - 1)/\lambda & \text{if } \lambda \neq 0 \\ \log(X) & \text{if } \lambda = 0 \end{cases}$$

Since we are transforming only X, we consider fitting models of the form

$$E(Y \mid X = x) = \alpha_0 + \alpha_1 \Psi_S(x, \lambda) \tag{3.5}$$

For a given value of λ, model (3.5) is just a simple linear regression model with the predictor variable given by $\Psi_S(x, \lambda)$ and the response variable given by y. As before we fit (3.5) by least squares for a range of values of λ and choosing the estimated optimal value of λ (which we will denote by $\hat{\lambda}$) as the value of λ that minimizes the residual sum of squares (which we will denote by RSS(λ)).

Alternatively, we could use a version of the Box-Cox transformation method that aims to make the distribution of the transformed version of X as normal as possible. Note that there is no regression model in this case and the Box-Cox method is modified to apply directly to X.

Important caution re-transformations

Transformations do not perform well in every situation. In particular, it may not be possible to develop a regression model for Y based on a single predictor X, no matter what transformations are considered for Y and/or X. This can occur for example when important predictors which interact with other are not included in the model. Secondly, the result of the Box-Cox transformation method may be a transformed variable that is not very close to normally distributed.

Transforming both the response and the predictor variables

When both X and Y are highly skewed and transformations of both variables are desirable the following two alternative approaches are suggested by Cook and Weisberg (1999b, p. 329):

Approach 1:

1. Transform X so that the distribution of the transformed version of X, $\Psi_S(x, \lambda_x)$ is as normal as possible. The univariate version of the Box-Cox transformation procedure is one way to do this.
2. Having transformed X to $\Psi_S(x, \lambda_x)$, consider a simple linear regression model of the form $Y = g(\beta_0 + \beta_1 \Psi_S(x, \lambda_x) + e)$. Then use an inverse response plot to decide on the transformation, g^{-1} for Y.

Approach 2:

Transform X and Y simultaneously to joint normality using the multivariate generalization of the Box-Cox method. We discuss this approach next.

Multivariate generalization of the Box-Cox transformation method

Velilla (1993) proposed the following extension of the Box-Cox method to transforming two or more variables towards joint normality. Here we consider just two variables (X and Y). Define the modified family of power transformations

$$(\Psi_M(Y,\lambda_Y), \Psi_M(X,\lambda_X))$$

where

$$\Psi_M(Y,\lambda_Y) = \Psi_S(Y,\lambda_Y) \times \mathrm{gm}(Y)^{1-\lambda_Y} = \begin{cases} \mathrm{gm}(Y)^{1-\lambda_Y}(Y^{\lambda_Y}-1)/\lambda_Y & \text{if } \lambda_Y \neq 0 \\ \mathrm{gm}(Y)\log(Y) & \text{if } \lambda_Y = 0. \end{cases}$$

Then we proceed by considering the log-likelihood function of $(\Psi_M(Y,\lambda_Y), \Psi_M(X,\lambda_X))$ and choosing λ_Y, λ_X to maximize it. The solution turns out to be obtained by minimizing the determinant of an estimated variance-covariance matrix (see Weisberg 2005, pp. 290–291 for details).

Example: Government salary data

This example is taken from Weisberg (2005, pp. 163–164). The data concern the maximum salary for 495 nonunionized job classes in a midwestern government unit in 1986. The data can be found in the R-package, alr3 in the file salarygov.txt. At present we shall focus on developing a regression model to predict

   MaxSalary = maximum salary (in \$) for employees in this job class
using just one predictor variable, namely,
   Score = score for job class based on difficulty, skill level, training requirements
   and level of responsibility as determined by a consultant to the government unit.

   We begin by considering a straight line model for the untransformed data, i.e., MaxSalary $= \beta_0 + \beta_1 \text{Score} + e$. Figure 3.32 contains a plot of the data, a plot of the standardized residuals against Score, and a plot of the square root of the absolute value of the standardized residuals against Score. There is clear evidence of non-linearity and nonconstant variance in these plots. Thus we consider transforming one or both of the variables.

   We next look at the shape of the distributions of MaxSalary and Score. Figure 3.33 contains box plots, normal Q–Q plots and Gaussian kernel density estimates (based on the Sheather-Jones bandwidth).

   It is evident from Figure 3.33 that the distribution of both MaxSalary and Score are skewed. We shall therefore consider transforming both variables.

Approach 2: Transforming both variables simultaneously

Given below is the output from R using the `bctrans` command from alr3 R-library.

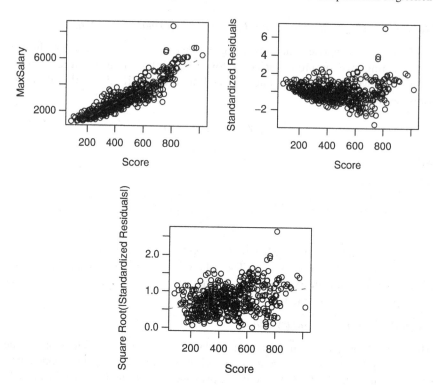

Figure 3.32 Plots associated with a straight line model to the untransformed data

Output from R

```
box.cox Transformations to Multinormality
              Est.Power    Std.Err.    Wald(Power=0)    Wald(Power=1)
MaxSalary      -0.0973      0.0770         -1.2627         -14.2428
Score           0.5974      0.0691          8.6405          -5.8240
                                       LRT df p.value
LR test, all lambda equal 0 125.0901   2       0
LR test, all lambda equal 1 211.0704   2       0
```

Using the Box-Cox method to transform the two variables simultaneously toward bivariate normality results in values of λ_Y, λ_X close to 0 and 0.5, respectively. Thus, we shall consider the following two variables, log(MaxSalary) and $\sqrt{\text{Score}}$.

Figure 3.34 shows a plot of log(MaxSalary) and $\sqrt{\text{Score}}$ with the least squares line of best fit added. It is evident from this figure that the relationship between the transformed variables is more linear than that between the untransformed variables.

Figure 3.35 contains box plots, normal Q–Q plots and Gaussian kernel density estimates (based on the Sheather-Jones bandwidth) of the transformed data. It is evident that the log and square root transformations have dramatically reduced skewness and produced variables which are more consistent with normally distributed data.

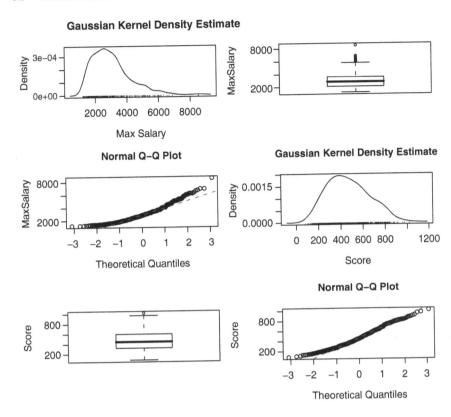

Figure 3.33 Plots of the untransformed data

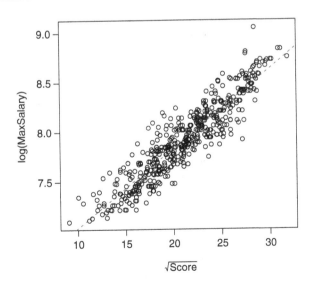

Figure 3.34 Plot of log(MaxSalary) and $\sqrt{\text{Score}}$ with the least squares line added

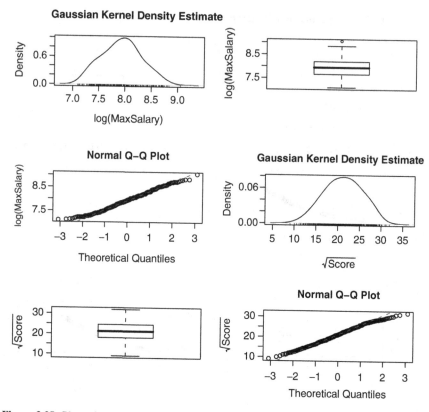

Figure 3.35 Plots of the transformed data

Next in Figure 3.36 we look at some diagnostic plots for the transformed data, namely, a plot of the standardized residuals against $\sqrt{\text{Score}}$, and a plot of the square root of the absolute value of the standardized residuals against $\sqrt{\text{Score}}$ for the regression model

$$\log(\text{MaxSalary}) = \beta_0 + \beta_1 \sqrt{\text{Score}} + e.$$

Approach 1: Transforming X first and then Y.

Finally in this section we consider what we previously called approach (1) to transforming both X and Y. In this approach we begin by transforming X. One way to do this is to use a variant of the Box-Cox transformation method. Define the modified family of power transformations

$$\Psi_M(X, \lambda_X)$$

where

$$\Psi_M(X, \lambda_X) = \Psi_S(X, \lambda_X) \times \text{gm}(X)^{1-\lambda_X} = \begin{cases} \text{gm}(X)^{1-\lambda_X} \left(X^{\lambda_X} - 1 \right) \big/ \lambda_X & \text{if } \lambda_X \neq 0 \\ \text{gm}(X)\log(X) & \text{if } \lambda_X = 0. \end{cases}$$

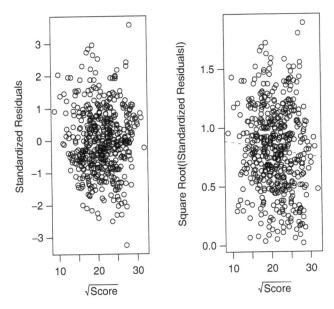

Figure 3.36 Diagnostic plots from the model based on the transformed data

Then we proceed by considering the log-likelihood function of $\Psi_M(X, \lambda_x)$ and choosing λ_x to maximize it. Note that there is no regression model here. Thus, this is an unconditional application of the Box-Cox procedure in that there is no regression model. The object is to make the distribution of the variable X as normal as possible.

Having transformed X to $\Psi_s(x, \lambda_x)$, consider a simple linear regression model of the form $Y = g(\beta_0 + \beta_1 \Psi_S(x, \lambda_x) + e)$. Then use an inverse response plot to decide on the transformation, g^{-1} for Y.

Example: Government salary data (cont.)

We apply approach (1) to the data. Recall that $Y = \text{MaxSalary}$ and $X = \text{Score}$. Given below is the output from R using the `bctrans` command from alr3.

Output from R

```
box.cox Transformations to Multinormality
        Est.Power   Std.Err.   Wald(Power=0)   Wald(Power=1)
Score     0.5481     0.0957          5.728         -4.7221
                                       LRT df          p.value
LR test, all lambda equal 0   35.16895   1     3.023047e-09
LR test, all lambda equal 1   21.09339   1     4.374339e-06
```

Using the Box-Cox method to transform the X variable, Score toward normality results in a value of λ_x close to 0.5. Thus, we shall again consider the following variable $\sqrt{\text{Score}}$. Thus we shall consider a model of the form

$$\text{MaxSalary} = g(\beta_0 + \beta_1 \sqrt{\text{Score}} + e) \tag{3.6}$$

and seek to find g^{-1}, the inverse of g, using an inverse response plot, since,

$$g^{-1}(\text{MaxSalary}) = \beta_0 + \beta_1 \sqrt{\text{Score}} + e.$$

Figure 3.37 contains an inverse response plot for the government salary data in the generated example. (Since the predictor variable $\sqrt{\text{Score}}$ has been transformed towards normality, we should be able to estimate g^{-1} from the inverse response plot.) Marked on the plot are three so-called power curves

$$\hat{y} = \text{MaxSalary}^{\lambda_y} \text{ for } \lambda_y = -0.19, 0, 1$$

where $\lambda = 0$ corresponds to natural logarithms. It is evident that among the three curves, the power curve

$$\hat{y} = \left(\hat{\beta}_0 + \hat{\beta}_1 \sqrt{\text{Score}}\right) = \text{MaxSalary}^{-0.19}$$

provides the closest fit to the data in Figure 3.37.

Rounding the estimated optimal value of λ_y to -0.25, we shall consider the following transformed Y-variable, $\text{MaxSalary}^{-0.25}$. Figure 3.38 contains a box plot, a normal Q–Q plot and Gaussian kernel density estimate (based on the Sheather-Jones bandwidth) for the variable $\text{MaxSalary}^{-0.25}$. Comparing Figure 3.39 with Figure 3.35 it seems that in terms of consistency with normality there is little to choose between the transformations $\text{MaxSalary}^{-0.25}$ and $\log(\text{MaxSalary})$.

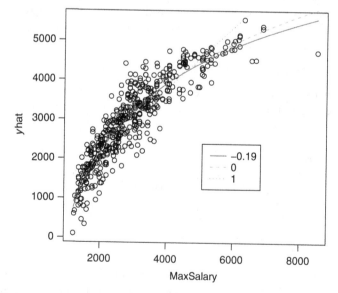

Figure 3.37 Inverse response plot based on model (3.6)

Gaussian Kernel Density Estimate

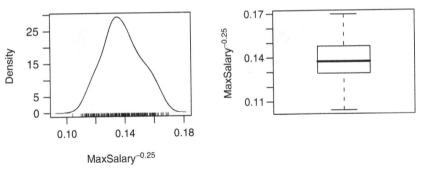

MaxSalary$^{-0.25}$

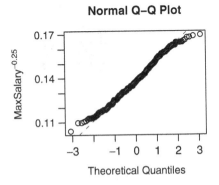

Figure 3.38   Plots of the transformed MaxSalary variable

Figure 3.39 shows a plot of MaxSalary$^{-0.25}$ and $\sqrt{\text{Score}}$ with the least squares line of best fit added. It is evident from this figure that the relationship between the transformed variables is more linear than that between the untransformed variables.

Figure 3.39 also contains some diagnostic plots for the transformed data, namely, a plot of the standardized residuals against $\sqrt{\text{Score}}$, and a plot of the square root of the absolute value of the standardized residuals against $\sqrt{\text{Score}}$ for the model

$$\text{MaxSalary}^{-0.25} = \beta_0 + \beta_1 \sqrt{\text{Score}} + e.$$

It is evident from the last two plots in Figure 3.39 that the variance of the standardized residuals decreases slightly as $\sqrt{\text{Score}}$ increases. Furthermore, comparing

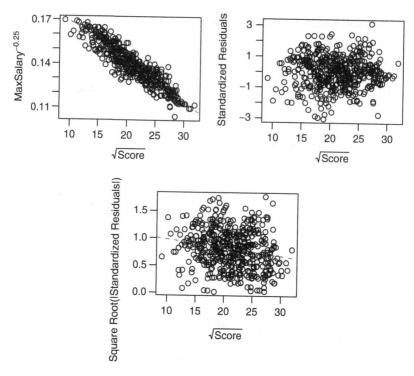

Figure 3.39 Plots associated with the model found using approach (1)

Figures 3.36 and 3.39, the model associated with Figure 3.36 is to be slightly preferred to the model associated with Figure 3.39 on the basis that the variance of the standardized residuals appears to be more constant in Figure 3.36. Thus, our preferred regression model is

$$\log(\text{MaxSalary}) = \beta_0 + \beta_1 \sqrt{\text{Score}} + e$$

Cook and Weisberg (1999, p. 329) point out that in many cases approaches (1) and (2) will lead to "essentially the same answers". In the government salary example the final models differed in terms of the transformation recommended for the response variable, MaxSalary. They also point out that the approach based on inverse fitted value plots (i.e. approach (1)) is more robust, as the Box-Cox method is susceptible to outliers, and as such approach (1) may give reasonable results when approach (2) fails.

Finally, Figure 3.40 contains a flow chart which summarizes the steps in developing a simple linear regression model.

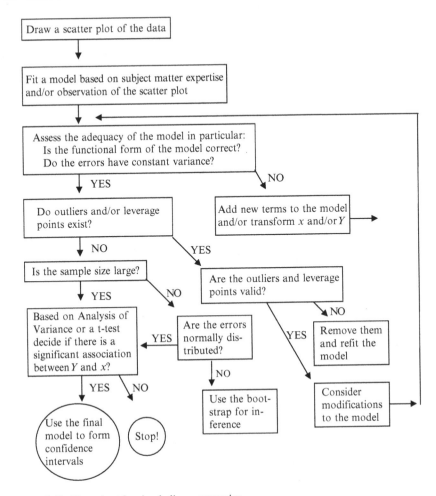

Figure 3.40 Flow chart for simple linear regression

3.4 Exercises

1. The data file airfares.txt on the book web site gives the one-way airfare (in US dollars) and distance (in miles) from city A to 17 other cities in the US. Interest centers on modeling airfare as a function of distance. The first model fit to the data was

$$\text{Fare} = \beta_0 + \beta_1 \text{Distance} + e \qquad (3.7)$$

(a) Based on the output for model (3.7) a business analyst concluded the following:

The regression coefficient of the predictor variable, Distance is highly statistically significant and the model explains 99.4% of the variability in the Y-variable, Fare. Thus model (1) is a highly effective model for both understanding the effects of Distance on Fare and for predicting future values of Fare given the value of the predictor variable, Distance.

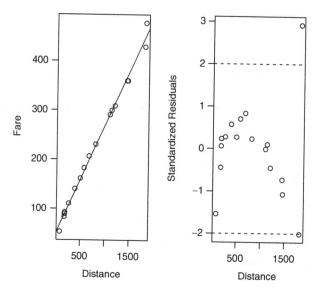

Figure 3.41 Output from model (3.7)

Provide a detailed critique of this conclusion.

(b) Does the ordinary straight line regression model (3.7) seem to fit the data well? If not, carefully describe how the model can be improved.

Given below and in Figure 3.41 is some output from fitting model (3.7).

Output from R

```
Call:
lm(formula = Fare ~ Distance)

Coefficients:
              Estimate    Std. Error    t value    Pr(>|t|)
(Intercept)   48.971770     4.405493      11.12    1.22e-08 ***
Distance       0.219687     0.004421      49.69     <2e-16  ***
---
Signif. codes: 0'***'0.001 '**' 0.01'*'0.05 '.' 0.1 ''1

Residual standard error: 10.41 on 15 degrees of freedom
Multiple R-Squared:0.994, Adjusted R-squared: 0.9936
F-statistic: 2469 on 1 and 15 DF, p-value: < 2.2e-16
```

2. Is the following statement true or false? If you believe that the statement is false, provide a brief explanation.

Suppose that a straight line regression model has been fit to bivariate data set of the form $(x_1, y_1), (x_2, y_2),..., (x_n, y_n)$. Furthermore, suppose that the distribution of X appears to be normal while the Y variable is highly skewed. A plot of standardized residuals from the least squares regression line produce a quadratic pattern with increasing variance when plotted against $(x_1, x_2, ..., x_n)$. In this case, one should consider adding a quadratic term in X to the regression model and thus consider a model of the form $Y = \beta_0 + \beta_1 x + \beta_2 x^2 + e$.

3. The price of advertising (and hence revenue from advertising) is different from one consumer magazine to another. Publishers of consumer magazines argue that magazines that reach more readers create more value for the advertiser. Thus, circulation is an important factor that affects revenue from advertising. In this exercise, we are going to investigate the effect of circulation on gross advertising revenue. The data are for the top 70 US magazines ranked in terms of total gross advertising revenue in 2006. In particular we will develop regression models to predict gross advertising revenue per advertising page in 2006 (in thousands of dollars) from circulation (in millions). The data were obtained from http://adage.com and are given in the file AdRevenue.csv which is available on the book web site. Prepare your answers to parts A, B and C in the form of a report.

Part A

(a) Develop a simple linear regression model based on least squares that predicts advertising revenue per page from circulation (i.e., feel free to transform either the predictor or the response variable or both variables). Ensure that you provide justification for your choice of model.

(b) Find a 95% prediction interval for the advertising revenue per page for magazines with the following circulations:
  (i) 0.5 million
  (ii) 20 million

(c) Describe any weaknesses in your model.

Part B

(a) Develop a polynomial regression model based on least squares that directly predicts the effect on advertising reve-nue per page of an increase in circulation of 1 million people (i.e., do not transform either the predictor nor the response variable). Ensure that you provide detailed justification for your choice of model. [Hint: Consider polynomial models of order up to 3.]

(b) Find a 95% prediction interval for the advertising page cost for magazines with the following circulations:
  (i) 0.5 million
  (ii) 20 million

(c) Describe any weaknesses in your model.

Part C

(a) Compare the model in Part A with that in Part B. Decide which provides a better model. Give reasons to justify your choice.

(b) Compare the prediction intervals in Part A with those in Part B. In each case, decide which interval you would recommend. Give reasons to justify each choice.

4. Tryfos (1998, p. 57) considers a real example involving the management at a Canadian port on the Great Lakes who wish to estimate the relationship between the volume of a ship's cargo and the time required to load and unload this cargo. It is envisaged that this relationship will be used for planning purposes as well as for making comparisons with the productivity of other ports. Records of the tonnage loaded and unloaded as well as the time spent in port by 31 liquid-carrying vessels that used the port over the most recent summer are available. The data are available on the book website in the file glakes.txt. The first model fit to the data was

$$\text{Time} = \beta_0 + \beta_1 \text{Tonnage} + e \tag{3.8}$$

On the following pages is some output from fitting model (3.8) as well as some plots of Tonnage and Time (Figures 3.42 and 3.43).

(a) Does the straight line regression model (3.8) seem to fit the data well? If not, list any weaknesses apparent in model (3.8).
(b) Suppose that model (3.8) was used to calculate a prediction interval for Time when Tonnage = 10,000. Would the interval be too short, too long or about right (i.e., valid)? Give a reason to support your answer.

The second model fitted to the data was

$$\log(\text{Time}) = \beta_0 + \beta_1 \text{Tonnage}^{0.25} + e \tag{3.9}$$

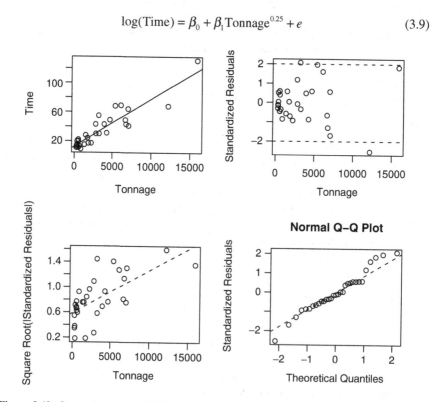

Figure 3.42 Output from model (3.8)

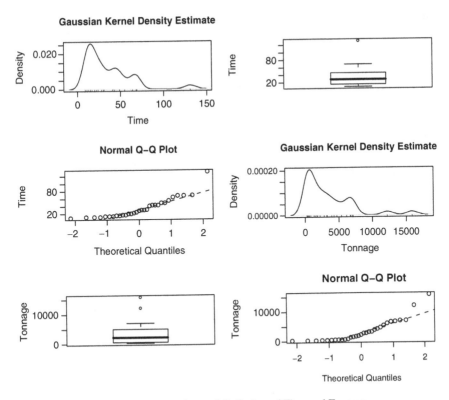

Figure 3.43 Density estimates, box plots and Q–Q plots of Time and Tonnage

Output from model (3.9) as well as some plots (Figures 3.44 and 3.45) appears on the following pages.

(a) Is model (3.9) an improvement over model (3.8) in terms of predicting Time? If so, please describe all the ways in which it is an improvement.

(b) List any weaknesses apparent in model (3.9).

Regression output from R for model (3.8)

```
Call:
lm(formula = Time ~ Tonnage)

Coefficients:
                Estimate Std. Error t value   Pr(>|t|)
(Intercept)   12.344707   2.642633    4.671   6.32e-05 ***
Tonnage        0.006518   0.000531   12.275   5.22e-13 ***
---
Residual standard error: 10.7 on 29 degrees of freedom
Multiple R-Squared: 0.8386, Adjusted R-squared: 0.833
F-statistic: 150.7 on 1 and 29 DF, p-value: 5.218e-13
```

Output from R

```
box.cox Transformations to Multinormality
          Est.Power    Std.Err.   Wald(Power=0)    Wald(Power=1)
Time        0.0228      0.1930         0.1183          -5.0631
Tonnage     0.2378      0.1237         1.9231          -6.1629
                                         LRT     df       p.value
LR test, all lambda equal     0     3.759605      2  1.526202e-01
LR test, all lambda equal     1    45.315290      2  1.445140e-10
```

Output from R for model (3.9)

```
Call:
lm(formula = log(Time) ~ I(Tonnage^0.25))

Coefficients:
                 Estimate   Std. Error  t value  Pr(>|t|)
(Intercept)       1.18842     0.19468     6.105   1.20e-06  ***
I(Tonnage^0.25)   0.30910     0.02728    11.332   3.60e-12  ***
---
Signif. codes: 0 '***' 0.001 '**' 0.01 '*' 0.05 '.' 0.1 ' ' 1

Residual standard error: 0.3034 on 29 degrees of freedom
Multiple R-Squared: 0.8158, Adjusted R-squared: 0.8094
F-statistic: 128.4 on 1 and 29 DF, p-value: 3.599e-12
```

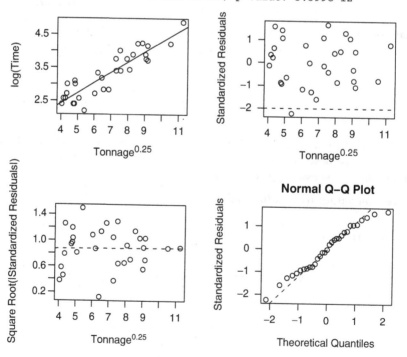

Figure 3.44 Output from model (3.9)

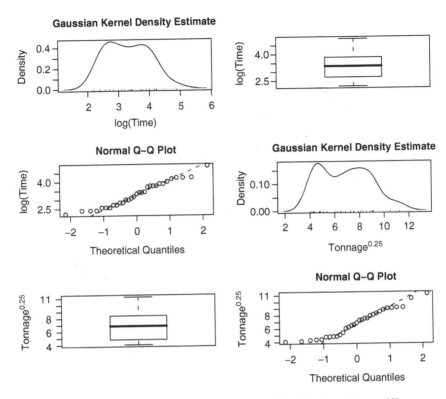

Figure 3.45 Density estimates, box plots and Q–Q plots of log(Time) and Tonnage$^{0.25}$

5. An analyst for the auto industry has asked for your help in modeling data on the prices of new cars. Interest centers on modeling suggested retail price as a function of the cost to the dealer for 234 new cars. The data set, which is available on the book website in the file cars04.csv, is a subset of the data from http://www.amstat.org/publications/jse/datasets/04cars.txt

(Accessed March 12, 2007)
The first model fit to the data was

$$\text{Suggested Retail Price} = \beta_0 + \beta_1 \text{ Dealer Cost} + e \qquad (3.10)$$

On the following pages is some output from fitting model (3.10) as well as some plots (Figure 3.46).

(a) Based on the output for model (3.10) the analyst concluded the following:

Since the model explains just more than 99.8% of the variability in Suggested Retail Price and the coefficient of Dealer Cost has a t-value greater than 412, model (1) is a highly effective model for producing prediction intervals for Suggested Retail Price.

Provide a detailed critique of this conclusion.

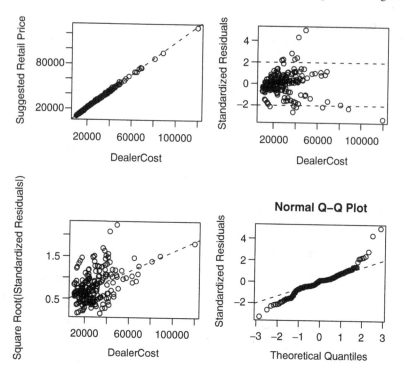

Figure 3.46 Output from model (3.10)

(b) Carefully describe all the shortcomings evident in model (3.10). For each short-coming, describe the steps needed to overcome the shortcoming.
   The second model fitted to the data was

$$\log(\text{Suggested Retail Price}) = \beta_0 + \beta_1\log(\text{Dealer Cost}) + e \qquad (3.11)$$

Output from model (3.11) and plots (Figure 3.47) appear on the following pages.

(c) Is model (3.11) an improvement over model (3.10) in terms of predicting Suggested Retail Price? If so, please describe all the ways in which it is an improvement.

(d) Interpret the estimated coefficient of log(Dealer Cost) in model (3.11).

(e) List any weaknesses apparent in model (3.11).

Regression output from R for model (3.10)

```
Call:
lm(formula = SuggestedRetailPrice ~ DealerCost)
```

```
Coefficients:
               Estimate    Std. Error   t value    Pr(>|t|)
(Intercept)   -61.904248    81.801381   -0.757       0.45
DealerCost      1.088841     0.002638   412.768    <2e-16 ***
---
Signif. codes: 0 '***' 0.001 '**' 0.01 '*' 0.05 '.' 0.1 '' 1

Residual standard error: 587 on 232 degrees of freedom
Multiple    R-Squared:0.9986,    Adjusted R-squared: 0.9986
F-statistic: 1.704e+05 on 1 and 232 DF, p-value: < 2.2e-16
```

Output from R for model (3.11)

```
Call:
lm(formula = log(SuggestedRetailPrice) ~ log(DealerCost))

Coefficients:
                 Estimate   Std. Error   t value    Pr(>|t|)
(Intercept)     -0.069459    0.026459    -2.625     0.00924 **
log(DealerCost)  1.014836    0.002616   387.942    <2e-16 ***
---
Signif.codes:0 '***' 0.001 '**' 0.01 '*' 0.05 '.' 0.1 '' 1

Residual standard error: 0.01865 on 232 degrees of freedom
Multiple R-Squared: 0.9985,    Adjusted R-squared: 0.9985
F-statistic: 1.505e+05 on 1 and 232 DF, p-value: < 2.2e-16
```

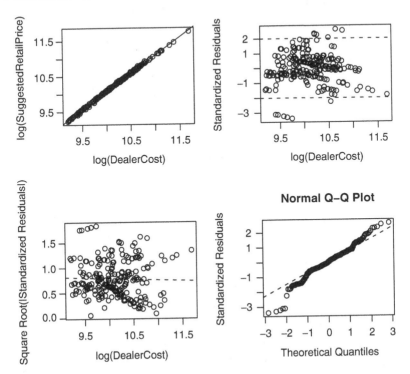

Figure 3.47 Output from model (3.11)

3. A sample of $n = 500$ data points were generated from the following model

$$Y = g(\beta_0 + \beta_1 x + e)$$

where $\beta_0 = 2.5, \beta_1 = 1, g(.) = \exp(.)$, the errors e being normally distributed and the distribution of x highly skewed. Figure 3.48 contains the inverse response plot. It produces the estimate $\hat{\lambda} = 0.61$. In this case, the correct transformation of Y corresponds to taking $\lambda = 0$ and hence using $g^{-1}(Y) = \log(Y) = \beta_0 + \beta_1 x + e$. Explain why the inverse response plot fails to produce an estimated value of λ close to the correct value of λ in this situation.

7. When Y has mean and variance both equal to μ we showed earlier in this chapter that the appropriate transformation of Y for stabilizing variance is the square root transformation. Now, suppose that Y has mean equal to μ and variance equal to μ^2 show that the appropriate transformation of Y for stabilizing variance is the log transformation.

8. Chu (1996) discusses the development of a regression model to predict the price of diamond rings from the size of their diamond stones (in terms of their weight in carats). Data on both variables were obtained from a full page advertisement placed in the *Straits Times* newspaper by a Singapore-based retailer of diamond jewelry. Only rings made with 20 carat gold and mounted with a single diamond stone were included in the data set. There were 48 such rings of varying designs. (Information on the designs was available but not used in the modeling.)

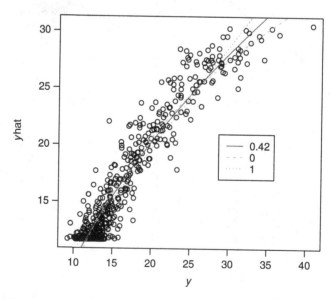

Figure 3.48 Inverse response plot

The weights of the diamond stones ranged from 0.12 to 0.35 carats (a one carat diamond stone weighs 0.2 gram) and were priced between $223 and $1086. The data are available on the course web site in the file diamonds.txt.

Part 1

(a) Develop a simple linear regression model based on least squares that directly predicts Price from Size (that is, do not transform either the predictor nor the response variable). Ensure that you provide justification for your choice of model.

(b) Describe any weaknesses in your model.

Part 2

(a) Develop a simple linear regression model that predicts Price from Size (i.e., feel free to transform either the predictor or the response variable or both variables). Ensure that you provide detailed justification for your choice of model.

(b) Describe any weaknesses in your model.

Part 3

Compare the model in Part A with that in Part B. Decide which provides a better model. Give reasons to justify your choice.

Chapter 4
Weighted Least Squares

In Chapter 3, we saw that it is sometimes possible to overcome nonconstant error variance by transforming Y and/or X. In this chapter we consider an alternative way of coping with nonconstant error variance, namely weighted least squares (WLS).

4.1 Straight-Line Regression Based on Weighted Least Squares

Consider the straight linear regression model

$$Y_i = \beta_0 + \beta_1 x_i + e_i$$

where the e_i have mean 0 but variance σ^2 / w_i. When w_i is very large then the variance of e_i is close to 0. In this situation, the estimates of the regression parameters β_0 and β_1 should be such that the fitted line at x_i should be very close to y_i. On the other hand, when w_i is very small then the variance of e_i is very large. In this situation, the estimates of the regression parameters β_0 and β_1 should take little account of the values (x_i, y_i). In the extreme situation that w_i is 0 then the variance of e_i is equal to infinity and the ith case (x_i, y_i) should be ignored in fitting the line: this is equivalent to deleting the ith case (x_i, y_i) from the data and fitting the line based on the other $n - 1$ cases.

Thus, we need to take account of the weights w_i when estimating the regression parameters β_0 and β_1. This is achieved by considering the following weighted version of the residual sum of squares

$$\text{WRSS} = \sum_{i=1}^{n} w_i (y_i - \hat{y}_{wi})^2 = \sum_{i=1}^{n} w_i (y_i - b_0 - b_1 x_i)^2.$$

WRSS is such that the larger the value of w_i the more the ith case (x_i, y_i) is taken into account. To obtain the weighted least squares estimates we seek the values of b_0 and b_1 that minimize WRSS. For WRSS to be a minimum with respect to b_0 and b_1 we require

S.J. Sheather, *A Modern Approach to Regression with R*,
DOI: 10.1007/978-0-387-09608-7_4, © Springer Science+Business Media LLC 2009

$$\frac{\partial \text{WRSS}}{\partial b_0} = -2\sum_{i=1}^{n} w_i(y_i - b_0 - b_1 x_i) = 0$$

$$\frac{\partial \text{WRSS}}{\partial b_1} = -2\sum_{i=1}^{n} w_i x_i(y_i - b_0 - b_1 x_i) = 0.$$

Rearranging terms in these last two equations gives

$$\sum_{i=1}^{n} w_i y_i = b_0 \sum_{i=1}^{n} w_i + b_1 \sum_{i=1}^{n} w_i x_i \qquad (4.1)$$

$$\sum_{i=1}^{n} w_i x_i y_i = b_0 \sum_{i=1}^{n} w_i x_i + b_1 \sum_{i=1}^{n} w_i x_i^2 \qquad (4.2)$$

These last two equations are called the **normal equations**. Multiplying equation (4.1) by $\sum_{i=1}^{n} w_i x_i$ and equation (4.2) by $\sum_{i=1}^{n} w_i$ gives

$$\sum_{i=1}^{n} w_i x_i \sum_{i=1}^{n} w_i y_i = b_0 \sum_{i=1}^{n} w_i \sum_{i=1}^{n} w_i x_i + b_1 \left(\sum_{i=1}^{n} w_i x_i\right)^2$$

and

$$\sum_{i=1}^{n} w_i \sum_{i=1}^{n} w_i x_i y_i = b_0 \sum_{i=1}^{n} w_i \sum_{i=1}^{n} w_i x_i + b_1 \sum_{i=1}^{n} w_i \sum_{i=1}^{n} w_i x_i^2.$$

Subtracting the first equation from the second and solving for b_1 gives the so-called **weighted least squares estimate** of the slope

$$\hat{\beta}_{1W} = \frac{\sum_{i=1}^{n} w_i \sum_{i=1}^{n} w_i x_i y_i - \sum_{i=1}^{n} w_i x_i \sum_{i=1}^{n} w_i y_i}{\sum_{i=1}^{n} w_i \sum_{i=1}^{n} w_i x_i^2 - \left(\sum_{i=1}^{n} w_i x_i\right)^2} = \frac{\sum_{i=1}^{n} w_i (x_i - \bar{x}_W)(y_i - \bar{y}_W)}{\sum_{i=1}^{n} w_i (x_i - \bar{x}_W)^2}$$

where $\bar{x}_W = \sum_{i=1}^{n} w_i x_i / \sum_{i=1}^{n} w_i$ and $\bar{y}_W = \sum_{i=1}^{n} w_i y_i / \sum_{i=1}^{n} w_i$. From which we can find the **weighted least squares estimate** of the intercept

$$\hat{\beta}_{0W} = \frac{\sum_{i=1}^{n} w_i y_i}{\sum_{i=1}^{n} w_i} - \hat{\beta}_{1W} \frac{\sum_{i=1}^{n} w_i x_i}{\sum_{i=1}^{n} w_i} = \bar{y}_W - \hat{\beta}_{1W} \bar{x}_W.$$

Example: Developing a bid on contract cleaning (cont.)

This example was first discussed in Chapter 3. Recall that the aim of the exercise was to develop a regression equation to model the relationship between the **number of rooms cleaned**, Y and the **number of crews**, X and predict the number of rooms that can be cleaned by 4 crews and by 16 crews. In this example, the x-variable (number of crews) is discrete with values 2, 4, 6, 8, 10, 12 and 16. Recall also that there are multiple measurements of the Y-variable (number of rooms cleaned) at each value of x. In *this special case*, it is possible to directly calculate the standard deviation of Y at each discrete value of x. Table 4.1 gives the value of each of these standard deviations along with the number of points each is based on.

Consider the simple linear regression model

$$Y_i = \beta_0 + \beta_1 x_i + e_i$$

where the e_i have mean 0 but variance σ^2 / w_i. In this case we take

$$w_i = \frac{1}{\left(\text{Standard Deviation}(Y_i)\right)^2}$$

then Y_i has variance σ^2 / w_i with $\sigma^2 = 1$. We shall use the estimated or sample standard deviations in Table 4.1 to produce the weights w_i. The data set with these weights included can be found on the book web site in the file cleaningwtd.txt. Given below is the weighted least squares regression output from R including 95% prediction intervals for Y when $x = 4$ and 16.

Regression output from R

```
Call:
lm(formula = Rooms ~ Crews, weights = 1/StdDev^2)

Residuals:
     Min       1Q     Median       3Q       Max
-1.43184  -0.82013   0.03909  0.69029   2.01030

Coefficients:
            Estimate  Std. Error  t value  Pr(>|t|)
(Intercept)  0.8095     1.1158     0.725    0.471
Crews        3.8255     0.1788    21.400    <2e-16   ***
---
Signif. codes: 0 `***' 0.001 `**' 0.01 `*' 0.05 `.' 0.1 ` ' 1

Residual standard error: 0.9648 on 51 degrees of freedom
Multiple R-Squared: 0.8998, Adjusted R-squared: 0.8978
F-statistic: 458 on 1 and 51 DF, p-value: < 2.2e-16

       fit       lwr       upr
1 16.11133  13.71210  18.51056
2 62.01687  57.38601  66.64773
```

Table 4.1 The standard deviation of Y for each value of x

x, Crews	N	Standard deviation(Y_i)
2	9	3.00
4	6	4.97
6	5	4.69
8	8	6.64
10	8	7.93
12	7	7.29
16	10	12.00

4.1.1   Prediction Intervals for Weighted Least Squares

According to Weisberg (2006, p. 42):[1]

> The predict helper function also works correctly for getting predictions and standard errors of fitted values, but it apparently does not always give the right answer for prediction intervals. Both R and S-Plus compute the standard error of prediction as

$$\sqrt{\left(\hat{\sigma}^2 + \text{sefit}(y \mid X = x^*)^2\right)},$$

> rather than

$$\sqrt{\left(\hat{\sigma}^2 / w^* + \text{sefit}(y \mid X = x^*)^2\right)}.$$

> The R/S-Plus formula assumes that the variance of the future observation is σ^2 rather than σ^2 / w^*, where w^* is the weight for the future value.

4.1.2   Leverage for Weighted Least Squares

The ith fitted or predicted value from weighted least squares is given by

$$\hat{y}_{Wi} = \hat{\beta}_{0W} + \hat{\beta}_{1W} x_i$$

where $\hat{\beta}_{0W} = \bar{y}_W - \hat{\beta}_{1W}\bar{x}_W$ and

$$\hat{\beta}_{1W} = \frac{\sum_{j=1}^{n} w_j \left(x_j - \bar{x}_W\right)\left(y_j - \bar{y}_W\right)}{\sum_{j=1}^{n} w_j \left(x_j - \bar{x}_W\right)^2} = \frac{\sum_{j=1}^{n} w_j \left(x_j - \bar{x}_W\right) y_j}{\sum_{j=1}^{n} w_j \left(x_j - \bar{x}_W\right)^2} = \frac{\sum_{j=1}^{n} w_j \left(x_j - \bar{x}_W\right) y_j}{WSXX}$$

---

[1] Weisberg, S. (2006) *Computing Primer for Applied Linear Regression, Third Edition Using R and S-Plus*, available at www.stat.umn.edu/alr/Links/RSprimer.pdf.

where $WSXX = \sum_{j=1}^{n} w_j \left(x_j - \bar{x}_W\right)^2$. So that, letting $w_j^s = w_j \Big/ \sum_{k=1}^{n} w_k$

$$
\begin{aligned}
\hat{y}_{Wi} &= \bar{y}_W - \hat{\beta}_{1W}\bar{x}_W + \hat{\beta}_{1W} x_i \\
&= \bar{y}_W + \hat{\beta}_{1W}\left(x_i - \bar{x}_W\right) \\
&= \sum_{j=1}^{n} w_j y_j \Big/ \sum_{k=1}^{n} w_k + \sum_{j=1}^{n} \frac{w_j\left(x_j - \bar{x}_W\right)}{WSXX} y_j \left(x_i - \bar{x}_W\right) \\
&= \sum_{j=1}^{n} \left[w_j^s + \frac{w_j\left(x_i - \bar{x}_W\right)\left(x_j - \bar{x}_W\right)}{WSXX} \right] y_j \\
&= \sum_{j=1}^{n} h_{Wij} y_j
\end{aligned}
$$

where

$$
h_{Wij} = \left[w_j^s + \frac{w_j\left(x_i - \bar{x}_W\right)\left(x_j - \bar{x}_W\right)}{WSXX} \right]
$$

Thus,

$$
\hat{y}_{Wi} = h_{Wii} y_i + \sum_{j \neq i} h_{Wij} y_j
$$

where $h_{Wii} = \left[w_i^s + \dfrac{w_i\left(x_i - \bar{x}_W\right)^2}{WSXX} \right]$.

Reality Check: All weights equal, then WLS = LS.
  Take $w_j = 1/n$ (then $w_j^s = 1/n$) and $h_{Wij} = h_{ij}$ as given in (3.1).

4.1.3 Using Least Squares to Calculate Weighted Least Squares

Consider the simple linear regression model

$$
Y_i = \beta_0 + \beta_1 x_i + e_i \tag{4.3}
$$

where the e_i have mean 0 but variance σ^2 / w_i. Notice that if we multiply both sides of the last equation by $\sqrt{w_i}$ we get

$$
\sqrt{w_i}\,Y_i = \beta_0 \sqrt{w_i} + \beta_1 \sqrt{w_i}\, x_i + \sqrt{w_i}\, e_i \tag{4.4}
$$

where the $\sqrt{w_i}\,e_i$ have mean 0 but variance $\left(\sqrt{w_i}\right)^2 \times \sigma^2/w_i = \sigma^2$. Thus, it is possible to calculate the weighted least squares fit of model (4.3) by calculating the least squares fit to model (4.4). Model (4.4) is a multiple linear regression with two predictors and no intercept. To see this, let

$$Y_{\text{NEW}i} = \sqrt{w_i}\,Y_i,\, x_{1\text{NEW}i} = \sqrt{w_i},\, x_{2\text{NEW}i} = \sqrt{w_i}\,x_i \text{ and } e_{\text{NEW}i} = \sqrt{w_i}\,e_i$$

then we can rewrite model (4.4) as

$$Y_{\text{NEW}i} = \beta_0 x_{1\text{NEW}i} + \beta_1 x_{2\text{NEW}i} + e_{\text{NEW}i} \tag{4.5}$$

Example: Developing a bid on contract cleaning (cont.)

Recall that in this case we take

$$w_i = \frac{1}{\left(\text{Standard deviation}(Y_i)\right)^2}.$$

We shall again use the estimated or sample standard deviations in Table 4.1 to produce the weights w_i. Given below is the least squares regression output from R for model (4.5) including 95% prediction intervals for Y when $x = 4$ and 16.

Comparing the output from R on the next page with that on page 4, we see that the results are the same down to and including Residual standard error.

Regression output from R

```
Call:
lm(formula = ynew ~ x1new + x2new - 1)

Residuals:
     Min        1Q     Median        3Q       Max
-1.43184  -0.82013    0.03909   0.69029   2.01030

Coefficients:
        Estimate Std.    Error   t value   Pr(>|t|)
x1new        0.8095      1.1158     0.725      0.471
x2new        3.8255      0.1788    21.400    <2e-16 ***
---
Signif. codes: 0 `***' 0.001 `**' 0.01 `*' 0.05 `.' 0.1 ` ' 1

Residual standard error: 0.9648 on 51 degrees of freedom
Multiple R-Squared: 0.9617, Adjusted R-squared: 0.9602
F-statistic: 639.6 on 2 and 51 DF, p-value: < 2.2e-16

        fit         lwr          upr
1   3.243965    1.286166     5.201763
2   5.167873    3.199481     7.136265
```

Table 4.2 Predictions and 95% prediction intervals for the number of crews

x, Crews	Prediction	Lower limit	Upper limit
4 (transformed data)	$16 = (4.003^2)$	$8 = (2.790^2)$	$27 = (5.217^2)$
4 (raw data WLS)	$16 = (3.244 \times 4.97)$	$6 = (1.286 \times 4.97)$	$26 = (5.202 \times 4.97)$
16 (transformed data)	$61 = (7.806^2)$	$43 = (6.582^2)$	$82 = (9.031^2)$
16 (raw data WLS)	$62 = (5.167 \times 12.00)$	$38 = (3.199 \times 12.00)$	$87 = (7.136 \times 12.00)$

The fits and prediction intervals above are for $Y_{\text{NEW}i} = \sqrt{w_i}\,Y_i$. To obtain the fits and prediction intervals for Y_i we need to multiply the values for $Y_{\text{NEW}i}$ by $1 \big/ \sqrt{w_i}$. Looking at Table 4.1 we see that when $x = 4, w_i = 1/4.97^2$ and when $x = 16, w_i = 1/12.00^2$. The results are given in Table 4.2. Also given, for comparison purposes, are results obtained from the transformed data in Chapter 3.

It is evident from Table 4.2 that the predictions are close for both methods. The prediction intervals are close when $x = 4$. However, when $x = 16$, the prediction interval is wider for weighted least squares. Can you think of a reason why this might be so?

4.1.4   Defining Residuals for Weighted Least Squares

In the case of weighted least squares, the residuals are defined by

$$\hat{e}_{Wi} = \sqrt{w_i}\left(y_i - \hat{y}_{Wi}\right)$$

where $\hat{y}_{Wi} = \hat{\beta}_{0W} + \hat{\beta}_{1W} x_i$. With this definition, the weighted least squares are obtained by minimizing the sum of the squared residuals

$$\text{WRSS} = \sum_{i=1}^{n} \hat{e}_{Wi}^2 = \sum_{i=1}^{n} w_i (y_i - \hat{y}_{Wi})^2$$

The second advantage of this choice is that the variance of the ith residual can be shown to depend on the weight w_i only through its leverage value.

4.1.5   The Use of Weighted Least Squares

The weighted least squares technique is commonly used in the important special case when Y_i is the average or the median of n_i observations so that $\text{Var}(Y_i) \propto \dfrac{1}{n_i}$. In this case we take $w_i = n_i$.

However, many situations exist in which the variance is not constant and in which it is not straightforward to determine the correct model for the variance. In these situations, the use of weighted least squares is problematic.

4.2 Exercises

1. A full professor of statistics at a major US university is interested in estimating the third quartile of salaries for full professors with 6 years of experience in that rank. Data, in the form of the 2005–2006 Salary Report of Academic Statisticians, are available at http://www.amstat.org/profession/salaryreport_acad2005-6.pdf (accessed March 12, 2007) (Table 4.3). Using weighted least squares, estimate the 2005–2006 third quartile for salary of full professors with 6 years of experience.

2. Consider regression though the origin (i.e., straight line regression with population intercept known to be zero) with $Var(e_i \mid x_i) = x_i^2 \sigma^2$. The corresponding regression model is $Y_i = \beta x_i + e_i \ (i = 1,...,n)$.

   Find an explicit expression for the weighted least squares estimate of β.

3. The Sunday April 15, 2007 issue of the Houston Chronicle included a section devoted to real estate prices in Houston. In particular, data are presented on the 2006 median price per square foot for 1922 subdivisions. The data (HoustonRealEstate.txt) can be found on the book web site. Interest centers on developing a regression model to predict

   $Y_i =$ 2006 median price per square foot

   from

   $x_{1i} =$ %NewHomes (i.e., of the houses that sold in 2006, the percentage that were built in 2005 or 2006)

   $x_{2i} =$ %Foreclosures (i.e., of the houses that sold in 2006, the percentage that were identified as foreclosures)

Table 4.3 Data on salaries

Years of experience as a full professor	Sample size, n_i	Third quartile ($)
0	17	101,300
2	33	111,303
4	19	98,000
6	25	124,000
8	18	128,475
12	60	117,410
17	58	115,825
22	31	134,300
28	34	128,066
34	19	164,700

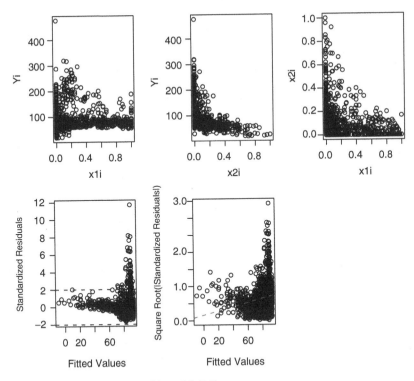

Figure 4.1 Plots associated with model (4.6)

for the $i = 1, \ldots 1922$ subdivisions.

The first model considered was

$$Y_i = \beta_0 + \beta_1 x_{1i} + \beta_2 x_{2i} + e \tag{4.6}$$

Model (4.6) was fit used weighted least squares with weights,

$$w_i = n_i$$

where

n_i = the number of homes sold in subdivision i in 2006.

Output from model (4.6), in the form of plots, appears in Figure 4.1.

(a) Explain it is necessary to use weighted least squares to fit model (4.6) and why $w_i = n_i$ is the appropriate choice for the weights.

(b) Explain why (4.6) is not a valid regression model.

(c) Describe what steps you would take to obtain a valid regression model (Figure 4.1).

Chapter 5
Multiple Linear Regression

It is common for more than one factor to influence an outcome. Fitting regression models to data involving two or more predictors is one of the most widely used statistical procedures. In this chapter we consider multiple linear regression problems involving modeling the relationship between a dependent variable, Y and two or more predictor variables x_1, x_2, x_3, etc. Throughout Chapter 5, we will assume that the multiple linear regression model under consideration is a valid model for the data. In the next chapter we will consider a series of tools to check model validity.

5.1 Polynomial Regression

We begin this chapter by looking at an important special case of multiple regression, known as polynomial regression. In this case the predictors are a single predictor, x, and its polynomial powers (x^2, x^3, etc.). In polynomial regression, we can display the results of our multiple regression on a single two-dimensional graph.

Example: Modeling salary from years of experience

This example is taken from Tryfos (1998, pp. 5–7). According to Tryfos:

> Professional organizations of accountants, engineers, systems analysts and others regularly survey their members for information concerning salaries, pensions, and conditions of employment. One product of these surveys is the so-called salary curve...which relates salary to years of experience.
> The salary curve is said to show the "normal" or "typical" salary of professionals with a given number of years experience. It is of considerable interest to members of the profession who like to know where they stand among their peers. It is also valuable to personnel departments of businesses considering salary adjustments or intending to hire new professionals.

We want to develop a regression equation to model the relationship between Y, salary (in thousands of dollars) and x, the number of years of experience and find a 95% prediction interval for Y when $x = 10$. The 143 data points are plotted in Figure 5.1 and can be found on the book web site in the file profsalary.txt.

It is clear from Figure 5.1 that the relationship between salary and years of experience is nonlinear. For illustrative purposes we will start by fitting a simple linear model

S.J. Sheather, *A Modern Approach to Regression with R*,
DOI: 10.1007/978-0-387-09608-7_5, © Springer Science + Business Media LLC 2009

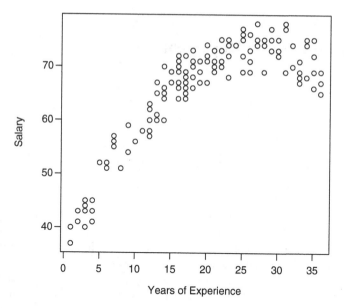

Figure 5.1 A plot of the professional salary data (prefsalary.txt)

and examining the associated regression diagnostics. The inadequacy of the straight-line model and potentially how to overcome it will be seen from this analysis.

We begin by considering the simple linear regression model

$$Y = \beta_0 + \beta_1 x + e \qquad (5.1)$$

where Y = salary and x = years of experience. Figure 5.2 shows a plot of the standardized residuals from model (5.1) against x.

A curved pattern resembling a quadratic is clearly evident in Figure 5.2. This suggests that we add a quadratic term in x to model (5.1) and thus consider the following polynomial regression model

$$Y = \beta_0 + \beta_1 x + \beta_2 x^2 + e \qquad (5.2)$$

where Y = salary and x = years of experience. Figure 5.3 contains a scatter plot of salary against years of experience. The curve in Figure 5.3 is just the least squares fit of model (5.2).

Figure 5.4 shows a plot of the standardized residuals from model (5.2) against x. The random pattern in Figure 5.4 indicates that model (5.2) is a valid model for the salary data.

Figure 5.5 shows a plot of leverage from model (5.2) against x. Marked on the plot as a horizontal dashed line is the cut-off value for a point of high leverage[1],

---

[1]In the next chapter we shall see that the cut-off is $2(p + 1)/n$ when there are p predictors.

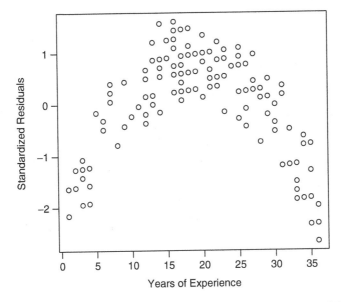

Figure 5.2 A plot of the standardized residuals from a straight-line regression model

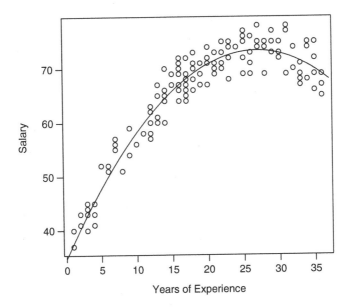

Figure 5.3 A plot of salary against experience with a quadratic fit added

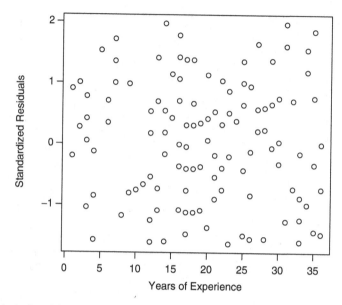

Figure 5.4 A plot of the standardized residuals from a quadratic regression model

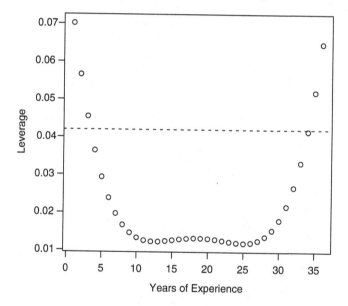

Figure 5.5 A plot of leverage against x, years of experience

namely, $6/n = 6/143 = 0.042$. It is evident from Figure 5.5 that the three smallest and the two largest x-values are leverage points.

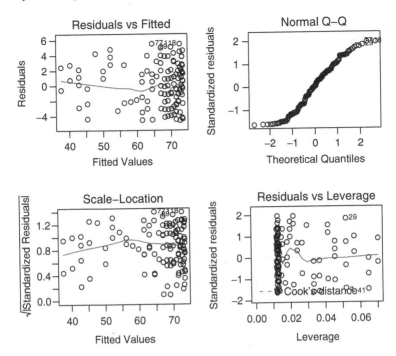

Figure 5.6 Diagnostic plots produced by R

Next we look at the four summary diagnostic plots produced by R in Figure 5.6. The bottom right plot shows that none of the points of high leverage have standardized residuals with absolute value of 2 or higher. The bottom left-hand plot is consistent with the errors in model (5.2) having constant variance. Thus (5.2) is a valid model.

The output from R associated with fitting model (5.2) is given below. The last part of the output gives the predicted salary (in thousands of dollars) and a 95% prediction interval for 10 years of experience. In this case the 95% prediction interval is ($52,505, $63,718).

Regression output from R

```
Call:

lm(formula = Salary ~ Experience + I(Experience^2))

Coefficients:
                 Estimate    Std. Error    t value    Pr(>|t|)
(Intercept)      34.720498     0.828724      41.90     <2e-16  ***
Experience        2.872275     0.095697      30.01     <2e-16  ***
I(Experience^2)  -0.053316     0.002477     -21.53     <2e-16  ***
---
```

```
Residual standard error: 2.817 on 140 degrees of freedom
Multiple R-Squared: 0.9247, Adjusted R-squared:0.9236
F-statistic: 859.3 on 2 and 140 DF, p-value:< 2.2e-16
         fit         lwr         upr
[1,]  58.11164   52.50481   63.71847
```

5.2 Estimation and Inference in Multiple Linear Regression

In the straight-line regression model, we saw that

$$E(Y \mid X = x) = \beta_0 + \beta_1 x$$

where the parameters β_0 and β_1 determine the intercept and the slope of a specific straight line, respectively.

Suppose in this case that $Y_1, Y_2, ..., Y_n$ are independent realizations of the random variable Y that are observed at the values $x_1, x_2, ..., x_n$ of a random variable X. Then for $i = 1, ..., n$

$$Y_i = E(Y_i \mid X_i = x_i) + e_i = \beta_0 + \beta_1 x_i + e_i$$

where

e_i = random fluctuation (or error) in Y_i such that $E(e_i \mid X) = 0$.

In this case the response variable Y is predicted from one predictor (or explanatory) variable X and the relationship between Y and X is linear in the parameters β_0 and β_1.

In the multiple linear regression model

$$E(Y \mid X_1 = x_1, X_2 = x_2, ..., X_p = x_p) = \beta_0 + \beta_1 x_1 + \beta_2 x_2 + ... + \beta_p x_p$$

Thus,

$$Y_i = \beta_0 + \beta_1 x_{1i} + \beta_2 x_{2i} + ... + \beta_p x_{pi} + e_i$$

where

e_i = random fluctuation (or error) in Y_i such that $E(e_i \mid X) = 0$. In this case the response variable Y is predicted from p predictor (or explanatory) variables X_1, X_2, ..., X_p and the relationship between Y and X_1, X_2, ..., X_p is linear in the parameters β_0, β_1, β_2, ..., β_p.

An example of a multiple linear regression model is between
Y = salary

and

$x_1 = x$, years of experience
$x_2 = x^2$, (years of experience)^2

Least squares estimates

The least squares estimates of $\beta_0, \beta_1, \beta_2, ..., \beta_p$ are the values of $b_0, b_1, b_2, ..., b_p$ for which the sum of the squared residuals,

$$\text{RSS} = \sum_{i=1}^{n} \hat{e}_i^2 = \sum_{i=1}^{n} (y_i - \hat{y}_i)^2 = \sum_{i=1}^{n} (y_i - b_0 - b_1 x_{1i} - b_2 x_{2i} - ... - b_p x_{pi})^2$$

is a minimum. For RSS to be a minimum with respect to $b_0, b_1, b_2, ..., b_p$ we require

$$\frac{\partial \text{RSS}}{\partial b_0} = -2 \sum_{i=1}^{n} (y_i - b_0 - b_1 x_{1i} - b_2 x_{2i} - ... - b_p x_{pi}) = 0$$

$$\frac{\partial \text{RSS}}{\partial b_1} = -2 \sum_{i=1}^{n} x_{1i} (y_i - b_0 - b_1 x_{1i} - b_2 x_{2i} - ... - b_p x_{pi}) = 0$$

$$\cdot$$
$$\cdot$$

$$\frac{\partial \text{RSS}}{\partial b_p} = -2 \sum_{i=1}^{n} x_{pi} (y_i - b_0 - b_1 x_{1i} - b_2 x_{2i} - ... - b_p x_{pi}) = 0$$

This gives a system of $(p + 1)$ equations in $(p + 1)$ unknowns. In practice, a computer package is needed to solve these equations and hence obtain the least squares estimates, $\hat{\beta}_0, \hat{\beta}_1, \hat{\beta}_2, ..., \hat{\beta}_p$.

Matrix formulation of least squares

A convenient way to study the properties of the least squares estimates, $\hat{\beta}_0, \hat{\beta}_1, \hat{\beta}_2, ..., \hat{\beta}_p$ is to use matrix and vector notation. Define the $(n \times 1)$ vector, $\mathbf{Y}$, the $n \times (p + 1)$ matrix, $\mathbf{X}$, the $(p + 1) \times 1$ vector, β of unknown regression parameters and the $(n \times 1)$ vector, $\mathbf{e}$ of random errors by

$$\mathbf{Y} = \begin{pmatrix} y_1 \\ y_2 \\ \vdots \\ y_n \end{pmatrix}, \mathbf{X} = \begin{pmatrix} 1 & x_{11} \cdots x_{1p} \\ 1 & x_{21} \cdots x_{2p} \\ \vdots & \vdots \\ 1 & x_{n1} \cdots x_{np} \end{pmatrix}, \beta = \begin{pmatrix} \beta_0 \\ \beta_1 \\ \vdots \\ \beta_p \end{pmatrix}, \mathbf{e} = \begin{pmatrix} e_1 \\ e_2 \\ \vdots \\ e_n \end{pmatrix}$$

We can write the multiple linear regression model in matrix notation as

$$\mathbf{Y} = \mathbf{X}\beta + \mathbf{e} \tag{5.3}$$

In addition, let $\mathbf{x}_i'$ denote the ith row of the matrix $\mathbf{X}$. Then

$$\mathbf{x}_i' = (1 \ x_{i1} \ x_{i2} \ ... \ x_{ip})$$

is a $1 \times (p + 1)$ row vector which allows us to write

$$E(Y \mid X = x) = \beta_0 + \beta_1 x_1 + \beta_2 x_2 + \ldots + \beta_p x_p = \mathbf{x}_i'\beta$$

The residual sum of squares as a function of β can be written in matrix form as

$$\text{RSS}(\beta) = (\mathbf{Y} - \mathbf{X}\beta)'(\mathbf{Y} - \mathbf{X}\beta) \tag{5.4}$$

Noting that $(\mathbf{AB})' = \mathbf{B}'\mathbf{A}'$ and that $\mathbf{B}'\mathbf{A} = \mathbf{A}'\mathbf{B}$ when the result is (1×1), expanding this last equation gives

$$\begin{aligned}
\text{RSS}(\beta) &= \mathbf{Y}'\mathbf{Y} + (\mathbf{X}\beta)'\,\mathbf{X}\beta - \mathbf{Y}'\mathbf{X}\beta - (\mathbf{X}\beta)'\,\mathbf{Y} \\
&= \mathbf{Y}'\mathbf{Y} + \beta'(\mathbf{X}'\mathbf{X})\beta - 2\mathbf{Y}'\mathbf{X}\beta
\end{aligned}$$

To find the least squares estimates we differentiate this last equation with respect to β, equate the result to zero and then cancel out the 2 common to both sides. This gives the following matrix form of the normal equations

$$(\mathbf{X}'\mathbf{X})\beta = \mathbf{X}'\mathbf{Y} \tag{5.5}$$

Assuming that the inverse of the matrix $(\mathbf{X}'\mathbf{X})$ exists, the least squares estimates are given by

$$\hat{\beta} = (\mathbf{X}'\mathbf{X})^{-1}\mathbf{X}'\mathbf{Y} \tag{5.6}$$

The fitted or predicted values are given by

$$\hat{\mathbf{Y}} = \mathbf{X}\hat{\beta} \tag{5.7}$$

and the residuals are given by

$$\hat{\mathbf{e}} = \mathbf{Y} - \hat{\mathbf{Y}} = \mathbf{Y} - \mathbf{X}\hat{\beta} \tag{5.8}$$

Special case: simple linear regression
Consider the regression model

$$\mathbf{Y} = \mathbf{X}\beta + \mathbf{e}.$$

For simple linear regression the matrix, $\mathbf{X}$ is given by

$$\mathbf{X} = \begin{pmatrix} 1 & x_1 \\ 1 & x_2 \\ \vdots & \vdots \\ 1 & x_n \end{pmatrix}$$

Thus,

$$\mathbf{X}'\mathbf{X} = \begin{pmatrix} 1 & 1 & \cdots & 1 \\ x_1 & x_2 & \cdots & x_n \end{pmatrix} \begin{pmatrix} 1 & x_1 \\ 1 & x_2 \\ \vdots & \vdots \\ 1 & x_n \end{pmatrix} = \begin{pmatrix} n & \sum_{i=1}^{n} x_i \\ \sum_{i=1}^{n} x_i & \sum_{i=1}^{n} x_i^2 \end{pmatrix} = n \begin{pmatrix} 1 & \bar{x} \\ \bar{x} & \dfrac{1}{n}\sum_{i=1}^{n} x_i^2 \end{pmatrix}$$

and

$$\mathbf{X}'\mathbf{Y} = \begin{pmatrix} 1 & 1 & \cdots & 1 \\ x_1 & x_2 & \cdots & x_n \end{pmatrix} \begin{pmatrix} y_1 \\ y_2 \\ \vdots \\ y_n \end{pmatrix} = \begin{pmatrix} \sum_{i=1}^{n} y_i \\ \sum_{i=1}^{n} x_i y_i \end{pmatrix}$$

Taking the inverse of $(\mathbf{X}'\mathbf{X})$ gives

$$(\mathbf{X}'\mathbf{X})^{-1} = \frac{1}{n\left(\dfrac{1}{n}\sum_{i=1}^{n} x_i^2 - (\bar{x})^2\right)} \begin{pmatrix} \dfrac{1}{n}\sum_{i=1}^{n} x_i^2 & -\bar{x} \\ -\bar{x} & 1 \end{pmatrix}$$

$$= \frac{1}{SXX} \begin{pmatrix} \dfrac{1}{n}\sum_{i=1}^{n} x_i^2 & -\bar{x} \\ -\bar{x} & 1 \end{pmatrix}$$

Putting all these pieces together gives

$$\hat{\beta} = (\mathbf{X}'\mathbf{X})^{-1}\mathbf{X}'\mathbf{Y}$$

$$= \frac{1}{SXX} \begin{pmatrix} \dfrac{1}{n}\sum_{i=1}^{n} x_i^2 & -\bar{x} \\ -\bar{x} & 1 \end{pmatrix} \begin{pmatrix} \sum_{i=1}^{n} y_i \\ \sum_{i=1}^{n} x_i y_i \end{pmatrix}$$

$$= \frac{1}{SXX} \begin{pmatrix} \dfrac{1}{n}\sum_{i=1}^{n} y_i \sum_{i=1}^{n} x_i^2 - \bar{x}\sum_{i=1}^{n} x_i y_i \\ \sum_{i=1}^{n} x_i y_i - \bar{x}\sum_{i=1}^{n} y_i \end{pmatrix}$$

$$= \begin{pmatrix} \dfrac{\bar{y}\left\{\sum_{i=1}^{n} x_i^2 - n\bar{x}^2\right\} - \bar{x}\left\{\sum_{i=1}^{n} x_i y_i - n\overline{xy}\right\}}{SXX} \\ \dfrac{SXY}{SXX} \end{pmatrix}$$

$$= \begin{pmatrix} \bar{y} - \dfrac{SXY}{SXX}\bar{x} \\ \dfrac{SXY}{SXX} \end{pmatrix}$$

matching the results in Chapter 2.

Properties of least squares estimates

Consider the regression model given by (5.3), i.e.,

$$\mathbf{Y} = \mathbf{X}\beta + \mathbf{e}$$

with $\text{Var}(\mathbf{e}) = \sigma^2 \mathbf{I}$ where $\mathbf{I}$ is the $(n \times n)$ identity matrix.

From (5.6), the least squares estimates are given

$$\hat{\beta} = (\mathbf{X}'\mathbf{X})^{-1}\mathbf{X}'\mathbf{Y}$$

We next derive the conditional mean and variance of the least squares estimates:

$$
\begin{aligned}
\text{E}\left(\hat{\beta} \mid \mathbf{X}\right) &= E\left((\mathbf{X}'\mathbf{X})^{-1}\mathbf{X}'\mathbf{Y} \mid \mathbf{X}\right) \\
&= (\mathbf{X}'\mathbf{X})^{-1}\mathbf{X}'E\left(\mathbf{Y} \mid \mathbf{X}\right) \\
&= (\mathbf{X}'\mathbf{X})^{-1}\mathbf{X}'\mathbf{X}\beta \\
&= \beta
\end{aligned}
$$

$$
\begin{aligned}
\text{Var}\left(\hat{\beta} \mid \mathbf{X}\right) &= \text{Var}\left((\mathbf{X}'\mathbf{X})^{-1}\mathbf{X}'\mathbf{Y} \mid \mathbf{X}\right) \\
&= (\mathbf{X}'\mathbf{X})^{-1}\mathbf{X}'\text{Var}\left(\mathbf{Y} \mid \mathbf{X}\right)\mathbf{X}(\mathbf{X}'\mathbf{X})^{-1} \\
&= (\mathbf{X}'\mathbf{X})^{-1}\mathbf{X}'\sigma^2\mathbf{I}\mathbf{X}(\mathbf{X}'\mathbf{X})^{-1} \\
&= \sigma^2(\mathbf{X}'\mathbf{X})^{-1}\mathbf{X}'\mathbf{X}(\mathbf{X}'\mathbf{X})^{-1} \\
&= \sigma^2(\mathbf{X}'\mathbf{X})^{-1}
\end{aligned}
$$

using the fact that $(\mathbf{X}'\mathbf{X})^{-1}$ is a symmetric matrix.

Reality check: simple linear regression

We saw earlier that for the case of simple linear regression

$$(\mathbf{X}'\mathbf{X})^{-1} = \frac{1}{SXX}\begin{pmatrix} \dfrac{1}{n}\sum_{i=1}^{n} x_i^2 & -\bar{x} \\ -\bar{x} & 1 \end{pmatrix}$$

Thus, for the slope of the fitted line

$$\text{Var}(\hat{\beta}_1) = \frac{\sigma^2}{SXX}$$

as we found in Chapter 2.

Residual sum of squares

From (5.4), the residual sum of squares as a function of the least squares estimates $\hat{\beta}$ can be written in matrix form as

$$\text{RSS} = \text{RSS}(\hat{\beta}) = \left(\mathbf{Y} - \mathbf{X}\hat{\beta}\right)'\left(\mathbf{Y} - \mathbf{X}\hat{\beta}\right) = \hat{\mathbf{e}}'\hat{\mathbf{e}} = \sum_{i=1}^{n}\hat{e}_i^2$$

Estimating the error variance

It can be shown that

$$S^2 = \frac{\text{RSS}}{n-p-1} = \frac{1}{n-p-1}\sum_{i=1}^{n}\hat{e}_i^2$$

is an unbiased estimate of σ^2.

Confidence intervals and tests of significance

Assuming that the errors are normally distributed with constant variance, it can be shown that for $i = 0, 1, ..., p$

$$T_i = \frac{\hat{\beta}_i - \beta_i}{\text{se}(\hat{\beta})} \sim t_{n-p-1}$$

where $\text{se}(\hat{\beta}_i)$ is the estimated standard deviation of $\hat{\beta}_i$ obtained by replacing σ by S. Note that $\text{se}(\hat{\beta}_i)$ can be obtained from R.

Notice that the degrees of freedom satisfy the formula:

Degrees of freedom = Sample size – Number of mean parameters estimated.

In this case we are estimating $p + 1$ such parameters, namely, $\beta_0, \beta_1, ..., \beta_p$.

Analysis of variance approach to testing whether there is a linear association between Y and a subset/all of the predictors

There is a linear association between Y and a subset/all of $X_1, X_2, ..., X_p$ if

$$Y = \beta_0 + \beta_1 x_1 + \beta_2 x_2 + ... + \beta_p x_p + e$$

and some/all of the $\beta_i \neq 0$ $(i = 1, 2, ..., p)$. Thus we wish to test

$$H_0 : \beta_1 = \beta_2 = ... = \beta_p = 0 \text{ against}$$

$$H_A : \text{at least some of the } \beta_i \neq 0 .$$

Once again we define the following terminology:

Total corrected sum of squares of the Y's,

$$\text{SST} = SYY = \sum_{i}^{n}(y_i - \overline{y})^2$$

Residual sum of squares, $\text{RSS} = \sum_{i}^{n}(y_i - \hat{y}_i)^2$

Regression sum of squares (i.e., sum of squares explained by the regression model),

$$\text{SSreg} = \sum_{i}^{n}(\hat{y}_i - \overline{y})^2$$

It can be shown that

$$\text{SST} = \text{SSreg} \qquad + \text{RSS}$$

$$\text{Total sample} = \text{Variability explained by} + \text{Unexplained (or error)}$$
$$\text{variability} \qquad \text{the model} \qquad\qquad \text{variability}$$

If

$$Y = \beta_0 + \beta_1 x_1 + \beta_2 x_2 + \ldots + \beta_p x_p + e \quad \text{and}$$

$$\text{at least one of the } \beta_i \neq 0$$

then RSS should be "small" and SSreg should be "close" to SST. But how small is "small" and how close is "close"? To test

$$H_0 : \beta_1 = \beta_2 = \ldots = \beta_p = 0 \quad \text{against}$$

$$H_A : \text{at least some of the } \beta_i \neq 0$$

we can use the test statistic

$$F = \frac{\text{SSreg} / p}{\text{RSS} / (n - p - 1)}$$

since RSS has $(n - p - 1)$ degrees of freedom and SSR has p degrees of freedom. Under the assumption that $e_1, e_2, \ldots, e_n$ are independent and normally distributed, it can be shown that F has an F distribution with p and $n - p - 1$ degrees of freedom when H_0 is true. The usual way of setting out this test is to use the following:

Analysis of variance table

Source of variation	Degrees of freedom (df)	Sum of squares (SS)	Mean square (MS)	F
Regression	p	SSreg	SSreg/p	$F = \dfrac{\text{SSreg} / p}{\text{RSS} / (n - p - 1)}$
Residual	$n - p - 1$	RSS	$S^2 = \text{RSS}/ (n - p - 1)$	
Total	$n - 1$	SST = SYY		

Notes:

1. R^2, the coefficient of determination of the regression line, is defined as the proportion of the total sample variability in the Y's explained by the regression model, that is,

$$R^2 = \frac{\text{SSreg}}{\text{SST}} = 1 - \frac{\text{RSS}}{\text{SST}}$$

Adding irrelevant predictor variables to the regression equation often increases R^2. To compensate for this one can define an adjusted coefficient of determination, R^2_{adj}

$$R^2_{adj} = 1 - \frac{RSS/(n-p-1)}{SST/(n-1)}$$

Note that $S^2 = \dfrac{RSS}{n-p-1}$ is an unbiased estimate of $\sigma^2 = Var(e_i) = Var(Y_i)$
while $SST/(n-1)$ is an unbiased estimate of $\sigma^2 = Var(Y_i)$ when $\beta_1 = ... = \beta_p = 0$.
Thus, when comparing models with different numbers of predictors one should use R^2_{adj} and not R^2.

2. The F-test is always used first to test for the existence of a linear association between Y and ANY of the p x-variables. If the F-test is significant then a natural question to ask is

For which of the p x-variables is there evidence of a linear association with Y?
To answer this question we could perform p separate t-tests of $H_0 : \beta_1 = 0$.
However, as we shall see later there are problems with interpreting these t-tests when the predictor variables are highly correlated.

Testing whether a specified subset of the predictors have regression coefficients equal to 0

Suppose that we are interested in testing

$$H_0 : \beta_1 = \beta_2 = ... = \beta_k = 0 \text{ where } k < p$$

i.e., $Y = \beta_0 + \beta_{k+1}x_{k+1} + ... + \beta_p x_p + e$ (reduced model)

against

$$H_A : H_0 \text{ is not true}$$

i.e., $Y = \beta_0 + \beta_1 x_1 + ... + \beta_k x_k + \beta_{k+1}x_{k+1} + ... + \beta_p x_p + e$ (full model)

This can be achieved using an F-test. Let RSS(Full) be the residual sum of squares under the full model (i.e., the model which includes all the predictors, i.e., H_A) and RSS(Reduced) be the residual sum of squares under the reduced model (i.e., the model which includes only the predictors thought to be non-zero, i.e., H_0). Then the F-statistic is given by

$$F = \frac{(RSS(reduced) - RSS(full))/(df_{reduced} - df_{full})}{RSS(full)/df_{full}}$$

$$= \frac{(RSS(reduced) - RSS(full))/k}{RSS(full)/(n-p-1)}$$

since the reduced model has $p + 1 - k$ predictors and

$$[n - (p + 1 - k)] - [n - (p + 1)] = k.$$

This is called a **partial F-test.**

Menu pricing in a new Italian restaurant in New York City (cont.)
Recall from Chapter 1 that you have been asked to produce a regression model to
predict the price of dinner. Data from surveys of customers of 168 Italian restaurants in the target area are available. The data are in the form of the average of
customer views on

Y = Price = the price (in $US) of dinner (including 1 drink & a tip)
x_1 = Food = customer rating of the food (out of 30)
x_2 = Décor = customer rating of the decor (out of 30)
x_3 = Service = customer rating of the service (out of 30)
x_4 = East = dummy variable = 1 (0) if the restaurant is east (west) of Fifth Avenue

The data are given on the book web site in the file nyc.csv. The source of the
data is the following restaurant guide book
Zagat Survey 2001: New York City Restaurants, Zagat, New York
In particular you have been asked to

(a) Develop a regression model that ***directly predicts*** the price of dinner (in dollars)
using a subset or all of the 4 potential predictor variables listed above.
(b) Determine which of the predictor variables Food, Décor and Service has the
largest estimated effect on Price? Is this effect also the most statistically
significant?
(c) If the aim is to choose the location of the restaurant so that the price achieved
for dinner is maximized, should the new restaurant be on the east or west of
Fifth Avenue?
(d) Does it seem possible to achieve a price premium for "setting a new standard
for high-quality service in Manhattan" for Italian restaurants?

Since interest centers on developing a regression model that directly predicts
price, we shall begin by considering the following model:

$$Y = \beta_0 + \beta_1 x_1 + \beta_2 x_2 + \beta_3 x_3 + \beta_4 x_4 + e \qquad (5.9)$$

At this point we shall assume that all the necessary assumptions hold. In particular, we shall assume that (5.9) is a valid model for the data. We shall check these
assumptions for this example in the next chapter and at that point identify any outliers. Given below is some output from R after fitting model (5.9):

Regression output from R

```
Call:
lm(formula =   Price  ~  Food  +  Decor  +  Service  +  East)
Coefficients:
                Estimate    Std. Error    t value    Pr(>|t|)
(Intercept)   -24.023800     4.708359      -5.102    9.24e-07   ***
Food            1.538120     0.368951       4.169    4.96e-05   ***
Decor           1.910087     0.217005       8.802    1.87e-15   ***
```

```
Service       -0.002727       0.396232      -0.007      0.9945
East           2.068050       0.946739       2.184      0.0304   *
---
Signif. codes:  `***' 0.001   `**' 0.01 `*' 0.05 `.'0.1 `  ` 1
Residual standard error: 5.738 on 163 degrees of freedom
Multiple R-Squared: 0.6279, Adjusted R-squared:0.6187
F-statistic: 68.76 on 4 and 163 DF, p-value: <2.2e-16
```

(a) The initial regression model is

$$Price = -24.02 + 1.54\ Food + 1.91\ Decor - 0.003\ Service + 2.07\ East$$

At this point we shall leave the variable Service in the model even though its regression coefficient is not statistically significant.

(b) The variable Décor has the largest effect on Price since its regression coefficient is largest. Note that Food, Décor and Service are each measured on the same 0 to 30 scale and so it is meaningful to compare regression coefficients. The variable Décor is also the most statistically significant since its p-value is the smallest of the three.

(c) In order that the price achieved for dinner is maximized, the new restaurant should be on the east of Fifth Avenue since the coefficient of the dummy variable is statistically significantly larger than 0.

(d) It does not seem possible to achieve a price premium for "setting a new standard for high quality service in Manhattan" for Italian restaurants since the regression coefficient of Service is not statistically significantly greater than zero.

Given below is some output from R after dropping the predictor Service from model (5.9):

Regression output from R

```
Call:
lm(formula  =    Price   ~     Food    +      Decor  + East)
Coefficients:
              Estimate    Std. Error   t value    Pr(>|t|)
(Intercept)   -24.0269      4.6727      -5.142    7.67e-07    ***
Food            1.5363      0.2632       5.838    2.76e-08    ***
Decor           1.9094      0.1900      10.049    <2e-16      ***
East            2.0670      0.9318       2.218    0.0279      *
---
Signif. codes:  0 `***' 0.001 `**' 0.01 `*' 0.05 `.' 0.1 `  ` 1
Residual standard error: 5.72 on 164 degrees of freedom
Multiple R-Squared:    0.6279, Adjusted  R-squared:0.6211
F-statistic:   92.24   on  3  and   164 DF, p-value:  < 2.2e-16
```

The final regression model is

$$\text{Price} = -24.03 + 1.54 \text{ Food} + 1.91 \text{ Decor} + 2.07 \text{ East}$$

Comparing the last two sets of output from R, we see that the regression coefficients for the variables in both models are very similar. This does **not** always occur. In fact, we shall see that dropping predictors from a regression model can have a dramatic effect on the coefficients of the remaining predictors. We shall discuss this and other issues related to choosing predictor variables for a "final" model in Chapter 7.

5.3 Analysis of Covariance

Consider the situation in which we want to model a response variable, Y based on a continuous predictor, x and a dummy variable, d. Suppose that the effect of x on Y is linear. This situation is the simplest version of what is commonly referred as **Analysis of Covariance**, since the predictors include both quantitative variables (i.e., x) and qualitative variables (i.e., d).

Coincident regression lines: The simplest model in the given situation is one in which the dummy variable has no effect on Y, that is,

$$Y = \beta_0 + \beta_1 x + e$$

and the regression line is exactly the same for both values of the dummy variable.

Parallel regression lines: Another model to consider for this situation is one in which the dummy variable produces only an additive change in Y, that is,

$$Y = \beta_0 + \beta_1 x + \beta_2 d + e = \begin{cases} Y = \beta_0 + \beta_1 x + e & \text{when } d=0 \\ Y = \beta_0 + \beta_2 + \beta_1 x + e & \text{when } d=1 \end{cases}$$

In this case, the regression coefficient β_2 measures the additive change in Y due to the dummy variable.

Regression lines with equal intercepts but different slopes: A third model to consider for this situation is one in which the dummy variable only changes the size of the effect of x on Y, that is,

$$Y = \beta_0 + \beta_1 x + \beta_3 d \times x + e = \begin{cases} Y = \beta_0 + \beta_1 x + e & \text{when } d=0 \\ Y = \beta_0 + (\beta_1 + \beta_3) x + e & \text{when } d=1 \end{cases}$$

Unrelated regression lines: The most general model is appropriate when the dummy variable produces an additive change in Y and also changes the size of the effect of x on Y. In this case the appropriate model is

$$Y = \beta_0 + \beta_1 x + \beta_2 d + \beta_3 d \times x + e = \begin{cases} Y = \beta_0 + \beta_1 x + e & \text{when } d=0 \\ Y = \beta_0 + \beta_2 + (\beta_1 + \beta_3) x + e & \text{when } d=1 \end{cases}$$

In the unrelated regression lines model, the regression coefficient β_2 measures the additive change in Y due to the dummy variable, while the regression coefficient β_3 measures the change in the size of the effect of x on Y due to the dummy variable.

Stylized example: Amount spent on travel

This stylized example is based on a problem in a text on business statistics. The background to the example is as follows:

> A small travel agency has retained your services to help them better understand two important customer segments. The first segment, which we will denote by A, consists of those customers who have purchased an adventure tour in the last twelve months. The second segment, which we will denote by C, consists of those customers who have purchased a cultural tour in the last twelve months. Data are available on 925 customers (i.e. on 466 customers from segment A and 459 customers from segment C). Note that the two segments are completely separate in the sense that there are no customers who are in both segments. Interest centres on *identifying any differences between the two segments in terms of the amount of money spent in the last twelve months*. In addition, data are also available on the age of each customer, since age is thought to have an effect on the amount spent.

The data in Figure 5.7 are given on the book web site in the file travel.txt. The first three and the last three rows of the data appear in Table 5.1.

It is clear from Figure 5.7 that the dummy variable for segment changes the size of the effect of Age, x on Amount Spent, Y. We shall also allow for the dummy variable for Segment to produce an additive change in Y. In this case the appropriate model is what we referred to above as *Unrelated regression lines*

$$Y = \beta_0 + \beta_1 x + \beta_2 C + \beta_3 C \times x + e = \begin{cases} Y = \beta_0 + \beta_1 x + e & \text{when } C=0 \\ Y = \beta_0 + \beta_2 + (\beta_1 + \beta_3) x + e & \text{when } C=1 \end{cases}$$

where

Y = amount spent; x = Age; and C is a dummy variable which is 1 when the customer is from Segment C and 0 otherwise (i.e., if the customer is in Segment A). The output from R is as follows:

Regression output from R

```
Call:
lm(formula   =   Amount    ~   Age    +      C +      C:Age)
Coefficients:
              Estimate    Std. Error    t value    Pr(>|t|)
(Intercept)   1814.5445      8.6011       211.0     <2e-16    ***
Age            -20.3175      0.1878      -108.2     <2e-16    ***
C            -1821.2337     12.5736      -144.8     <2e-16    ***
Age:C           40.4461      0.2724       148.5     <2e-16    ***

Residual standard error: 47.63 on 921 degrees of freedom
Multiple  R-Squared:    0.9601, Adjusted R-squared:0.9599
F-statistic:    7379  on  3  and  921    DF,  p-value:<2.2e-16
```

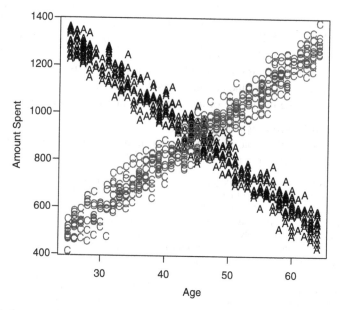

Figure 5.7 A scatter plot of Amount Spent versus Age for segments A and C

Table 5.1 Amount spent on travel for two market segments (A & C)

Amount	Age	Segment	C
997	44	A	0
997	43	A	0
951	41	A	0
.	.	.	.
1,111	57	C	1
883	43	C	1
1,038	53	C	1

Notice that all the regression coefficients are highly statistically significant. Thus, we shall use as a final model

$$Y = \hat{\beta}_0 + \hat{\beta}_1 x + \hat{\beta}_2 C + \hat{\beta}_3 C \times x = \begin{cases} Y = \hat{\beta}_0 + \hat{\beta}_1 x & \text{when } C=0 \\ Y = \hat{\beta}_0 + \hat{\beta}_2 + \left(\hat{\beta}_1 + \hat{\beta}_3\right)x & \text{when } C=1 \end{cases}$$

For customers in segment A (i.e., $C = 0$) our model predicts

$$\text{Amount Spent} = \$1814.54 - \$20.32 \times \text{Age}$$

while for customers in segment C (i.e., $C = 1$) our model predicts

$$\text{Amount Spent} = -\$6.69 + \$20.13 \times \text{Age}$$

since

$$1814.5445 + (-1821.2337) = -6.69$$

and

$$-20.3175 + 40.4461 = 20.13.$$

Thus, in segment A (i.e., those customers who have purchased an adventure tour) the amount spent decreases with Age while in Segment C (i.e., those customers who have purchased a cultural tour) the amount spent increases with Age.

Finally, imagine that we are interested in an overall test of

$$H_0 : \beta_2 = \beta_3 = 0$$

i.e., $Y = \beta_0 + \beta_1 x_1 + e$ (reduced model: *coincident regression lines:*)

against

$$H_A : H_0 \text{ is not true}$$

i.e., $Y = \beta_0 + \beta_1 x + \beta_2 C + \beta_3 C \times x + e$ (full model: *unrelated lines*)

The fit under the full model is given on a previous page, while the fit under the reduced model appears next:

Regression output from R

```
Call:
lm(formula = Amount ~ Age)
Coefficients:
               Estimate    Std. Error    t value    Pr(>|t|)
(Intercept)    957.9103      31.3056      30.599     <2e-16    ***
Age             -1.1140       0.6784      -1.642      0.101
---
Residual standard error: 237.7 on 923 degrees of freedom
Multiple R-Squared: 0.002913, Adjusted R-squared: 0.001833
F-statistic: 2.697 on 1 and 923 DF, p-value: 0.1009
```

The test can be achieved using a partial F-test. Let RSS(Full) be the residual sum of squares under the full model (i.e., the model which includes all the predictors, i.e., H_A) and RSS(Reduced) be the residual sum of squares under the reduced model (i.e., the model which includes only the predictors thought to be nonzero, i.e., H_0). Then the F-statistic is given by

$$F = \frac{\left(\text{RSS(reduced)} - \text{RSS(full)}\right) / \left(\text{df}_{\text{reduced}} - \text{df}_{\text{full}}\right)}{\text{RSS(full)} / \text{df}_{\text{full}}}$$

Given below is the output from R associated with this F-statistic:

Regression output from R

```
Analysis of Variance Table
Model 1: Amount ~ Age
Model 2: Amount ~ Age + C + C:Age
  Res. Df       RSS   Df   Sum of Sq        F      Pr(>F)
1     923  52158945
2     921   2089377    2    50069568    11035    <2.2e-16    ***
---
Signif.codes:    0 `***' 0.001 `**' 0.01 `*' 0.05 `.' 0.1 ` '1
```

Calculating by hand we find that

$$F = \frac{(52158945 - 2089377)/(923 - 921)}{2089377/921} = \frac{25034784}{2268.6} = 11035$$

which agrees with the result from the R-output.

As expected there is very strong evidence against the reduced model in favour of the full model. Thus, we prefer the unrelated regression lines model to the coincident lines model.

Menu pricing in a new Italian restaurant in New York City (cont.)

Recall from Chapter 1 and earlier in Chapter 5 that the data are in the form of the average of customer views on

Y = Price = the price (in \$US) of dinner (including one drink and a tip)
x_1 = Food = customer rating of the food (out of 30)
x_2 = Décor = customer rating of the decor (out of 30)
x_3 = Service = customer rating of the service (out of 30)
x_4 = East = dummy variable = 1 (0) if the restaurant is east (west) of Fifth Avenue

Earlier in Chapter 5 we obtained what we referred to as the "final regression model," namely

$$\text{Price} = -24.03 + 1.54 \text{ Food} + 1.91 \text{ Decor} + 2.07 \text{ East}$$

In particular, the variable Service had very little effect on Price and was omitted from the model.

When the young New York City chef, who plans to create the new Italian restaurant in Manhattan, is shown your regression model there is much discussion of the implications of the model. In particular, heated debate focuses on the statistical insignificance of the predictor variable Service, since the stated aims of the restaurant are to provide the highest quality Italian food utilizing state-of-the-art décor while setting a new standard for high-quality service in Manhattan. The general consensus from the other members of the team supporting the chef is that the model is too simple to reflect the reality of Italian restaurants in Manhattan. In particular, there is general consensus amongst the team that restaurants on the east of Fifth Avenue are very different from those on the west side with service and décor thought to be more important on the east of Fifth Avenue. As such you have been asked to consider different models for the East and West.

In order to investigate whether the effect of the predictors depends on the dummy variable East, we shall consider the following model which is an extension of the *unrelated regression lines* model to more than one predictor variable:

$$Y = \beta_0 + \beta_1 x_1 + \beta_2 x_2 + \beta_3 x_3 + \beta_4 \text{East}$$
$$+ \beta_5 x_1 \times \text{East} + \beta_6 x_2 \times \text{East} + \beta_7 x_3 \times \text{East} + e \quad \text{(Full)}$$

where x_1 = Food, x_2 = Décor and x_3 = Service. Regression output from R showing the fit of this model appears next:

Regression output from R

```
Call:
lm(formula = Price ~ Food + Decor + Service + East +
Food:East + Decor:East + Service:East)
Coefficients:
              Estimate   Std. Error   t value   Pr(>|t|)
(Intercept)   -26.9949     8.4672      -3.188    0.00172    **
Food            1.0068     0.5704       1.765    0.07946    .
Decor           1.8881     0.2984       6.327    2.40e-09   ***
Service         0.7438     0.6443       1.155    0.25001
East            6.1253    10.2499       0.598    0.55095
Food:East       1.2077     0.7743       1.560    0.12079
Decor:East     -0.2500     0.4570      -0.547    0.58510
Service:East   -1.2719     0.8171      -1.557    0.12151
---
Signif. codes:  0 '***' 0.001 '**' 0.01 '*' 0.05 '.'0.1 ' ' 1
Residual standard error: 5.713 on 160 degrees of freedom
Multiple R-Squared: 0.6379,       Adjusted R-squared:0.622
F-statistic: 40.27 on 7 and 160 DF, p-value: <2.2e-16
```

Notice how none of the regression coefficients for the interaction terms are statistically significant. However, both the interactions between Food and East and Service and East have *p*-values equal to 0.12.

We next compare the full model above with what we previously called the "final" model. We consider an overall test of

$$H_0 : \beta_3 = \beta_5 = \beta_6 = \beta_7 = 0$$

i.e., $Y = \beta_0 + \beta_1 x_1 + \beta_2 x_2 + \beta_4 \text{ East} + e$ (Reduced)

against

$$H_A : H_0 \text{ is not true}$$

i.e.,

$$Y = \beta_0 + \beta_1 x_1 + \beta_2 x_2 + \beta_3 x_3 + \beta_4 \text{East}$$
$$+ \beta_5 x_1 \times \text{East} + \beta_6 x_2 \times \text{East} + \beta_7 x_3 \times \text{East} + e \quad \text{(Full)}$$

The fit under the full model is given above, while the fit under the reduced model appears next.

Regression output from R

```
Call:
lm(formula = Price ~ Food + Decor + East)
Coefficients:
               Estimate   Std. Error   t value   Pr(>|t|)
(Intercept)    -24.0269       4.6727    -5.142   7.67e-07   ***
Food             1.5363       0.2632     5.838   2.76e-08   ***
Decor            1.9094       0.1900    10.049    < 2e-16   ***
East             2.0670       0.9318     2.218     0.0279   *
---

Residual standard error: 5.72 on 164 degrees of freedom
Multiple R-Squared: 0.6279, Adjusted R-squared:0.6211
F-statistic: 92.24 on 3 and 164 DF,  p-value:<2.2e-16
```

The test of whether the effect of the predictors depends on the dummy variable East can be achieved using the following partial F-test:

$$F = \frac{\left(\text{RSS(reduced)} - \text{RSS(full)}\right)/\left(\text{df}_{\text{reduced}} - \text{df}_{\text{full}}\right)}{\text{RSS(full)}/\text{df}_{\text{full}}}$$

Given below is the output from R associated with this F-statistic:

Regression output from R

```
Analysis of Variance Table
Model 1: Price ~ Food + Decor + East
Model 2:  Price ~ Food + Decor + Service + East + Food:East +
Decor:East      + Service:East
    Res.Df       RSS    Df   Sum of Sq         F    Pr(>F)
1      164    5366.5
2      160    5222.2     4       144.4    1.1057    0.3558
```

Given the *p*-value equals 0.36, there is little, if any, evidence to support the alternative hypothesis (i.e., the need for different models for the East and West). This means that we are happy to adopt the reduced model:

$$Y = \beta_0 + \beta_1 x_1 + \beta_2 x_2 + \beta_4 \text{ East} + e \quad \text{(Reduced)}$$

5.4 Exercises

1. This problem is based on CASE 32 – Overdue Bills from Bryant and Smith (1995). Quick Stab Collection Agency (QSCA) is a bill-collecting agency that specializes in collecting small accounts. To distinguish itself from competing collection agencies, the company wants to establish a reputation for collecting delinquent accounts quickly. The marketing department has just suggested that QSCA adopt the slogan: "Under 60 days or your money back!!!!"

   You have been asked to look at account balances. In fact, you suspect that the number of days to collect the payment is related to the size of the bill. If this is the case, you may be able to estimate how quickly certain accounts are likely to be collected, which, in turn, may assist the marketing department in determining an appropriate level for the money-back guarantee.

To test this theory, a random sample of accounts closed out during the months of January through June has been collected. The data set includes the initial size of the account and the total number of days to collect payment in full. Because QSCA deals in both household and commercial accounts in about the same proportion, an equal number have been collected from both groups. The first 48 observations in the data set are residential accounts and the second 48 are commercial accounts. The data can be found on the book web site in the file named overdue.txt. In this data set, the variable LATE is the number of days the payment is overdue, BILL is the amount of the overdue bill in dollars and TYPE identifies whether an account is RESIDENTIAL or COMMERCIAL.

Develop a regression model to predict LATE from BILL.

2. On July 23, 2006, the *Houston Chronicle* published an article entitled "Reading: First-grade standard too tough for many". The article claimed in part that "more students (across Texas) are having to repeat first grade. Experts attribute the increase partially to an increase in poverty." The article presents data for each of 61 Texas counties on

Y = Percentage of students repeating first grade
x = Percentage of low-income students

for both 2004–2005 and 1994–1995. The data can be found on the book web site in the file HoustonChronicle.csv. Use analysis of covariance to decide whether:

(a) An increase in the percentage of low income students is associated with an increase in the percentage of students repeating first grade.
(b) There has been an increase in the percentage of students repeating first grade between 1994–1995 and 2004–2005
(c) Any association between the percentage of students repeating first grade and the percentage of low-income students differs between 1994–1995 and 2004–2005.

3. Chateau Latour is widely acknowledged as one of the world's greatest wine estates with a rich history dating back to at least 1638. The Grand Vin de Chateau Latour is a wine of incredible power and longevity. At a tasting in New York in April 2000, the 1863 and 1899 vintages of Latour were rated alongside the 1945 and the 1961 vintages as the best in a line-up of 39 vintages ranging from 1863 to 1999 (*Wine Spectator*, August 31, 2000). Quality of a particular vintage of Chateau Latour has a huge impact on price. For example, in March 2007, the 1997 vintage of Chateau Latour could be purchased for as little as $159 per bottle while the 2000 vintage of Chateau Latour costs as least $700 per bottle (www.wine-searcher.com).

While many studies have identified that the timing of the harvest of the grapes has an important effect on the quality of the vintage, with quality improving the earlier the harvest. A less explored issue of interest is the effect of unwanted rain at vintage time on the quality of icon wine like Chateau Latour. This question addresses this issue.

The Chateau Latour web site (www.chateau-latour.com) provides a rich source of data. In particular, data on the quality of each vintage, harvest dates and weather at harvest time were obtained from the site for the vintages from 1961 to 2004. An example of the information on weather at harvest time is given below for the 1994 vintage:

Harvest began on the 13th September and lasted on the 29th, frequently interrupted by storm showers. But quite amazingly the dilution effect in the grapes was very limited .... (http://www.chateau-latour.com/commentaires/1994uk.html"; Accessed: March 16, 2007)

Each vintage was classified as having had "unwanted rain at harvest" (e.g., the 1994 vintage) or not (e.g., the 1996 vintage) on the basis of information like that reproduced above. Thus, the data consist of:

Vintage = year the grapes were harvested
Quality – on a scale from 1 (worst) to 5 (best) with some half points
End of harvest – measured as the number days since August 31
Rain – a dummy variable for unwanted rain at harvest = 1 if yes.
The data can be found on the book web site in the file latour.csv.
The first model considered was:

$$\text{Quality} = \beta_0 + \beta_1 \text{End of Harvest} + \beta_2 \text{Rain}$$
$$+ \beta_3 \text{End of Harvest} \times \text{Rain} + e \qquad (5.10)$$

A plot of the data and the two regression lines from model (5.10) can be found in Figure 5.8. In addition, numerical output appears below.

(a) Show that the coefficient of the interaction term in model (5.10) is statistically significant. In other words, show that the rate of change in quality rating depends on whether there has been any unwanted rain at vintage.
(b) Estimate the number of days of delay to the end of harvest it takes to decrease the quality rating by 1 point when there is:

(i) No unwanted rain at harvest
(ii) Some unwanted rain at harvest

Regression output from R

```
Call:
lm(formula = Quality ~ EndofHarvest + Rain +
Rain:EndofHarvest)
Coefficients:
                   Estimate   Std. Error   t value   Pr(>|t|)
(Intercept)         5.16122      0.68917     7.489   3.95e-09 ***
EndofHarvest       -0.03145      0.01760    -1.787   0.0816 .
Rain                1.78670      1.31740     1.356   0.1826
EndofHarvest:Rain  -0.08314      0.03160    -2.631   0.0120 *
---

Residual  standard  error: 0.7578  on 40 degrees  of freedom
Multiple  R-Squared:   0.6848,        Adjusted R-squared: 0.6612
F-statistic:   28.97 on 3 and    40 DF, p-value: 4.017e-10
```

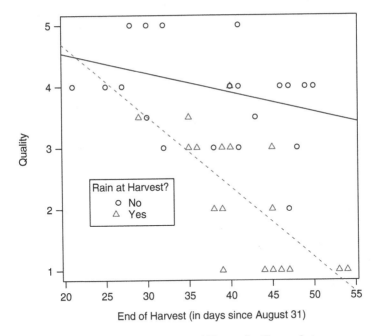

Figure 5.8 A scatter plot of Quality versus End of Harvest for Chateau Latour

```
Call:
lm(formula = Quality ~ EndofHarvest + Rain)
Coefficients:
              Estimate   Std. Error   t value   Pr(>|t|)
(Intercept)    6.14633      0.61896     9.930   1.80e-12   ***
EndofHarvest  -0.05723      0.01564    -3.660   0.000713   ***
Rain          -1.62219      0.25478    -6.367   1.30e-07   ***
---
Residual standard error: 0.8107 on 41 degrees of freedom
Multiple R-Squared: 0.6303,      Adjusted R-squared: 0.6123
F-statistic: 34.95 on 2 and 41 DF,  p-value: 1.383e-09

Analysis of Variance Table
Model 1: Quality ~ EndofHarvest + Rain
Model 2: Quality ~ EndofHarvest + Rain + Rain:EndofHarvest
    Res.Df      RSS    Df   Sum of Sq       F    Pr(>F)
1       41  26.9454
2       40  22.9705     1      3.9749  6.9218   0.01203   *
```

Chapter 6
Diagnostics and Transformations for Multiple Linear Regression

In the previous chapter we studied multiple linear regression. Throughout Chapter 5, we assumed that the multiple linear regression model was a valid model for the data. Thus, we implicitly made a series of assumptions. In this chapter we consider a series of tools known as regression diagnostics to check each of these assumptions. Having used these tools to diagnose potential problems with the assumptions, we look at how to first identify and then overcome or deal with common problems such as nonlinearity and nonconstant variance.

6.1 Regression Diagnostics for Multiple Regression

We next look at regression diagnostics in order to check the validity of all aspects of a regression model. When fitting a multiple regression model we will discover that it is important to:

1. Determine whether the proposed regression model is a valid model (i.e., determine whether it provides an adequate fit to the data). The main tools we will use to validate regression assumptions are plots involving **standardized residuals** and/ or **fitted values**. We shall see that these plots enable us to assess visually whether the assumptions are being violated and, *under certain conditions*, point to what should be done to overcome these violations. We shall also consider a tool, called **marginal model plots**, which have wider application than residual plots.
2. Determine which (if any) of the data points have predictor values that have an unusually large effect on the estimated regression model. (Recall that such points are called **leverage points**.)
3. Determine which (if any) of the data points are **outliers**, that is, points which do not follow the pattern set by the bulk of the data, when one takes into account the given model.
4. Assess the effect of each predictor variable on the response variable, having adjusted for the effect of other predictor variables using **added variable plots**.
5. Assess the extent of **collinearity** among the predictor variables using **variance inflation factors**.

S.J. Sheather, *A Modern Approach to Regression with R*,
DOI: 10.1007/978-0-387-09608-7_6, © Springer Science+Business Media LLC 2009

6. Examine whether the assumption of constant error variance is reasonable. If not, decide how can we overcome this problem.
7. If the data are collected over time, examine whether the data are correlated over time.

We shall begin by looking at the second item of the above list, leverage points, as once again, these will be needed in the definition of standardized residuals. However, before we begin let's briefly review some material from Chapter 5.

Matrix formulation of least squares regression

Define the $(n \times 1)$ vector, $\mathbf{Y}$ and the $n \times (p+1)$ matrix, $\mathbf{X}$ by

$$\mathbf{Y} = \begin{pmatrix} y_1 \\ y_2 \\ \vdots \\ y_n \end{pmatrix} \quad \mathbf{X} = \begin{pmatrix} 1 & x_{11} \cdots x_{1p} \\ 1 & x_{21} \cdots x_{2p} \\ \vdots & \vdots \\ 1 & x_{n1} \cdots x_{np} \end{pmatrix}$$

Also define the $(p+1) \times 1$ vector, β of unknown regression parameters and the $(n \times 1)$ vector, $\mathbf{e}$ of random errors

$$\beta = \begin{pmatrix} \beta_0 \\ \beta_1 \\ \vdots \\ \beta_p \end{pmatrix} \quad \mathbf{e} = \begin{pmatrix} e_1 \\ e_2 \\ \vdots \\ e_n \end{pmatrix}$$

As in (5.3), we can write the multiple linear regression model in matrix notation as

$$\mathbf{Y} = \mathbf{X}\beta + \mathbf{e} \qquad (6.1)$$

where $\text{Var}(\mathbf{e}) = \sigma^2 \mathbf{I}$ and $\mathbf{I}$ is the $(n \times n)$ identity matrix. Recall from (5.6) to (5.8) that the fitted values are given by

$$\hat{\mathbf{Y}} = \mathbf{X}\hat{\beta} \qquad (6.2)$$

where

$$\hat{\beta} = (\mathbf{X}'\mathbf{X})^{-1}\mathbf{X}'\mathbf{Y} \qquad (6.3)$$

and the residuals are given by

$$\hat{\mathbf{e}} = \mathbf{Y} - \hat{\mathbf{Y}} = \mathbf{Y} - \mathbf{X}\hat{\beta} \qquad (6.4)$$

6.1.1 Leverage Points in Multiple Regression

Recall that data points which exercise considerable influence on the fitted model are called *leverage points*. Recall also that leverage measures the extent to which the fitted regression model is attracted by the given data point. We are therefore interested in the relationship between the fitted values $\hat{\mathbf{Y}}$ and $\mathbf{Y}$.

From (6.2) and (6.3)

$$\hat{\mathbf{Y}} = \mathbf{X}\hat{\beta} = \mathbf{X}(\mathbf{X}'\mathbf{X})^{-1}\mathbf{X}'\mathbf{Y} = \mathbf{H}\mathbf{Y}$$

where

$$\mathbf{H} = \mathbf{X}(\mathbf{X}'\mathbf{X})^{-1}\mathbf{X}' \tag{6.5}$$

The $(n \times n)$ matrix $\mathbf{H}$ is commonly called the **hat matrix** since pre-multiplying $\mathbf{Y}$ by $\mathbf{H}$ changes $\mathbf{Y}$ into $\hat{\mathbf{Y}}$. According to Hoaglin and Welsh (1978, p. 17) the term hat matrix is due to John Tukey, who coined the term in the 1960s.

Let h_{ij} denote the (i,j)th element of $\mathbf{H}$, then

$$\hat{Y}_i = h_{ii}Y_i + \sum_{j \neq i} h_{ij}Y_j$$

where h_{ii} denotes the ith diagonal element of $\mathbf{H}$. Thus, as we saw in Chapter 3, h_{ii} measures the extent to which the fitted regression model $\hat{Y}_i$ is attracted by the given data point, Y_i.

Special case: Simple linear regression

Consider the simple linear regression model in matrix form

$$\mathbf{Y} = \mathbf{X}\beta + \mathbf{e}.$$

where $\mathbf{X}$ is given by

$$\mathbf{X} = \begin{pmatrix} 1 & x_1 \\ 1 & x_2 \\ \vdots & \vdots \\ 1 & x_n \end{pmatrix}$$

We found in Chapter 5 that

$$(\mathbf{X}'\mathbf{X})^{-1} = \frac{1}{SXX} \begin{pmatrix} \dfrac{1}{n}\sum_{i=1}^{n} x_i^2 & -\bar{x} \\ -\bar{x} & 1 \end{pmatrix}$$

Putting all the pieces together we find that

$$\mathbf{H} = \mathbf{X}(\mathbf{X}'\mathbf{X})^{-1}\mathbf{X}'$$

$$= \begin{pmatrix} 1 & x_1 \\ 1 & x_2 \\ \vdots & \vdots \\ 1 & x_n \end{pmatrix} \frac{1}{SXX} \begin{pmatrix} \dfrac{1}{n}\sum_{i=1}^{n} x_i^2 & -\bar{x} \\ -\bar{x} & 1 \end{pmatrix} \begin{pmatrix} 1 & 1 & \cdots & 1 \\ x_1 & x_2 & \cdots & x_n \end{pmatrix}$$

After multiplication and simplification we find that the (i,j) th element of **H** is given by

$$h_{ij} = \frac{1}{n} + \frac{(x_i - \bar{x})(x_j - \bar{x})}{SXX} = \frac{1}{n} + \frac{(x_i - \bar{x})(x_j - \bar{x})}{\sum_{j=1}^{n}(x_j - \bar{x})^2}$$

as we found in Chapter 3.

Rule for identifying leverage points

A popular rule, which we shall adopt, is to classify the ith point as a point of high leverage (i.e., a leverage point) in a multiple linear regression model with p predictors if

$$h_{ii} > 2 \times \text{average}\,(h_{ii}) = 2 \times \frac{(p+1)}{n}$$

6.1.2   Properties of Residuals in Multiple Regression

Recall from (6.4) and (6.5) that the vector of residuals is given by

$$\hat{\mathbf{e}} = \mathbf{Y} - \hat{\mathbf{Y}} = \mathbf{Y} - \mathbf{HY} = (\mathbf{I} - \mathbf{H})\mathbf{Y}$$

where $\mathbf{H} = \mathbf{X}(\mathbf{X}'\mathbf{X})^{-1}\mathbf{X}'$.

The **expected value** of the vector of **residuals** is

$$\begin{aligned}
E(\hat{\mathbf{e}} \mid \mathbf{X}) &= (\mathbf{I} - \mathbf{H})E(\mathbf{Y}) \\
&= (\mathbf{I} - \mathbf{H})\mathbf{X}\beta \\
&= \mathbf{X}\beta - \mathbf{X}(\mathbf{X}'\mathbf{X})^{-1}\mathbf{X}'\mathbf{X}\beta \\
&= \mathbf{X}\beta - \mathbf{X}\beta \\
&= 0
\end{aligned}$$

The **variance** of the vector of **residuals** is

$$\begin{aligned}
\text{Var}(\hat{\mathbf{e}} \mid \mathbf{X}) &= (\mathbf{I} - \mathbf{H})\,\text{Var}(\mathbf{Y})(\mathbf{I} - \mathbf{H})' \\
&= (\mathbf{I} - \mathbf{H})\sigma^2 \mathbf{I}(\mathbf{I} - \mathbf{H})' \\
&= \sigma^2 (\mathbf{I} - \mathbf{H})(\mathbf{I}' - \mathbf{H}') \\
&= \sigma^2 (\mathbf{II}' - \mathbf{IH}' - \mathbf{HI}' + \mathbf{HH}') \\
&= \sigma^2 (\mathbf{I} - \mathbf{H} - \mathbf{H} + \mathbf{H}) \\
&= \sigma^2 (\mathbf{I} - \mathbf{H})
\end{aligned}$$

since $\mathbf{HH}' = \mathbf{H}^2 = \mathbf{X}(\mathbf{X}'\mathbf{X})^{-1}\mathbf{X}'\mathbf{X}(\mathbf{X}'\mathbf{X})^{-1}\mathbf{X}' = \mathbf{X}(\mathbf{X}'\mathbf{X})^{-1}\mathbf{X}' = \mathbf{H}$

Standardized residuals

The ith least squares residual has variance given by

$$\mathrm{Var}\,(\hat{e}_i) = \sigma^2 \left[1 - h_{ii} \right]$$

where h_{ii} is the ith diagonal element of **H**. Thus, the ith *standardized residual, r_i* is given by

$$r_i = \frac{\hat{e}_i}{s\sqrt{1 - h_{ii}}}$$

where $s = \sqrt{\dfrac{1}{n-(p+1)} \sum_{j=1}^{n} \hat{e}_j^2}$ is the usual estimate of σ.

We shall follow the common practice of labelling points as **outliers** in small to moderate size data sets if the standardized residual for the point falls outside the interval from **-2 to 2**. In very large data sets, we shall change this rule to **-4 to 4**. (Otherwise, many points will be flagged as potential outliers.) Recall, however, that a point can only be declared to be an outlier, only after we are convinced that the model under consideration is a valid one.

Using residuals and standardized residuals for model checking

In its simplest form, a multiple linear regression model is a **valid model** for the data if the conditional mean of Y given X is a linear function of X and the conditional variance of Y given X is constant. In other words,

$$\mathrm{E}(Y \mid X = x) = \beta_0 + \beta_1 x_1 + \beta_2 x_2 + ... + \beta_p x_p \quad \text{and} \quad \mathrm{Var}\,(Y \mid X = x) = \sigma^2.$$

When a **valid model** has been fit, a plot of standardized residuals, r_i against any predictor or any linear combination of the predictors (such as the fitted values) will have the following features:

- A random scatter of points around the horizontal axis, since the mean function of the e_i is zero when a correct model has been fit
- Constant variability as we look along the horizontal axis

An implication of these features is that **any pattern in a plot of standardized residuals is indicative that an invalid model has been fit to the data**.

In multiple regression, plots of residuals or standardized residuals provide direct information on the way on which the model is misspecified when the following two conditions hold:

$$\mathrm{E}(Y \mid X = x) = g(\beta_0 + \beta_1 x_1 + \beta_2 x_2 + ... + \beta_p x_p) \qquad (6.6)$$

and

$$\mathrm{E}(X_i \mid X_j) \approx \alpha_0 + \alpha_1 X_j \qquad (6.7)$$

This finding is based on the work of Li and Duan (1989). The linearity condition (6.7) is just another way to say that the distribution of the predictors follow an elliptically symmetric distribution. Note that if the X's follow a multivariate normal distribution then this is stronger than condition (6.7).

Furthermore, when (6.6) and (6.7) hold, then the **plot of Y against fitted values,** $\hat{Y}$ provides direct information about g. In particular, in the usual multiple regression model g is the identity function (i.e., $g(x) = x$). In this case the plot of Y against $\hat{Y}$ should produce points scattered around a straight line.

If either condition (6.6) or (6.7) does not hold, then a pattern in a residual plot indicates that an incorrect model has been fit, but the pattern itself does not provide direct information on how the model is misspecified. For example, we shall see shortly that in these circumstances it is possible for the standardized residuals to display nonconstant variance when the errors in fact have constant variance but the conditional mean is modelled incorrectly. Cook and Weisberg (1999a, p. 36) give the following advice for this situation:

> Using residuals to guide model development will often result in misdirection, or at best more work than would otherwise be necessary.

To understand how (6.7) affects the interpretability of residual plots we shall consider the following stylized situation in which the true model is

$$y_i = \beta_0 + \beta_1 x_{1i} + \beta_2 x_{2i} + \beta_3 x_{3i} + e_i$$

and that x_1 and x_3 and x_2 and x_3 are nonlinearly related (i.e., that (6.7) does not hold). Suppose that we fit a model without the predictor x_3 and obtain the following least squares fitted values

$$\hat{y}_i = \hat{\beta}_0 + \hat{\beta}_1 x_{1i} + \hat{\beta}_2 x_{2i}$$

Thus, the residuals are given by

$$\hat{e}_i = y_i - \hat{y}_i = (\beta_0 - \hat{\beta}_0) + (\beta_1 - \hat{\beta}_1)x_{1i} + (\beta_2 - \hat{\beta}_2)x_{2i} + \beta_3 x_{3i} + e_i$$

Then, due to the term $\beta_3 x_3$ and the fact that x_1 and x_3 are nonlinearly related, the residuals plotted against x_1 would show a potentially misleading nonrandom nonlinear pattern in x_1. Similarly, residuals plotted against x_2 would show a potentially misleading nonrandom nonlinear pattern in x_2. In summary, in this situation the residual plots would show nonrandom patterns indicating that an invalid model has been fit to the data. However, the nonrandom patterns do not provide direct information on how the model is misspecified.

Menu pricing in a new Italian restaurant in New York City (cont.)

Recall from Chapter 1 that you have been asked to produce a regression model to predict the price of dinner. Data from surveys of customers of 168 Italian restaurants in the target area are available. The data are in the form of the average of customer views on

Y = Price = the price (in \$US) of dinner (including one drink and a tip)

x_1 = Food = customer rating of the food (out of 30)

x_2 = Décor = customer rating of the decor (out of 30)

x_3 = Service = customer rating of the service (out of 30)

x_4 = East = dummy variable = 1 (0) if the restaurant is east (west) of Fifth Avenue

The data are given on the book web site in the file nyc.csv.

Recall further that interest centers on developing a regression model that directly predicts Price and so we began by considering the following model:

$$Y = \beta_0 + \beta_1 x_1 + \beta_2 x_2 + \beta_3 x_3 + \beta_4 x_4 + e \qquad (6.8)$$

We begin by looking at the validity of condition (6.7) above. Figure 6.1 shows a scatter plot matrix of the three continuous predictors. The predictors seem to be related linearly at least approximately.

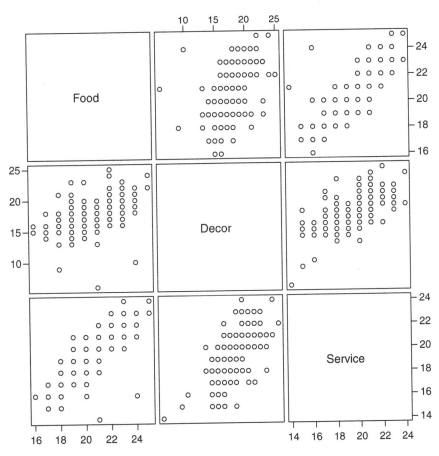

Figure 6.1 Scatter plot matrix of the three continuous predictor variables

Assuming that condition (6.6) holds we next look at plots of standardized residuals against each predictor (see Figure 6.2). The random nature of these plots is indicative that model (6.8) is a valid model for the data.

Finally, Figure 6.3 shows a plot of Y, price against fitted values, $\hat{Y}$. We see from this figure that Y and $\hat{Y}$ appear to be linearly related, i.e., that condition (6.6) appears

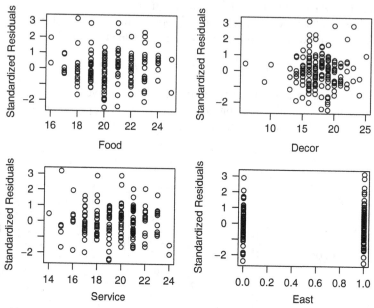

Figure 6.2 Plots of standardized residuals against each predictor variable

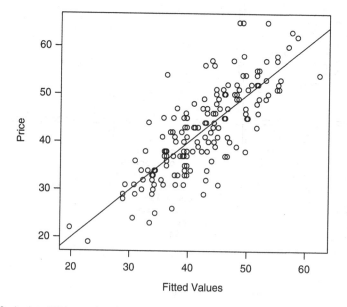

Figure 6.3 A plot of Price against fitted values

to hold with g equal to the identity function. This provides a further indication that (6.8) is a valid model.

Generated example for which condition (6.6) does not hold

In this example we look at a generated data set for which condition (6.6) does not hold. The example is taken from Cook and Weisberg (1999a, p. 36). The data are available in the R-library, alr3 in the file called caution. According to Cook and Weisberg (1999, p. 36), "... the $p = 2$ predictors were sampled from a Pearson type II distribution on the unit disk, which is an elliptical distribution," which satisfies condition (6.7). The mean function was chosen to be ·

$$E(Y \mid X) = \frac{\mid x_1 \mid}{2 + (1.5 + x_2)^2} = \frac{g_1(x_1)}{g_2(x_2)}$$

This clearly does not satisfy condition (6.6) since two functions, rather than one, are needed to model $Y|X$. The errors were chosen to be normally distributed with mean 0 and variance 1 and the data set consists of 100 cases.

Figure 6.4 shows a scatter plot matrix of the data. The predictor variables x_1 and x_2 are close to being uncorrelated with $\text{corr}(x_1, x_2) = -0.043$ and there is no evidence of a nonlinear relationship between x_1 and x_2.

We begin by considering the following model:

$$Y = \beta_0 + \beta_1 x_1 + \beta_2 x_2 + e$$

even though condition (6.6) does not hold.

We next look at plots of standardized residuals against each predictor and the fitted values (see Figure 6.5). The nonrandom nature of these plots is indicative that model is not a valid model for the data. The usual interpretations of the plots of standardized residuals against x_2 and fitted values are indicative of nonconstant error variance. However, this is not true in this case, as the error variance is constant. Instead the mean function of the model is misspecified. Since condition (6.6) does not hold, all we can say in this case is that an invalid model has been fit to the data. *Based on residual plots, we cannot say anything about what part of the model is misspecified.*

Finally, Figure 6.6 shows a plot of Y against fitted values, $\hat{Y}$. We see from this figure that Y does not seem to be a single function of $\hat{Y}$, i.e., that condition (6.6) appears not to hold.

Next steps:

In this example, since there are just two predictor variables, a three-dimensional plot of Y against the predictors will reveal the shape of the mean function. In a situation with more than two predictor variables, methods exist to directly

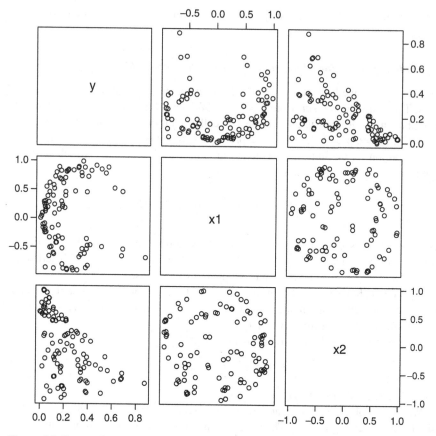

Figure 6.4 Scatter plot matrix of the response and the two predictor variables

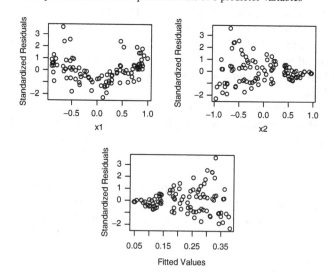

Figure 6.5 Plots of standardized residuals against each predictor and the fitted values

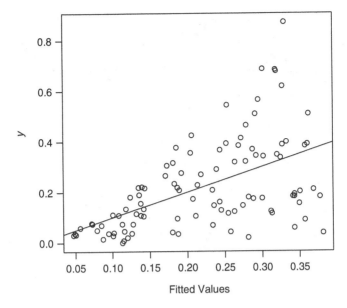

Figure 6.6 A plot of Y against fitted values

estimate more than one g-function and to determine the number of g-functions required. One such method, which was developed by Li (1991) is called Sliced Inverse Regression (or SIR). Unfortunately, a discussion of SIR is beyond the scope of this book.

Generated example for which condition (6.7) does not hold

In this example we look at a generated data set for which condition (6.7) does not hold. The data consist of $n = 601$ points generated from the following model

$$Y = x_1 + 3x_2^2 + e$$

where

$$E(x_2 \mid x_1) = \sin(x_1)$$

x_1 is equally spaced from -3 to 3 and the errors are normally distributed with standard deviation equal to 0.1. The data can be found on the book website in the file called nonlinear.txt. Figure 6.7 shows scatter plots of the data. There nonlinear relationship between x_1 and x_2 is clearly evident in Figure 6.7.

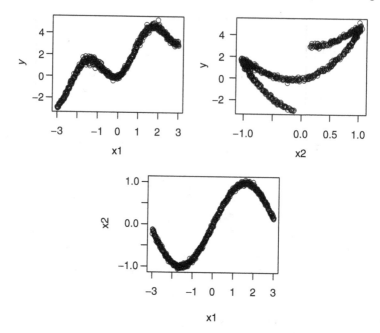

Figure 6.7 Scatter plots of the response and the two predictor variables

We begin by considering the following model:

$$Y = \beta_0 + \beta_1 x_1 + \beta_2 x_2 + e$$

even though condition (6.7) does not hold. Figure 6.8 shows plots of standardized residuals against each predictor and the fitted values. The nonrandom nature of these plots is indicative that model is not a valid model for the data. The usual interpretation of the plot of standardized residuals against x_1 is that a periodic function of x_1 is missing from the model. However, this is not true in this case. The highly nonlinear relationship between the two predictors has produced the nonrandom pattern in the plot of standardized residuals against x_1. Since (6.7) does not hold, all we can say in this case is that model fit to the data is invalid. ***Based on residual plots, we cannot say anything about what part of the model is misspecified.***

6.1.3 Added Variable Plots

Added-variable plots enable us to visually assess the effect of each predictor, having adjusted for the effects of the other predictors.

Throughout this section we shall assume that our current regression model is the multiple linear regression model,

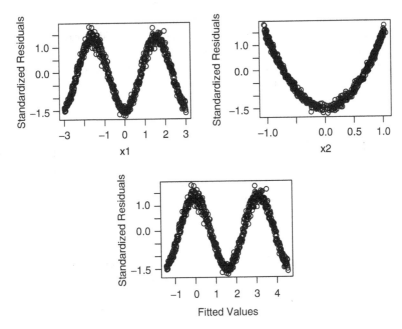

Figure 6.8   Plots of standardized residuals against each predictor and the fitted values

$$\mathbf{Y} = \mathbf{X}\beta + \mathbf{e} \tag{6.9}$$

where $\mathrm{Var}(\mathbf{e}) = \sigma^2 \mathbf{I}$ and $\mathbf{I}$ is the $(n \times n)$ identity matrix and the $(n \times 1)$ vector, $\mathbf{Y}$ the $n \times (p + 1)$ matrix, $\mathbf{X}$ and the $(n \times 1)$ vector β are given by

$$\mathbf{Y} = \begin{pmatrix} y_1 \\ y_2 \\ \vdots \\ y_n \end{pmatrix} \quad \mathbf{X} = \begin{pmatrix} 1 & x_{11} \cdots x_{1p} \\ 1 & x_{21} \cdots x_{2p} \\ \vdots & \vdots \\ 1 & x_{n1} \cdots x_{np} \end{pmatrix} \quad \beta = \begin{pmatrix} \beta_0 \\ \beta_1 \\ \vdots \\ \beta_p \end{pmatrix}$$

Further, suppose that we are considering the introduction of an additional predictor variable $\mathbf{Z}$ to model (6.9). In other words, we are considering the model

$$\mathbf{Y} = \mathbf{X}\beta + \mathbf{Z}\alpha + \mathbf{e} \tag{6.10}$$

where

$$\mathbf{Z} = \begin{pmatrix} z_1 \\ z_2 \\ \vdots \\ z_n \end{pmatrix}$$

In particular, we are interested in α, the regression coefficient measuring the effect of $\mathbf{Z}$ on $\mathbf{Y}$, having adjusted for the effect of $\mathbf{X}$ on $\mathbf{Y}$. The **added-variable plot** for predictor variable $\mathbf{Z}$ enables us to visually estimate α. The added-variable plot is obtained by plotting on the vertical axis the residuals from model (6.9) against on the horizontal axis the residuals from model

$$\mathbf{Z} = \mathbf{X}\delta + \mathbf{e} \qquad (6.11)$$

Notice that the residuals from model (6.9) give that part of $\mathbf{Y}$ that is not predicted by $\mathbf{X}$ while the residuals from model (6.11) give that part of $\mathbf{Z}$ that is not predicted by $\mathbf{X}$. Thus, the added-variable plot for predictor variable $\mathbf{Z}$ shows that part of $\mathbf{Y}$ that is not predicted by $\mathbf{X}$ against that part of $\mathbf{Z}$ that is not predicted by $\mathbf{X}$ (i.e., the effects due to $\mathbf{X}$ are removed from both axes). The added-variable plot was introduced by Mosteller and Tukey (1977).

Mathematical justification for added-variable plots

In what follows, we follow the approach taken by Chatterjee and Hadi (1988, pp. 54–56). The vector of residuals from model (6.9) is given by

$$\hat{\mathbf{e}}_{Y.X} = \mathbf{Y} - \hat{\mathbf{Y}} = \mathbf{Y} - \mathbf{H}_X\mathbf{Y} = \left(\mathbf{I} - \mathbf{H}_X\right)\mathbf{Y}$$

where $\mathbf{H}_X = \mathbf{X}(\mathbf{X}'\mathbf{X})^{-1}\mathbf{X}'$. Multiplying equation (6.10) by $\left(\mathbf{I} - \mathbf{H}_X\right)$ results in

$$\begin{aligned}
\left(\mathbf{I} - \mathbf{H}_X\right)\mathbf{Y} &= \left(\mathbf{I} - \mathbf{H}_X\right)\mathbf{X}\beta + \left(\mathbf{I} - \mathbf{H}_X\right)\mathbf{Z}\alpha + \left(\mathbf{I} - \mathbf{H}_X\right)\mathbf{e} \\
&= \left(\mathbf{I} - \mathbf{H}_X\right)\mathbf{Z}\alpha + \left(\mathbf{I} - \mathbf{H}_X\right)\mathbf{e} \qquad (6.12)
\end{aligned}$$

since $\left(\mathbf{I} - \mathbf{H}_X\right)\mathbf{X} = \left(\mathbf{I} - \mathbf{X}(\mathbf{X}'\mathbf{X})^{-1}\mathbf{X}'\right)\mathbf{X} = \mathbf{X} - \mathbf{X} = 0$. Notice that $\left(\mathbf{I} - \mathbf{H}_X\right)\mathbf{Y} = \hat{\mathbf{e}}_{Y.X}$ is just the vector of residuals from model (6.9).

We next consider the first term on the right-hand side of (6.12), namely, $\left(\mathbf{I} - \mathbf{H}_X\right)\mathbf{Z}\alpha$. Notice that the multiple linear regression model (6.11) has residuals given by

$$\hat{\mathbf{e}}_{Z.X} = \mathbf{Z} - \hat{\mathbf{Z}} = \mathbf{Z} - \mathbf{H}_X\mathbf{Z} = \left(\mathbf{I} - \mathbf{H}_X\right)\mathbf{Z}$$

where $\mathbf{H}_X = \mathbf{X}(\mathbf{X}'\mathbf{X})^{-1}\mathbf{X}'$.

Thus (6.12) can be rewritten as

$$\left(\mathbf{I} - \mathbf{H}_X\right)\mathbf{Y} = \left(\mathbf{I} - \mathbf{H}_X\right)\mathbf{Z}\alpha + \left(\mathbf{I} - \mathbf{H}_X\right)\mathbf{e}$$

i.e. $\hat{\mathbf{e}}_{Y.X} = \hat{\mathbf{e}}_{Z.X}\alpha + \mathbf{e}*$ (Added-variable plot model)

where $\mathbf{e}* = \left(\mathbf{I} - \mathbf{H}_X\right)\mathbf{e}$. Thus, α is the slope parameter in a regression of $\hat{\mathbf{e}}_{Y.X}$ (i.e., the residuals from the regression of $\mathbf{Y}$ on $\mathbf{X}$) on $\hat{\mathbf{e}}_{Z.X}$ (i.e., the residuals from the regression of $\mathbf{Z}$ on $\mathbf{X}$). Let $\hat{\alpha}_{AVP}$ denote the least squares estimate of the slope parameter in a regression of $\hat{\mathbf{e}}_{Y.X}$ (i.e., the residuals from the regression of $\mathbf{Y}$ on $\mathbf{X}$) on $\hat{\mathbf{e}}_{Z.X}$ (i.e., the residuals from the regression of $\mathbf{Z}$ on $\mathbf{X}$). It can be shown that $\hat{\alpha}_{AVP}$ is equal to $\hat{\alpha}_{LS}$, the least squares estimate of α in model (6.10). Furthermore, assuming that (6.10) is a valid model for the data, then the added-variable plot should produce points randomly scattered around a line through the origin with slope $\hat{\alpha}_{LS}$. This plot will also enable the user to identify any data points which have undue influence on the least squares estimate of α.

Menu pricing in a new Italian restaurant in New York City (cont.)

Recall from Chapter 1 that you have been asked to produce a regression model to predict the price of dinner. The data are in the form of the average of customer views on

Y = Price = the price (in $US) of dinner (including one drink and a tip)
x_1 = Food = customer rating of the food (out of 30)
x_2 = Décor = customer rating of the decor (out of 30)
x_3 = Service = customer rating of the service (out of 30)
x_4 = East = 1 (0) if the restaurant is east (west) of Fifth Avenue

The data are given on the book web site in the file nyc.csv.

Recall further that interest centers on developing a regression model that directly predicts Price and so we began by considering the following model:

$$Y = \beta_0 + \beta_1 x_1 + \beta_2 x_2 + \beta_3 x_3 + \beta_4 x_4 + e \qquad (6.13)$$

Figure 6.9 contains a plot of Y, Price against each predictor. Added to each plot is the least squares line of best fit for a simple linear regression of Price on that predictor variable.

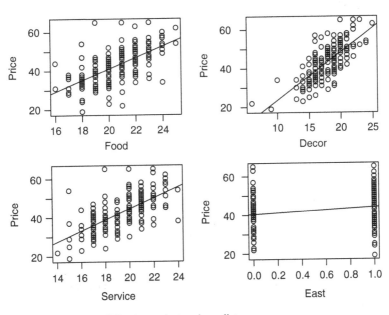

Figure 6.9 A scatter plot of Y, price against each predictor

A shortcoming of each plot in Figure 6.9 is that it looks at the effect of a given predictor on Y, Price, ignoring the effects of the other predictors on Price. This shortcoming is overcome by looking at added-variable plots (see Figure 6.10). The lack of statistical significance of the regression coefficient associated with the variable Service is clearly evident in the bottom left-hand plot of Figure 6.10. Thus, having adjusted for the effects of the other predictors, the variable Service adds little to the prediction of Y, Price. Two points are identified in the top left-hand plot as having a large influence on the least squares estimate of the regression coefficient for Food. These points correspond to cases 117 and 168 and should be investigated. Case 117 corresponds to a restaurant called Veronica which has very low scores for Décor and Service, namely 6 and 14, respectively while achieving a relatively high food score of 21 given a price of $22. Case 168 corresponds to a restaurant called Gennaro, which has low scores for Décor and Service, namely 10 and 16, respectively while achieving a high food score of 24 for a relatively low price of $34. Gennaro, still in existence at the end of 2007, is described in the *2008 Zagat Guide to New York City Restaurants* as follows:

Upper Westsiders gennar-ally gush over this "unassuming" cash-only Italian "gem", citing "sophisticated" preparations at "bargain" prices; to cope with "awful lines", "crapshoot" service and a room "packed tighter than a box of pasta", go at off-hours.

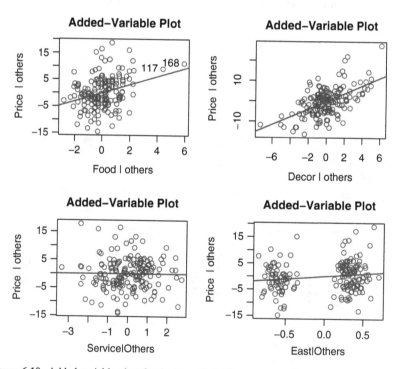

Figure 6.10 Added-variable plots for the New York City restaurant data

6.2   Transformations

In this section we shall see how transformations in multiple regression can be used to:

- Overcome problems due to nonlinearity
- Estimate percentage effects

6.2.1   Using Transformations to Overcome Nonlinearity

In this section we consider the following two general methods for transforming the response variable Y and/or the predictor variables $X_1, X_2, ..., X_p$ to overcome problems due to nonlinearity:

- Inverse response plots
- Box-Cox procedure

There are three situations we need to consider (a) only the response variable needs to be transformed; (b) only the predictor variables needs to be transformed; and (c) both the response and predictor variables need to be transformed. We begin by looking at the first situation.

Transforming only the response variable Y using inverse regression

Suppose that the true regression model between Y and $X_1, X_2, ..., X_p$ is given by

$$Y = g(\beta_0 + \beta_1 x_1 + \beta_2 x_2 + ... + \beta_p x_p + e)$$

where g is a function which is generally unknown. The previous model can be turned into a multiple linear regression model by transforming Y by g^{-1}, the inverse of g, since,

$$g^{-1}(Y) = \beta_0 + \beta_1 x_1 + \beta_2 x_2 + ... + \beta_p x_p + e$$

For example suppose that

$$Y = \exp(\beta_0 + \beta_1 x_1 + \beta_2 x_2 + ... + \beta_p x_p + e)$$

then,

$$g(Y) = \exp(Y) \text{ and so } g^{-1}(Y) = \log(Y).$$

We next look at methods for estimating g^{-1}.

Example: Modelling defective rates

This example is adapted from Siegel (1997, pp. 509–510). According to Siegel:

> Everybody seems to disagree about just why so many parts have to be fixed or thrown away after they are produced. Some say that it's the standard deviation of the temperature of the production process, which needs to be minimised. Others claim it is clearly the density of

the product, and that the problems would disappear if the density is increased. Then there is Ole, who has been warning everyone forever to take care not to push the equipment beyond its limits. This problem would be easiest to fix, simply by slowing down the production rate; however, this would increase some costs. The table below gives the average number of defects per 1,000 parts produced (denoted by Defective) along with values of the other variables described above for 30 independent production runs.

The data are given in Table 6.1 and can be found on the book web site in the file defects.txt.

Interest centres on developing a model for Y, Defective, based on the predictors x_1, Temperature; x_2, Density and x_3, Rate.

Figure 6.11 contains a scatter plot matrix of the response variable, Defective and the predictors Temperature, Density and Rate. The three predictors appear to be linearly related. However, the separate relationships between the response variable and each of the predictor variables do not appear to be linear.

We begin by fitting the regression model

$$Y = \beta_0 + \beta_1 x_1 + \beta_2 x_2 + \beta_3 x_3 + e \tag{6.14}$$

Figure 6.12 contains scatter plots of the standardized residuals against each predictor and the fitted values for model (6.14). Each of the plots in Figure 6.12 shows a curved rather than a random pattern. Thus, model (6.14) does not appear to be a valid model for the data.

Figure 6.13 contains a plot of Y, Defective against the fitted values, $\hat{Y}$. The straight-line fit to this plot (displayed as a dashed line) provides a poor fit. This provides further evidence that model (6.14) is not a valid model for the data. The solid line in Figure 6.11 is a quadratic fit and it follows the points more closely than the straight line.

In summary, model (6.14) does not provide a valid model to the data since:

Table 6.1 Data on defective rates (defects.txt)

Temperature	Density	Rate	Defective	Temperature	Density	Rate	Defective
0.97	32.08	177.7	0.2	2.76	21.58	244.7	42.2
2.85	21.14	254.1	47.9	2.36	26.3	222.1	13.4
2.95	20.65	272.6	50.9	1.09	32.19	181.4	0.1
2.84	22.53	273.4	49.7	2.15	25.73	241	20.6
1.84	27.43	210.8	11	2.12	25.18	226	15.9
2.05	25.42	236.1	15.6	2.27	23.74	256	44.4
1.5	27.89	219.1	5.5	2.73	24.85	251.9	37.6
2.48	23.34	238.9	37.4	1.46	30.01	192.8	2.2
2.23	23.97	251.9	27.8	1.55	29.42	223.9	1.5
3.02	19.45	281.9	58.7	2.92	22.5	260	55.4
2.69	23.17	254.5	34.5	2.44	23.47	236	36.7
2.63	22.7	265.7	45	1.87	26.51	237.3	24.5
1.58	27.49	213.3	6.6	1.45	30.7	221	2.8
2.48	24.07	252.2	31.5	2.82	22.3	253.2	60.8
2.25	24.38	238.1	23.4	1.74	28.47	207.9	10.5

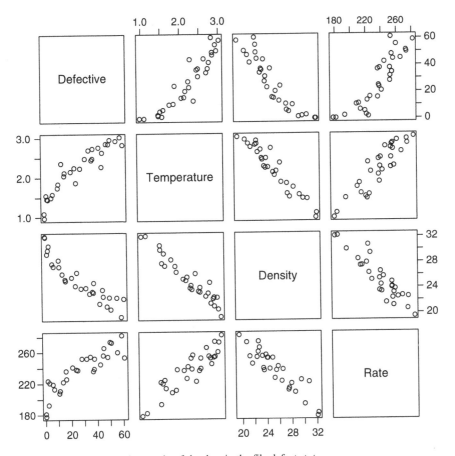

Figure 6.11 A scatter plot matrix of the data in the file defects.txt

- The plots of standardized residuals against each predictor and the fitted values do not produce random scatters
- The summary plot of Y against $\hat{Y}$ shows a quadratic rather than a linear trend, that is, $Y \cong g(\hat{Y})$ where g is a quadratic function

In view of the last point, it is natural to consider a transformation of Y.

Inverse response plots

Suppose that the true regression model between Y and X_1, X_2, ..., X_p is given by

$$Y = g(\beta_0 + \beta_1 x_1 + \beta_2 x_2 + \ldots + \beta_p x_p + e)$$

where g is an unknown function. Recall that the previous model can be turned into a multiple linear regression model by transforming Y by g^{-1}, the inverse of g, since,

$$g^{-1}(Y) = \beta_0 + \beta_1 x_1 + \beta_2 x_2 + \ldots + \beta_p x_p + e.$$

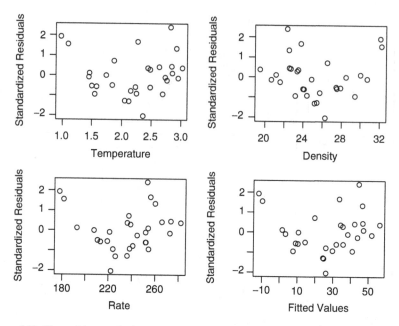

Figure 6.12   Plots of the standardized residuals from model (6.14)

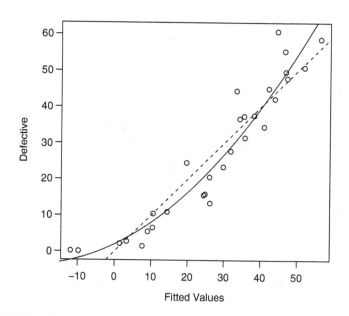

Figure 6.13   Plot of Y against fitted values with a straight line and a quadratic curve

Thus, if we knew $\beta_0, \beta_1, \ldots \beta_p$ we could discover the shape of g^{-1} by plotting Y on the horizontal axis and $\beta_0 + \beta_1 x_1 + \ldots \beta_p x_p$ on the vertical axis.

Based on the results of Li and Duan (1989), Cook and Weisberg (1994) showed that if conditions (6.6) and (6.7) hold then g^{-1} can be estimated from the scatter plot of Y (on the horizontal axis) and $\hat{y} = \hat{\beta}_0 + \hat{\beta}_1 x_1 + \ldots + \hat{\beta}_p x_p$ (on the vertical axis). Such a plot is commonly referred to as **inverse response plot** (since the usual axis for Y is the vertical axis). It can be shown that the assumption that the predictors X_1, X_2, ..., X_p have a multivariate normal distribution is much stronger than the assumption than the predictors X_1, X_2, ..., X_p are linearly related at least approximately.

Example: Modelling defective rates (cont.)

Figure 6.14 contains an inverse response plot for the data in the example involving defective rates. Since we found that the predictors are linearly at least approximately (see Figure 6.11), we should be able to estimate g^{-1} from the inverse response plot. Marked on the plot are three so-called power curves

$$\hat{y} = y^\lambda \text{ for } \lambda = 0, 0.44, 1$$

where, as we shall see below, $\lambda = 0$ corresponds to natural logarithms. It is evident that among the three curves, the power curve

$$\hat{y} = \left(\hat{\beta}_0 + \hat{\beta}_1 x_1 + \hat{\beta}_2 x_2 + \hat{\beta}_3 x_3 \right) = y^{0.44}$$

provides the closest fit to the data in Figure 6.14. This is not unexpected since we found in Figure 6.13 that a quadratic provided an approximation to g. Rounding

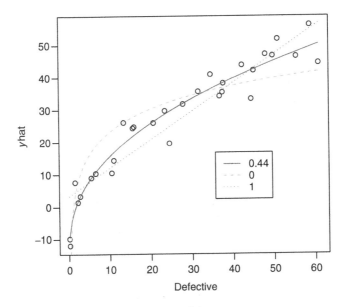

Figure 6.14   Inverse response plot for the data set defects.txt

0.44 to the nearest reasonable value, namely, 0.5 we shall transform Y by taking the square root and thus consider the following model

$$g^{-1}(Y) = Y^{0.5} = \beta_0 + \beta_1 x_1 + \beta_2 x_2 + \beta_3 x_3 + e \qquad (6.15)$$

Transforming only the response variable Y using Box-Cox method

Box and Cox (1964) provide a general method for transforming a strictly positive response variable Y. The Box-Cox procedure aims to find a transformation that makes the transformed response variable close to normally distributed having taken into account the regression model under consideration. Box and Cox (1964) considered the modified family of power transformations

$$\Psi_M(Y,\lambda) = \psi_S(Y,\lambda) \times gm(Y)^{1-\lambda} = \begin{cases} gm(Y)^{1-\lambda}(Y^\lambda - 1)/\lambda & \text{if } \lambda \neq 0 \\ gm(Y)\log(Y) & \text{if } \lambda = 0 \end{cases}$$

where $gm(Y) = \prod_{i=1}^{n} Y_i^{1/n} = \exp\left(\dfrac{1}{n}\sum_{i=1}^{n}\log(Y_i)\right)$ is the geometric mean of Y. The Box-Cox method is based on the notion that for some value of λ the transformed version of Y, namely, $\Psi_M(Y,\lambda)$ is normally distributed. The method is based on choosing λ that maximizes the log-likelihood function for $\beta_0, \beta_1, \ldots, \beta_p, \sigma^2, \lambda \mid \Psi_M(Y,\lambda)$.

Example: Modelling defective rates (cont.)

Figure 6.15 provides plots of the log-likelihood against λ for the data in the example involving defective rates. The value of λ that maximizes the log-likelihood and 95% confidence limits for λ are marked on the plot. The value of λ that maximizes the log-likelihood is 0.45. Thus, in this case, both the inverse response plot and the Box-Cox transformation method point to using a square root transformation of Y. Thus, we next consider the multiple linear regression model given by (6.15).

Figure 6.16 contains plots of $Y^{0.5}$ against each predictor. It is evident from Figure 6.16 that the relationship between $Y^{0.5}$ and each predictor is more linear than the relationship between Y and each predictor.

Figure 6.17 contains scatter plots of the standardized residuals against each predictor and the fitted values for model (6.15). Each of the plots in Figure 6.15 shows a random pattern. Thus, model (6.15) appears to be a valid model for the data.

Figure 6.18 contains a plot of $Y^{0.5} = \sqrt{\text{Defective}}$ against the fitted values. The straight-line fit to this plot provides a reasonable fit. This provides further evidence that model (6.15) is a valid model for the data.

The diagnostic plots provided by R for model (6.15) shown in Figure 6.19 further confirm that it is a valid model for the data.

We finish by considering the following theories put forward regarding the causes of the defects (taken from Siegel, 1997, p. 509):

Others claim it is clearly the density of the product, and that the problems would disappear if the density is increased. Then there is Ole, who has been warning everyone forever to take care not to push the equipment beyond its limits. This problem would be easiest to fix, simply by slowing down the production rate ....

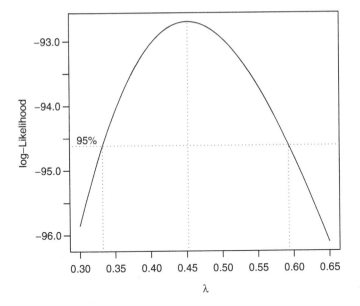

Figure 6.15 Log-likelihood for the Box-Cox transformation method

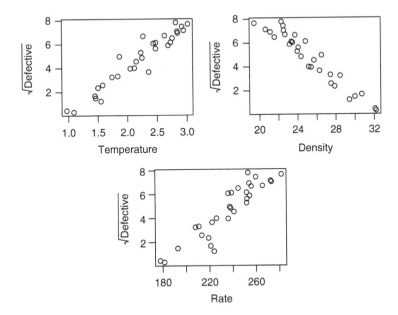

Figure 6.16 Plots of $Y^{0.5}$ against each predictor

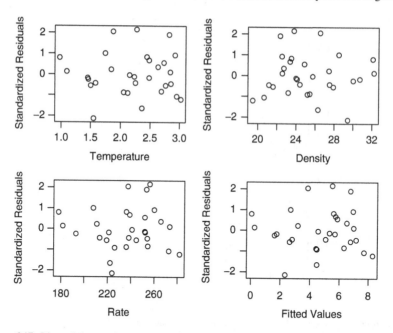

Figure 6.17   Plots of the standardized residuals from model (6.15)

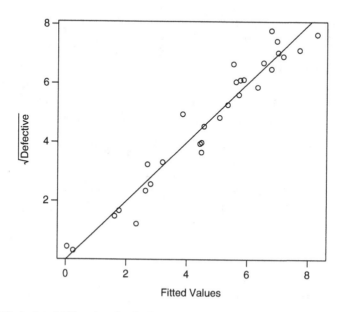

Figure 6.18   A plot of $Y^{0.5}$ against fitted values with a straight line added

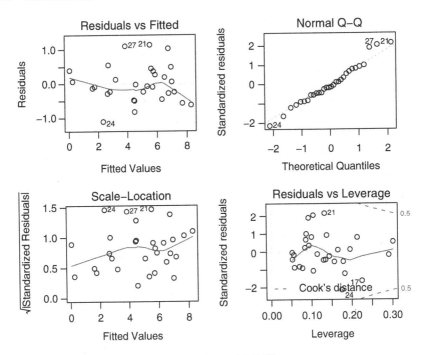

Figure 6.19 Diagnostic plots provided by R for model (6.15)

Regression Output from R from model (6.15)

```
Call:
lm(formula = sqrt(Defective) ~ Temperature + Density + Rate)

Coefficients:
              Estimate  Std. Error  t value  Pr(>|t|)
(Intercept)    5.59297     5.26401    1.062    0.2978
Temperature    1.56516     0.66226    2.363    0.0259 *
Density       -0.29166     0.11954   -2.440    0.0218 *
Rate           0.01290     0.01043    1.237    0.2273
---
Residual standard error: 0.5677 on 26 degrees of freedom
Multiple R-Squared: 0.943,        Adjusted R-squared: 0.9365
F-statistic: 143.5 on 3 and 26 DF,  p-value: 2.713e-16
```

The variable Rate is not statistically significant, thus not supporting Ole's theory above. On the other hand, the coefficient of Density is statistically significantly less than zero in line with the theory above of increasing the density as a way of lowering the defect rate. However, the value of the variable Rate still needs to be considered when adjustments are made to one or both of the statistically significant predictors, since from Figure 6.11 Rate is clearly related to the other two predictors. Finally, we show in Figure 6.20 the added-variable plots associated with model (6.15). The lack of statistical significance of the predictor Rate is evident in the bottom left-hand plot of Figure 6.18.

Transforming both the response and the predictor variables

When some, or all, of the predictors and the response are highly skewed and transformations of these variables are desirable, the following two alternative approaches are suggested by Cook and Weisberg (1999b, p. 329):

Approach 1:

1. Transform X_1, X_2, ..., X_p so that the distribution of the transformed versions $\Psi_s(x_1, \lambda_{X_1}), \Psi_s(x_2, \lambda_{X_2}), ... \Psi_s(x_p, \lambda_{X_p})$ are as jointly normal as possible. The multivariate version of the Box-Cox transformation procedure is one way to do this.

2. Having transformed X_1, X_2, ..., X_p to $\Psi_s(x_1, \lambda_{X_1}), \Psi_s(x_2, \lambda_{X_2}), ... \Psi_s(x_p, \lambda_{X_p})$, consider a multivariate linear regression model of the form

$$Y = g(\beta_0 + \beta_1 \Psi_s(x_1, \lambda_{X_1}) + ... + \beta_p \Psi_s(x_p, \lambda_{X_p}) + e).$$

Then use an inverse response plot to decide on the transformation, g^{-1} for Y.

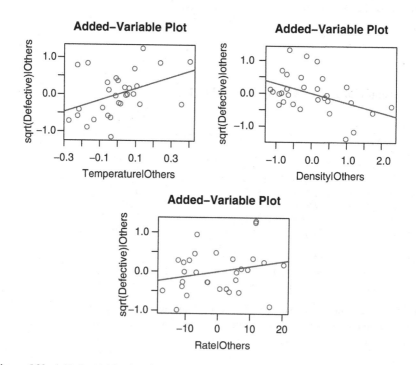

Figure 6.20 Added-variable plots for model (6.15)

Approach 2:

Transform X_1, X_2, ..., X_p and Y simultaneously to joint normality using the multivariate generalization of the Box-Cox method.

Example: Magazine revenue

An analyst is interested in understanding the relationship between revenue from magazine sales and that from advertising. The analyst has obtained some US data from *Advertising Age's* 14th annual Magazine 300 report (http://www.adage.com) which was released in September 2003. Data are available for 204 US magazines for the following variables:

Y = AdRevenue = Revenue from advertising (in thousands of $)
X_1 = AdPages = Number of pages of paid advertising
X_2 = SubRevenue = Revenue from paid subscriptions (in thousands of $)
X_3 = NewsRevenue = Revenue from newsstand sales (in thousands of $)

The data are on the book web site in the file magazines.csv. Interest centers on building a regression model to predict Ad Revenue from Ad Pages, Sub Revenue and News Revenue.

Figure 6.21 shows a scatter plot matrix of the response variable and the three predictor variables. The response variable and the three predictor variables are each highly skewed. In addition, the predictors do not appear to be linearly related. Thus, we need to consider transformations of the response and the three predictor variables.

Approach 1: Transforming the predictors first and then the response

Given below is the R output using the bctrans command from alr3.

Output from R

```
box.cox Transformations to Multinormality
```

	Est.Power	Std.Err.	Wald(Power=0)	Wald(Power=1)
AdPages	0.1119	0.1014	1.1030	-8.7560
SubRevenue	-0.0084	0.0453	-0.1864	-22.2493
NewsRevenue	0.0759	0.0333	2.2769	-27.7249

					LRT	df	p.value
LR test, all	lambda	equal	0		6.615636	3	0.08521198
LR test, all	lambda	equal	1		1100.018626	3	0.00000000

Using the Box-Cox method to transform the predictor variables toward normality, results values of λ close to 0. Thus, we shall log transform all three predictors and consider a model of the form

AdRevenue =

$$g(\beta_0 + \beta_1 \log(\text{AdPages}) + \beta_2 \log(\text{SubRevenue}) + \beta_3 \log(\text{NewsRevenue}) + e)$$

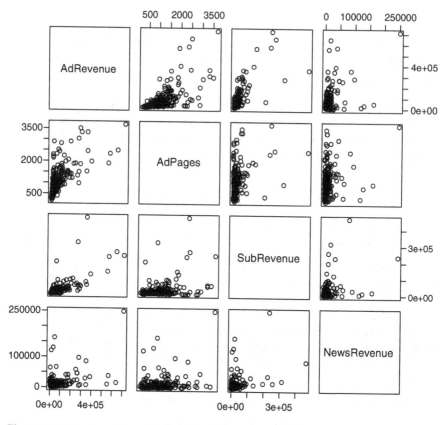

Figure 6.21 A scatter plot matrix of the data in file magazines.csv

and seek to find g^{-1} using an inverse response plot, since,

$$g^{-1}(\text{AdRevenue}) = \beta_0 + \beta_1 \log(\text{AdPages}) + \beta_2 \log(\text{SubRevenue})$$
$$+ \beta_3 \log(\text{NewsRevenue}) + e$$

Figure 6.22 contains an inverse response plot. (Since the predictor variables have been transformed towards normality, we should be able to estimate g^{-1} from the inverse response plot.) Marked on the plot are three so-called power curves

$$\hat{y} = \text{AdRevenue}^{\lambda_Y} \text{ for } \lambda_Y = 0, 0.23, 1$$

where $\lambda = 0$ corresponds to natural logarithms. It is evident that among the three curves, the power curve

$$\hat{y} = \text{AdRevenue}^{0.23}$$

provides the closest fit to the data in Figure 6.22. However, the curve based on $\lambda = 0$ also seems to provide an adequate fit, especially for small to moderate values of

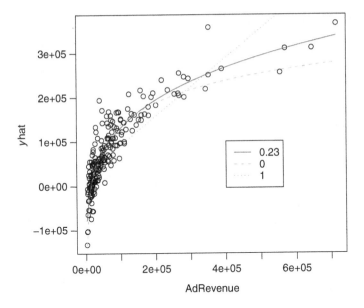

Figure 6.22 Inverse response plot

AdRevenue. In addition, notice that two of the predictors and the response are all measured in the same units (i.e., $). Thus, a second argument in favor of the log transformation for the response variable is that it means that all the variables measured in dollars are transformed in the same way.

Approach 2: Transforming all variables simultaneously

Given below is the output from R using bctrans from alr3.

Output from R

```
box.cox Transformations to Multinormality
              Est.Power Std.Err. Wald(Power=0) Wald(Power=1)
AdRevenue        0.1071   0.0394        2.7182      -22.6719
AdPages          0.0883   0.0836        1.0566      -10.9068
SubRevenue      -0.0153   0.0362       -0.4217      -28.0413
NewsRevenue      0.0763   0.0330        2.3087      -27.9682
                                         LRT df       p.value
LR test, all  lambda equal    0     13.87021   4   0.007721018
LR test, all  lambda equal    1   1540.50928   4   0.000000000
```

Using the Box-Cox method to transform the predictor and response variables simultaneously toward multivariate normality, results in values of each λ close to 0. Thus, the two approaches agree in that they suggest that each variable be transformed using the log transformation.

Figure 6.23 shows a scatter plot matrix of the log-transformed response and predictor variables. The pair-wise relationships in Figure 6.23 are much more linear than those in Figure 6.21. The least linear relationship appears to be between log(AdRevenue) and log(NewsRevenue).

We next consider a multiple linear regression model based on the log-transformed data, namely,

$$\log(\text{AdRevenue}) = \beta_0 + \beta_1 \log(\text{AdPages}) + \beta_2 \log(\text{SubRevenue})$$
$$+ \beta_3 \log(\text{NewsRevenue}) + e \qquad (6.16)$$

Figure 6.24 contains scatter plots of the standardized residuals against each predictor and the fitted values for model (6.16). Each of the plots in Figure 6.24 shows a random pattern. Thus, model (6.16) appears to be a valid model for the data.

The plot of log(AdRevenue) against the fitted values in Figure 6.25 provides further evidence that model (6.16) is a valid model.

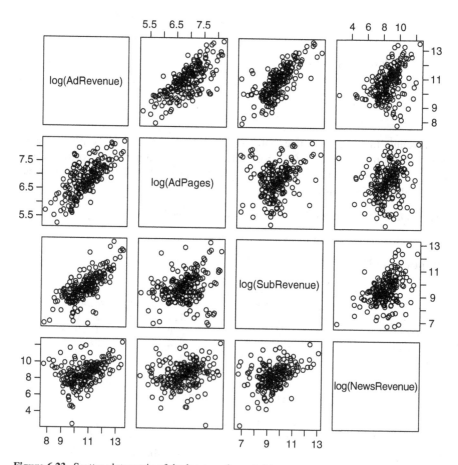

Figure 6.23 Scatter plot matrix of the log-transformed data

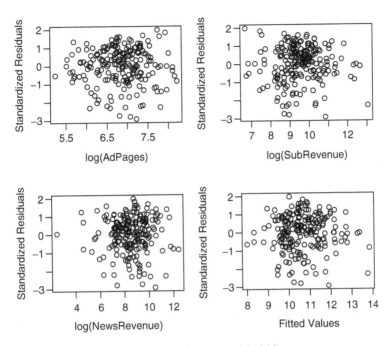

Figure 6.24   Plots of the standardized residuals from model (6.16)

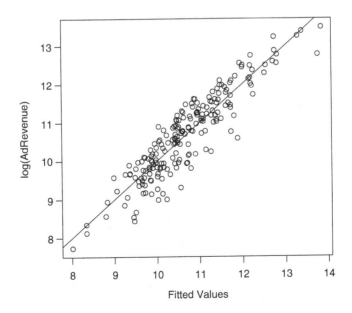

Figure 6.25   A plot of log(adrevenue) against fitted values with a straight line added

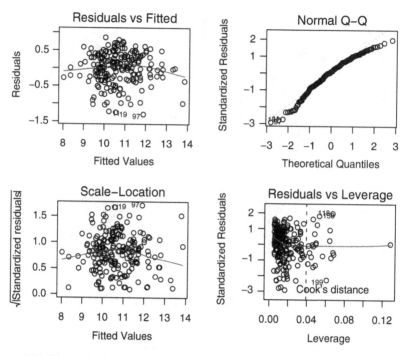

Figure 6.26 Diagnostic plots provided by R for model (6.16)

Figure 6.26 shows the diagnostic plots provided by R for model (6.16). These plots further confirm that model (6.16) is a valid model for the data. The dashed vertical line in the bottom right-hand plot of Figure 6.26 is the usual cut-off for declaring a point of high leverage (i.e., $2(p+1)/n = 8/204 = 0.039$). Thus, there is a bad leverage point (i.e., case 199) that requires further investigation.

Given below is the output from R associated with fitting model (6.16). The variable log(NewsRevenue) is not statistically significant, while the other two predictors are. Because both the predictor and response variables have been log-transformed the usual interpretation of regression coefficients as percentages holds. Thus, for example, holding all else constant, the model (6.16) predicts:

- A 1.03% increase in AdRevenue for every 1% increase in AdPages
- A 0.56% increase in AdRevenue for every 1% increase in SubRevenue

Finally, we show in Figure 6.27 the added-variable plots associated with model (6.16). The lack of statistical significance of the predictor Log(NewsRevenue) is evident in the bottom left-hand plot of Figure 6.27.

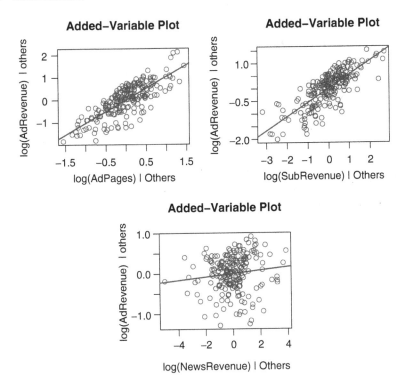

Figure 6.27 Added-variable plots for model (6.16)

Regression output from R

```
Call:
lm(formula = log(AdRevenue) ~ log(AdPages) + log(SubRevenue) +
log(NewsRevenue))
Coefficients:
                 Estimate Std. Error t value Pr(>|t|)
(Intercept)      -2.02894    0.41407  -4.900 1.98e-06 ***
log(AdPages)      1.02918    0.05564  18.497  < 2e-16 ***
log(SubRevenue)   0.55849    0.03159  17.677  < 2e-16 ***
log(NewsRevenue)  0.04109    0.02414   1.702   0.0903 .
---
Residual standard error: 0.4483 on 200 degrees of freedom
Multiple R-Squared: 0.8326,     Adjusted R-squared: 0.8301
F-statistic: 331.6 on 3 and 200 DF, p-value: < 2.2e-16
```

6.2.2 Using Logarithms to Estimate Percentage Effects: Real Valued Predictor Variables

In this section we illustrate how logarithms can be used to estimate percentage change in Y based on a one unit change in a given predictor variable. In particular, we consider the regression model

$$\log(Y) = \beta_0 + \beta_1 \log(x_1) + \beta_2 x_2 + e \tag{6.17}$$

where log refers to log to the base e or natural logarithms and x_2 is a predictor variable taking numerical values (and hence x_2 is allowed to be a dummy variable). In this situation the slope

$$\begin{aligned}
\beta_2 &= \frac{\Delta \log(Y)}{\Delta x_2} \\
&= \frac{\log(Y_2) - \log(Y_1)}{\Delta x_2} \\
&= \frac{\log(Y_2/Y_1)}{\Delta x_2} \\
&\approx \frac{Y_2/Y_1 - 1}{\Delta x_2} \quad (\text{using } \log(1+z) \approx z) \\
&= \frac{100(Y_2/Y_1 - 1)}{100\Delta x_2} \\
&= \frac{\%\Delta Y}{100\Delta x_2}
\end{aligned}$$

So that, for small β_2

$$\%\Delta Y \approx \beta_2 \times 100\Delta x_2$$

Thus for every 1 unit change in x_2 (i.e., $\Delta x_2 = 1$) the model predicts a $100 \times \beta_2 \%$ change in Y.

Example: Newspaper circulation

Recall from Chapter 1 that the company that publishes a weekday newspaper in a mid size American city has asked for your assistance in an investigation into the feasibility of introducing a Sunday edition of the paper. The current circulation of the company's weekday newspaper is 210,000. Interest focuses on developing a regression model that enables you to predict the Sunday circulation of a newspaper with a weekday circulation of 210,000. Circulation data from September 30, 2003 are available for 89 US newspapers that publish both weekday and Sunday editions. The data are available on the book website, in the file circulation.txt.

The situation is further complicated by the fact that in some cities there is more than one newspaper In particular, in some cities there is a tabloid newspaper along with a so called "serious" newspaper as a competitor. As such the data contains a dummy variable, which takes value 1 when the newspaper is a tabloid with a serious competitor in the same city and value 0 othervise.

Figure 6.28 is a repeat of Figure 1.3, which is a plot of log(Sunday Circulation) versus log(Weekday Circulation) with the dummy variable Tabloid identified. On the basis of Figure 6.28 we consider model (6.17) with

$Y = $log(Sunday Circulation)
$X_1 = $log(Weekday Circulation)
$X_2 = $Tabloid.with.a.Serious.Competitor (a dummy variable)

Thus we consider the following multiple linear regression model:

$$\log(\text{SundayCirculation}) = \beta_0 + \beta_1 \log(\text{WeekdayCirculation})$$
$$+ \beta_3 \text{ Tabloid.with.a.Serious.Competitor} + e \qquad (6.18)$$

Figure 6.29 contains scatter plots of the standardized residuals against each predictor and the fitted values for model (6.18). Each of the plots in Figure 6.29 shows a random pattern. Thus, model (6.18) appears to be a valid model for the data.

Figure 6.30 contains a plot of log(Sunday Circulation) against the fitted values. The straight-line fit to this plot provides a reasonable fit. This provides further evidence that model (6.18) is a valid model for the data.

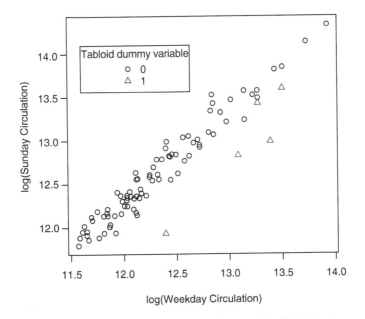

Figure 6.28 A plot of log(Sunday Circulation) against log(Weekday Circulation)

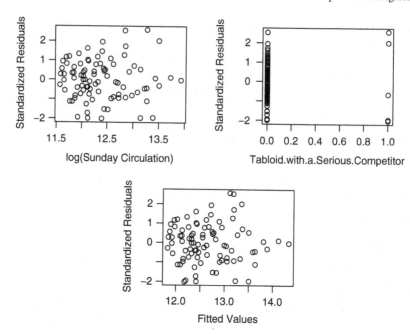

Figure 6.29   Plots of the standardized residuals from model (6.17)

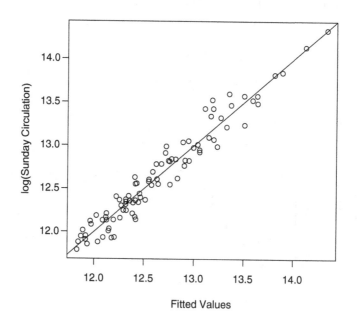

Figure 6.30   A plot of log(Sunday Circulation) against fitted values

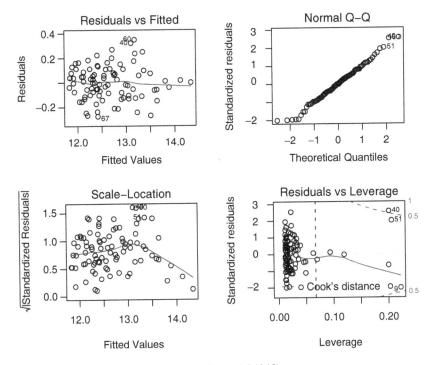

Figure 6.31 Diagnostic plots provided by R for model (6.18)

Figure 6.31 shows the diagnostic plots provided by R for model (6.18). These plots further confirm that model (6.18) is a valid model for the data.

The dashed vertical line in the bottom right-hand plot of Figure 6.31 is the usual cut-off for declaring a point of high leverage (i.e., $2 \times (p+1)/n = 6/89 = 0.067$). The points with the largest leverage correspond to the cases where the dummy variable is 1.

The output from R associated with fitting model (6.18) shows that both predictor variables are highly statistically significant. Because of the log transformation model (6.18) predicts:

- A 1.06% increase in Sunday Circulation for every 1% increase in Weekday Circulation
- A 53.1% decrease in Sunday Circulation if the newspaper is a tabloid with a serious competitor

Regression output from R

```
Call:
lm(formula = log(Sunday) ~ log(Weekday) + Tabloid.with.a.Serious.
Competitor)

Coefficients:
                 Estimate  Std. Error  t value  Pr(>|t|)
(Intercept)      -0.44730    0.35138    -1.273    0.206
log(Weekday)      1.06133    0.02848    37.270   < 2e-16  ***
Tabloid.with.
a.Serious.
Competitor       -0.53137    0.06800    -7.814  1.26e-11  ***
---
Residual standard error: 0.1392 on 86 degrees of freedom
Multiple R-Squared: 0.9427,       Adjusted R-squared: 0.9413
F-statistic: 706.8 on 2 and 86 DF, p-value: < 2.2e-16 ---
Signif. codes: 0 '***' 0.001 '**' 0.01 '*' 0.05 '.' 0.1 ' ' 1
```

Figure 6.32 contains the added-variable plots associated with model (6.18). The fact that both predictor variables are highly statistically significant is evident from the added variable plots.

Finally, we are now able to predict the Sunday circulation of a newspaper with a weekday circulation of 210,000. There are the following two cases to consider corresponding to whether the newspaper is a tabloid with a serious competitor or not. Given below are the prediction intervals obtained from R for log(Sunday Circulation):

Output from R

```
Tabloid.with.a.Serious.Competitor=1
         fit       lwr       upr
[1,] 12.02778 11.72066 12.33489

Tabloid.with.a.Serious.Competitor=0
         fit       lwr       upr
[1,] 12.55915 12.28077 12.83753
```

Back transforming these results by exponentiating them produces the numbers in Table 6.2.

Can you think of a way of improving model (6.18)?

Table 6.2 Predictions of Sunday circulation

Tabloid with a serious competitor	Weekday circulation	Prediction	95% Prediction interval
Yes	210000	167340	(123089, 227496)
No	210000	284668	(215512, 376070)

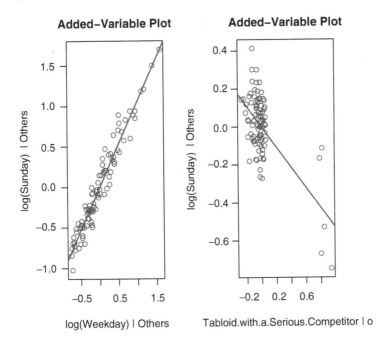

Figure 6.32 Added-variable plots for model (6.18)

6.3 Graphical Assessment of the Mean Function Using Marginal Model Plots

We begin by briefly considering simple linear regression. In this case, we wish to visually assess whether

$$Y = \beta_0 + \beta_1 x + e \qquad (6.19)$$

models $E(Y|x)$ adequately. One way to assess this is to compare the fit from (6.19) with a fit from a general or nonparametric regression model (6.20) where

$$Y = f(x) + e \qquad (6.20)$$

There are many ways to estimate f nonparametrically. We shall use a popular estimator called loess, which is based on local linear or locally quadratic regression fits. Further details on nonparametric regression in general and loess in particular can be found in Appendix A.2.

Under model (6.19), $E_{M_1}(Y \mid x) = \beta_0 + \beta_1 x$, while under model (6.20), $E_{F_1}(Y \mid x) = f(x)$. Thus, we shall decide that model (6.19) is an adequate model if $\hat{\beta}_0 + \hat{\beta}_1 x$ and $\hat{f}(x)$ agree well.

Example: Modeling salary from years of experience (cont.)

Recall from Chapter 5 that we wanted to develop a regression equation to model the relationship between Y, salary (in thousands of $) and x, the number of years of experience. The 143 data points can be found on the book web site in the file prof-salary.txt.

For illustrative purposes we will start by considering the model

$$Y = \beta_0 + \beta_1 x + e \tag{6.21}$$

and compare this with nonparametric regression model (6.22) where

$$Y = f(x) + e \tag{6.22}$$

Figure 6.33 includes the least squares fit for model (6.21) and as a solid curve, the loess fit (with $\alpha = 2/3$) for model (6.22). The two fits differ markedly indicating that model (6.21) is not an adequate model for the data.

We next consider a quadratic regression model for the data

$$Y = \beta_0 + \beta_1 x + \beta_2 x^2 + e \tag{6.23}$$

Figure 6.34 includes the least squares fit for model (6.23) and as a solid curve loess fit (with $\alpha = 2/3$) for model (6.22). The two fits are virtually indistinguishable. This implies that model (6.23) models $E(Y|x)$ adequately.

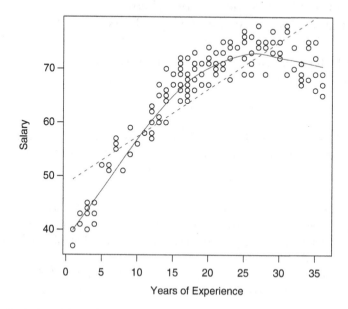

Figure 6.33 A plot of the professional salary data with straight line and loess fits

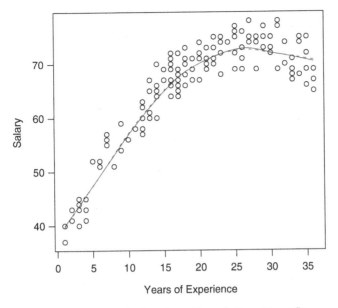

Figure 6.34 A plot of the professional salary data with quadratic and loess fits

The challenge for the approach we have just taken is how to extend it to regression models based on more than one predictor. In what follows we shall describe the approach proposed and developed by Cook and Weisberg (1997).

Marginal Model Plots

Consider the situation when there are just two predictors x_1 and x_2. We wish to visually assess whether

$$Y = \beta_0 + \beta_1 x_1 + \beta_2 x_2 + e \tag{M1}$$

models $E(Y|x)$ adequately. Again we wish to compare the fit from (M1) with a fit from a nonparametric regression model (F1) where

$$Y = f(x_1, x_2) + e \tag{F1}$$

Under model (F1), we can estimate $E_{F_1}(Y \mid x_1)$ by adding a nonparametric fit to the plot of Y against x_1. We want to check that the estimate of $E_{F_1}(Y \mid x_1)$ is close to the estimate of $E_{M_1}(Y \mid x_1)$.

Under model (M1)

$$E_{M_1}(Y \mid x_1) = E(\beta_0 + \beta_1 x_1 + \beta_2 x_2 + e \mid x_1) = \beta_0 + \beta_1 x_1 + \beta_2 E(x_2 \mid x_1)$$

Notice that this last equation includes the unknown $E_{M_1}(x_2 \mid x_1)$ and that in general there would be $(p-1)$ unknowns, where p is the number of predictor variables in model (M1). Cook and Weisberg (1997) overcome this problem by utilizing the following result:

$$E_{M_1}(Y \mid x_1) = E\left[E_{M_1}(Y \mid x) \mid x_1\right] \qquad (6.24)$$

The result follows from the well-known general result re conditional expectations. However, it is easy and informative to demonstrate the result in this special case. First, note that

$$E_{M_1}(Y \mid x) = E_{M_1}(\beta_0 + \beta_1 x_1 + \beta_2 x_2 + e \mid x) = \beta_0 + \beta_1 x_1 + \beta_2 x_2$$

so that

$$E\left[E_{M_1}(Y \mid x) \mid x_1\right] = E(\beta_0 + \beta_1 x_1 + \beta_2 x_2 \mid x_1) = \beta_0 + \beta_1 x_1 + \beta_2 E(x_2 \mid x_1)$$

matching what we found on the previous page for $E_{M_1}(Y \mid x)$.

Under model (M1), we can estimate $E_{M_1}(Y \mid x) = \beta_0 + \beta_1 x_1 + \beta_2 x_2$ by the fitted values $\hat{Y} = \hat{\beta}_0 + \hat{\beta}_1 x_1 + \hat{\beta}_2 x_2$. Utilizing (6.24) we can therefore estimate $E_{M_1}(Y \mid x_1) = E\left[E_{M_1}(Y \mid x) \mid x_1\right]$ by estimating $E\left[E_{M_1}(Y \mid x) \mid x_1\right]$ with an estimate of $E\left[\hat{Y} \mid x_1\right]$.

In summary, we wish to compare estimates under models (F1) and (M1) by comparing nonparametric estimates of $E(Y \mid x_1)$ and $E\left[\hat{Y} \mid x_1\right]$. If the two nonparametric estimates agree then we conclude that x_1 is modelled correctly by model (M1). If **not** then we conclude that x_1 is **not** modelled correctly by model (M1).

Example: Modelling defective rates (cont.)

Recall from earlier in Chapter 6 that interest centres on developing a model for Y, Defective, based on the predictors x_1, Temperature; x_2, Density and x_3, Rate. The data can be found on the book web site in the file defects.txt.

The first model we considered was the following:

$$Y = \beta_0 + \beta_1 x_1 + \beta_2 x_2 + \beta_3 x_3 + e \qquad (6.25)$$

The left-hand plot in Figure 6.35 is a plot of Y against x_1, Temperature with the loess estimate of $E(Y \mid x_1)$ included. The right-hand plot in Figure 6.35 is a plot of $\hat{Y}$ against x_1, Temperature with the loess estimate of $E\left[\hat{Y} \mid x_1\right]$ included.

The two curves in Figure 6.35 do not agree with the fit in the left-hand plot showing distinct curvature, while the fit in the right-hand plot is close to a straight line. Thus, we decide that x_1 is **not** modelled correctly by model (6.25).

In general, it is difficult to compare curves in different plots. Thus, following Cook and Weisberg (1997) we shall from this point on include both nonparametric curves on the plot of Y against x_1. The plot of Y against x_1 with the loess fit for Y against x_1 and the loess fit for $\hat{Y}$ against x_1 both marked on it is called a **marginal model plot** for Y and x_1.

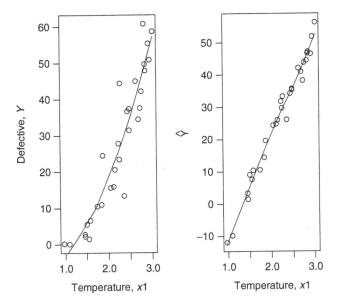

Figure 6.35 Plots of Y and $\hat{Y}$ against x_1, Temperature

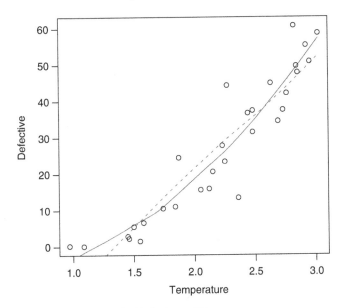

Figure 6.36 A marginal mean plot for Defective and Temperature

Figure 6.36 contains a **marginal model plot** for Y and x_1. The solid curve is the loess estimate of $E(Y \mid x_1)$ while the dashed curve is the loess estimate of $E\left[\hat{Y} \mid x_1\right]$. It is once again clear that these two curves do not agree well.

It is recommended in practice that marginal model plots be drawn for each pre-dictor (except dummy variables) and for $\hat{Y}$. Figure 6.37 contains these recommended marginal model plots for model (6.25) in the current example. The two fits in each of the plots in Figure 6.37 differ markedly. In particular, each of the non-parametric estimates in Figure 6.37 (marked as solid curves) show distinct curvature which is not present in the smooths of the fitted values (marked as dashed curves). Thus, we again conclude that (6.25) is not a valid model for the data.

We found earlier that in this case, both the inverse response plot and the Box-Cox transformation method point to using a square root transformation of Y. Thus, we next consider the following multiple linear regression model

$$Y^{0.5} = \beta_0 + \beta_1 x_1 + \beta_2 x_2 + \beta_3 x_3 + e \tag{6.26}$$

Figure 6.38 contains the recommended marginal model plots for model (6.26) in the current example. These plots again point to the conclusion that (6.26) is a valid model for the data.

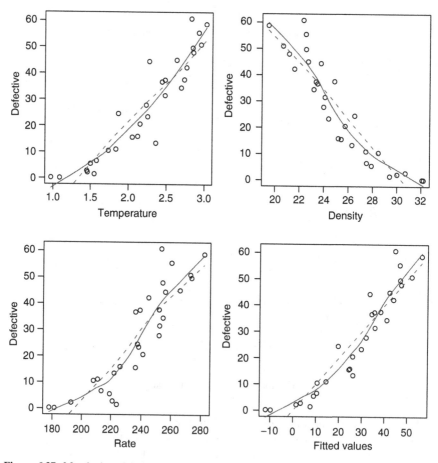

Figure 6.37 Marginal model plots for model (6.25)

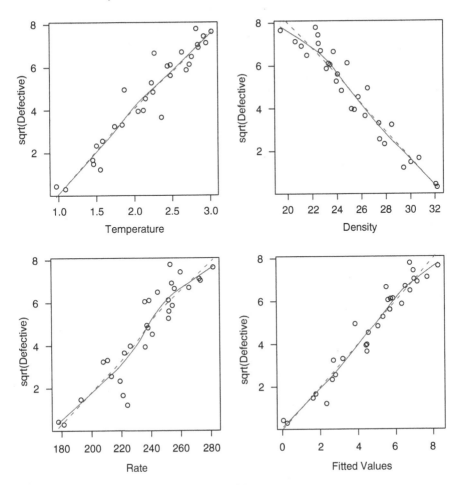

Figure 6.38 Marginal model plots for model (6.26)

6.4 Multicollinearity

A number of important issues arise when strong correlations exist among the predictor variables (often referred to as multicollinearity). In particular, in this situation regression coefficients can have the wrong sign and/or many of the predictor variables are not statistically significant when the overall F-test is highly significant. We shall use the following example to illustrate these issues.

Example: Bridge construction

The following example is adapted from Tryfos (1998, pp. 130–1). According to Tryfos:

> Before construction begins, a bridge project goes through a number of stages of production, one of which is the design stage. This phase is composed of various activities,

each of which contributes directly to the overall design time. ....In short, predicting the design time is helpful for budgeting and internal as well as external scheduling purposes.

Information from 45 bridge projects was compiled for use in this study. The data are partially listed in Table 6.3 below and can be found on the book web site in the file bridge.txt. The response and predictor variables are as follows:

Y = Time = design time in person-days
x_1 = DArea = Deck area of bridge (000 sq ft)
x_2 = CCost = Construction cost ($000)
x_3 = Dwgs = Number of structural drawings
x_4 = Length = Length of bridge (ft)
x_5 = Spans = Number of spans

We begin by plotting the data. Figure 6.39 contains a scatter plot matrix of response variable and the five predictor variables. The response variable and a number of the predictor variables are highly skewed. There is also evidence of nonconstant variance in the top row of plots. Thus, we need to consider transformations of the response and the five predictor variables.

The multivariate version of the Box-Cox transformation method can be used to transform all variables simultaneously. Given below is the output from R using the bctrans command from alr3.

Output from R

```
box.cox Transformations to Multinormality
            Est.Power    Std.Err.    Wald(Power=0)    Wald(Power=1)
Time         -0.1795      0.2001        -0.8970          -5.8951
DArea        -0.1346      0.0893        -1.5073         -12.7069
CCost        -0.1762      0.0942        -1.8698         -12.4817
Dwgs         -0.2507      0.2402        -1.0440          -5.2075
Length       -0.1975      0.1073        -1.8417         -11.1653
Spans        -0.3744      0.2594        -1.4435          -5.2991
                                        LRT      df    p.value
LR test, all lambda equal 0         8.121991     6   0.2293015
LR test, all lambda equal 1       283.184024     6   0.0000000
```

Using the Box-Cox method to transform the predictor and response variables simultaneously toward multivariate normality, results in values of each λ close to 0. Thus,

Table 6.3 Partial listing of the data on bridge construction (bridge.txt)

Case	TIME	DAREA	CCOST	DWGS	LENGTH	SPANS
1	78.8	3.6	82.4	6	90	1
2	309.5	5.33	422.3	12	126	2
3	184.5	6.29	179.8	9	78	1
.	.	.	.	.	.	.
45	87.2	3.24	70.2	6	90	1

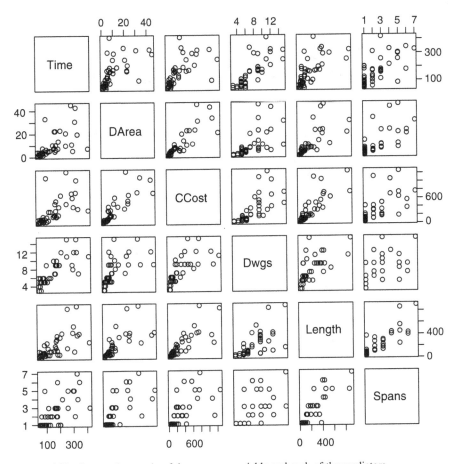

Figure 6.39 Scatter plot matrix of the response variable and each of the predictors

we shall transform each variable using the log transformation. Figure 6.40 shows a scatter plot matrix of the log-transformed response and predictor variables. The pairwise relationships in Figure 6.40 are much more linear than those in Figure 6.39. There is no longer any evidence of nonconstant variance in the top row of plots.

We next consider a multiple linear regression model based on the log-transformed data, namely,

$$\log(Y) = \beta_0 + \beta_1 \log(x_1) + \beta_2 \log(x_2) + \beta_3 \log(x_3) + \beta_4 \log(x_4) \qquad (6.28)$$
$$+ \beta_5 \log(x_5) + e$$

Figure 6.41 contains scatter plots of the standardized residuals against each predictor and the fitted values for model (6.28). Each of the plots in Figure 6.41 shows a random pattern. Thus, model (6.28) appears to be a valid model for the data.

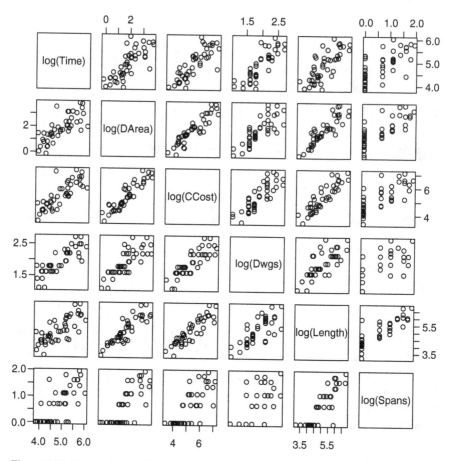

Figure 6.40 Scatter plot matrix of the log-transformed data

Figure 6.42 contains a plot of log(Time) against the fitted values. The straight-line fit to this plot provides a reasonable fit. This provides further evidence that model (6.28) is a valid model for the data.

Figure 6.43 shows the diagnostic plots provided by R for model (6.28). These plots further confirm that model (6.28) is a valid model for the data.

The dashed vertical line in the bottom right-hand plot of Figure 6.43 is the usual cut-off for declaring a point of high leverage (i.e., $2 \times (p + 1)/n = 12/45 = 0.267$). Thus, there is a bad leverage point (i.e., case 22) that requires further investigation.

Figure 6.44 contains the recommended marginal model plots for model (6.28). The nonparametric estimates of each pair-wise relationship are marked as solid curves, while the smooths of the fitted values are marked as dashed curves. There is some curvature present in the top three plots which is not present in the smooths of the fitted values. However, at this stage we shall continue under the assumption that (6.28) is a valid model.

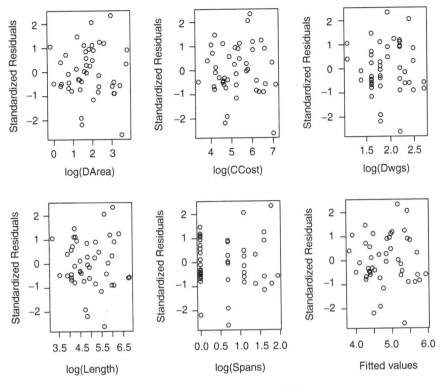

Figure 6.41 Plots of the standardized residuals from model (6.28)

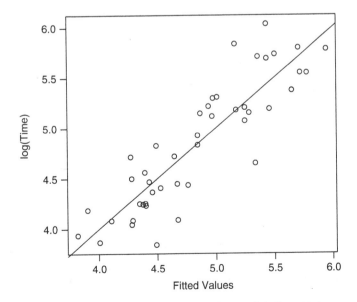

Figure 6.42 A plot of log(Time) against fitted values with a straight line added

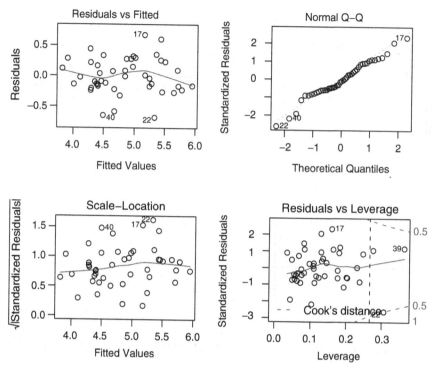

Figure 6.43 Diagnostic plots from R for model (6.28)

Given below is the output from R associated with fitting model (6.28).

Regression output from R

```
Call:
lm(formula = log(Time) ~ log(DArea) + log(CCost) + log(Dwgs) +
log(Length) + log(Spans))

Coefficients:
             Estimate   Std. Error   t value   Pr(>|t|)
(Intercept)   2.28590     0.61926     3.691    0.00068    ***
log(DArea)   -0.04564     0.12675    -0.360    0.72071
log(CCost)    0.19609     0.14445     1.358    0.18243
log(Dwgs)     0.85879     0.22362     3.840    0.00044    ***
log(Length)  -0.03844     0.15487    -0.248    0.80530
log(Spans)    0.23119     0.14068     1.643    0.10835
---

Residual standard error: 0.3139 on 39 degrees of freedom
Multiple R-Squared: 0.7762,  Adjusted R-squared: 0.7475
F-statistic: 27.05 on 5 and 39 DF,  p-value: 1.043e-11
```

Notice that while the overall F-test for model (6.28) is highly statistically significant (i.e., has a very small *p*-value), only one of the estimated regression

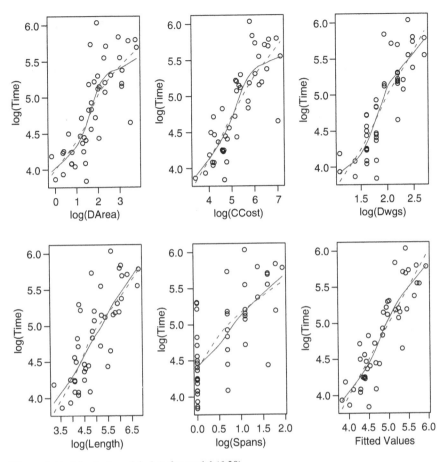

Figure 6.44 Marginal model plots for model (6.28)

coefficients is statistically significant (i.e., log(Dwgs) with a p-value < 0.001). Even more troubling is the fact that the estimated regression coefficients for log(DArea) and log(Length) are of the wrong sign (i.e., negative), since longer bridges or bridges with larger area should take a longer rather than a shorter time to design.

Finally, we show in Figure 6.45 the added-variable plots associated with model (6.28). The lack of statistical significance of the predictor variables other than log(Dwgs) is evident from Figure 6.45.

When two or more highly correlated predictor variables are included in a regression model, they are effectively carrying very similar information about the response variable. Thus, it is difficult for least squares to distinguish their separate effects on the response variable. In this situation the overall F-test will be highly statistically significant but very few of the regression coefficients may be statistically significant. Another consequence of highly correlated predictor variables is that some of the coefficients in the regression model are of the opposite sign than expected.

The output from R below gives the correlations between the predictors in model (6.28). Notice how most of the correlations are greater than 0.8.

Output from R: Correlations between the predictors in (6.28)

	logDArea	logCCost	logDwgs	logLength	logSpans
logDArea	1.000	0.909	0.801	0.884	0.782
logCCost	0.909	1.000	0.831	0.890	0.775
logDwgs	0.801	0.831	1.000	0.752	0.630
logLength	0.884	0.890	0.752	1.000	0.858
logSpans	0.782	0.775	0.630	0.858	1.000

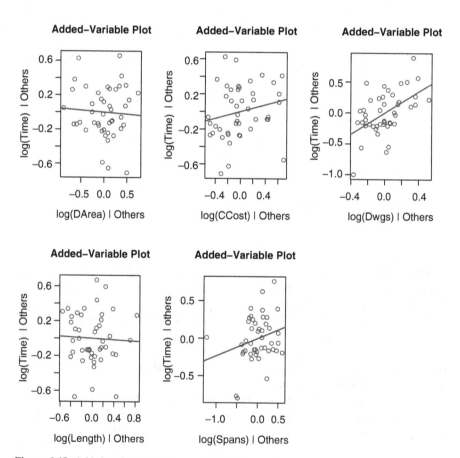

Figure 6.45 Added-variable plots for model (6.28)

6.4.1   Multicollinearity and Variance Inflation Factors

First, consider a multiple regression model with two predictors

$$Y = \beta_0 + \beta_1 x_1 + \beta_2 x_2 + e$$

Let r_{12} denote the correlation between x_1 and x_2 and S_{x_j} denote the standard deviation of x_j. Then it can be shown that

$$\text{Var}(\hat{\beta}_j) = \frac{1}{1-r_{12}^2} \times \frac{\sigma^2}{(n-1)S_{x_j}^2} \quad j = 1, 2$$

Notice how the variance of $\hat{\beta}_j$ gets larger as the absolute value of r_{12} increases. Thus, *correlation amongst the predictors increases the variance of the estimated regression coefficients.* For example, when $r_{12}^2 = 0.99$ the variance of $\hat{\beta}_j$ is

$$\frac{1}{1-r_{12}^2} = \frac{1}{1-0.99^2} = 50.25 \text{ times larger than it would be if } r_{12}^2 = 0 . \text{ The term } \frac{1}{1-r_{12}^2}$$

is called a variance inflation factor (VIF).

Next consider the general multiple regression model

$$Y = \beta_0 + \beta_1 x_1 + \beta_2 x_2 + ... + \beta_p x_p + e$$

Let R_j^2 denote the value of R^2 obtained from the regression of x_j on the other x's (i.e., the amount of variability explained by this regression). Then it can be shown that

$$\text{Var}(\hat{\beta}_j) = \frac{1}{1-R_j^2} \times \frac{\sigma^2}{(n-1)S_{x_j}^2} \quad j = 1, ..., p$$

The term $1/(1-R_j^2)$ is called the jth **variance inflation factor (VIF)**.

The variance inflation factors for the bridge construction example are as follows:

log(DArea)	log(CCost)	log(Dwgs)	log(Length)	log(Spans)
7.164619	8.483522	3.408900	8.014174	3.878397

A number of these variance inflation factors exceed 5, the cut-off often used, and so the associated regression coefficients are poorly estimated due to multicollinearity.

We shall return to this example in Chapter 7.

6.5   Case Study: Effect of Wine Critics' Ratings on Prices of Bordeaux Wines

We next answer the questions in Section 1.1.4. In particular, we are interested in the effects of an American wine critic, Robert Parker and an English wine critic, Clive Coates on the London auction prices of Bordeaux wines from the 2000 vintage.

Part (a)

Since interest centres on estimating the percentage effect on price of a 1% increase in ParkerPoints and a 1% increase in CoatesPoints we consider the following model

$$\log(Y) = \beta_0 + \beta_1 \log(x_1) + \beta_2 \log(x_2) + \beta_3 x_3 + \beta_4 x_4 + \beta_5 x_5$$
$$+ \beta_6 x_6 + \beta_7 x_7 + e \tag{6.29}$$

where

Y = Price = the price (in pounds sterling) of 12 bottles of wine
x_1 = ParkerPoints = Robert Parker's rating of the wine (out of 100)
x_2 = CoatesPoints = Clive Coates' rating of the wine (out of 20)
x_3 = P95andAbove = 1 (0) if the wine scores 95 or above from Robert Parker (otherwise)
x_4 = FirstGrowth = 1 (0) if the wine is a First Growth (otherwise)
x_5 = CultWine = 1 (0) if the wine is a cult wine (otherwise)
x_6 = Pomerol = 1 (0) if the wine is from Pomerol (otherwise)
x_7 = VintageSuperstar = 1 (0) if the wine is a vintage superstar (otherwise)

Recall from Chapter 1 that Figure 1.9 contains a matrix plot of log(Price), log(Parker's ratings) and log(Coates ratings), while Figure 1.10 shows box plots of log(Price) against each of the dummy variables.

Figure 6.46 contains plots of the standardized residuals against each predictor and the fitted values for model (6.29). The plots are in the form of scatter plots for real valued predictors and box plots for predictors in the form of dummy variables. Each of the scatter plots in Figure 6.46 shows a random pattern. In addition, the box plots show that the variability of the standardized residuals is relatively constant across both values of each dummy predictor variable. Thus, model (6.29) appears to be a valid model for the data.

Figure 6.47 contains a plot of log(Price) against the fitted values. The straight-line fit to this plot provides a reasonable fit. This provides further evidence that model (6.29) is a valid model for the data.

Figure 6.48 shows the diagnostic plots provided by R for model (6.29). These plots further confirm that model (6.29) is a valid model for the data. The dashed vertical line in the bottom right-hand plot of Figure 6.48 is the usual cut-off for declaring a point of high leverage (i.e., $2 \times (p+1)/n = 16/72 = 0.222$). Case 67, Le Pin is a bad leverage point.

Figure 6.49 contains the recommended marginal model plots for model (6.29). Notice that the nonparametric estimates of each pair-wise relationship are marked as solid curves, while the smooths of the fitted values are marked as dashed curves. The two curves in each plot match very well thus providing further evidence that (6.29) is a valid model.

Given below is the output from R associated with fitting model (6.29). Notice that the overall F-test for model (6.29) is highly statistically significant and the only estimated regression coefficient that is not statistically significant is P95andAbove.

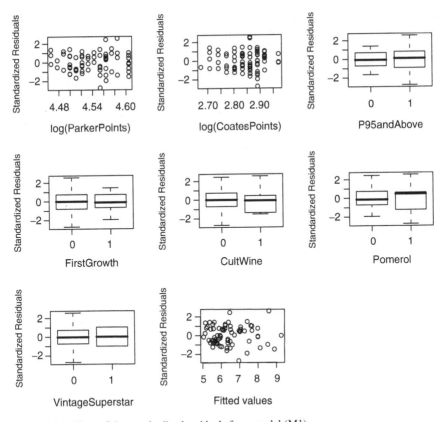

Figure 6.46 Plots of the standardized residuals from model (M1)

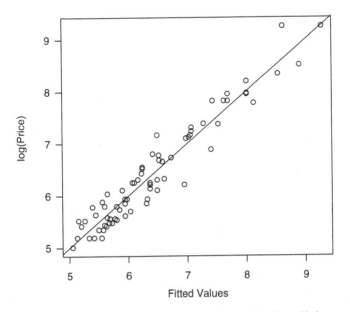

Figure 6.47 A plot of log(Price) against fitted values with a straight line added

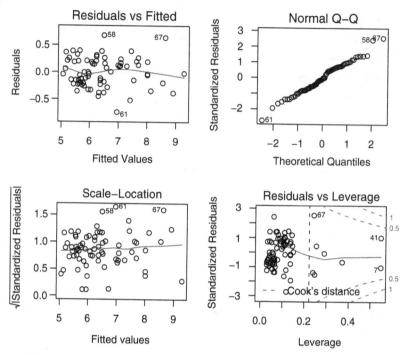

Figure 6.48 Diagnostic plots from R for model (6.29)

Regression output from R

```
Call:
lm(formula  =  log(Price)   ~   log(ParkerPoints)   +   log(Coates
Points)  +  P95andAbove  +  FirstGrowth  +  CultWine  +  Pomerol  +
VintageSuperstar)
Coefficients:
                     Estimate   Std. Error   t value   Pr(>|t|)
(Intercept)          -51.14156    8.98557    -5.692   3.39e-07  ***
log(ParkerPoints)     11.58862    2.06763     5.605   4.74e-07  ***
log(CoatesPoints)      1.62053    0.61154     2.650    0.01013  *
P95andAbove            0.10055    0.13697     0.734    0.46556
FirstGrowth            0.86970    0.12524     6.944   2.33e-09  ***
CultWine               1.35317    0.14569     9.288   1.78e-13  ***
Pomerol                0.53644    0.09366     5.727   2.95e-07  ***
VintageSuperstar       0.61590    0.22067     2.791    0.00692  **
---
Residual standard error: 0.2883 on 64 degrees of freedom
Multiple R-Squared: 0.9278,  Adjusted R-squared: 0.9199
F-statistic: 117.5 on 7 and 64 DF,  p-value: < 2.2e-16
```

Figure 6.50 shows the added-variable plots associated with model (6.29). Case 53 (Pavie) appears to be highly influential in the added variable plot for log(CoatesPoints), and, as such, it should be investigated. Other outliers are evident from the added variable plots in Figure 6.50. We shall continue under the assumption that (6.29) is a valid model.

The variance inflation factors for the training data set are as follows:

```
log(ParkerPoints)  log(CoatesPoints)     P95andAbove      FirstGrowth
         5.825135           1.410011        4.012792         1.625091
CultWine      Pomerol   VintageSuperstar
1.188243    1.124300           1.139201
```

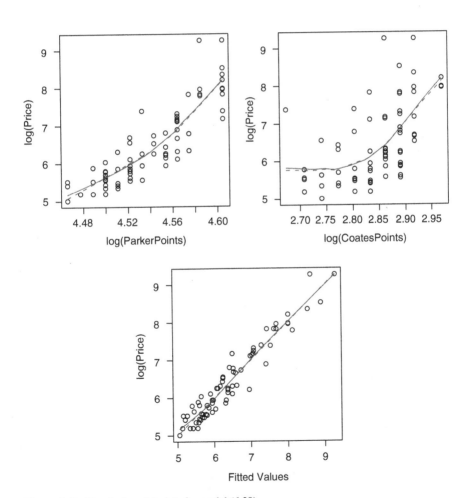

Figure 6.49   Marginal model plots for model (6.29)

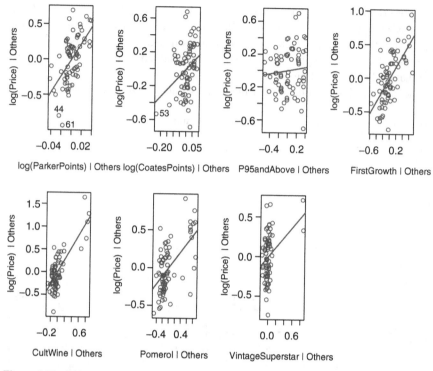

Figure 6.50 Added-variable plots for model (6.29)

Only one of the variance inflation factors exceeds 5 and so multicollinearity is only a minor issue.

Since (6.29) is a valid model and the only estimated regression coefficient that is not statistically significant is x_3, P95andAbove we shall drop it from the model and consider the reduced model

$$\log(Y) = \beta_0 + \beta_1 \log(x_1) + \beta_2 \log(x_2) + \beta_4 x_4 + \beta_5 x_5$$
$$+ \beta_6 x_6 + \beta_7 x_7 + e \tag{6.30}$$

Given below is the output from R associated with fitting model (6.30).

Regression output from R

```
Call:
lm(formula = log(Price) ~ log(ParkerPoints) + log(CoatesPoints) +
        FirstGrowth + CultWine + Pomerol + VintageSuperstar)
Coefficients:
                     Estimate   Std. Error   t value   Pr(>|t|)
(Intercept)         -56.47547      5.26798   -10.721   5.20e-16 ***
log(ParkerPoints)    12.78432      1.26915    10.073   6.66e-15 ***
```

```
log(CoatesPoints)     1.60447      0.60898      2.635    0.01052  *
FirstGrowth           0.86149      0.12430      6.931    2.30e-09 ***
CultWine              1.33601      0.14330      9.323    1.34e-13 ***
Pomerol               0.53619      0.09333      5.745    2.64e-07 ***
VintageSuperstar      0.59470      0.21800      2.728    0.00819  **
---
Signif. codes: 0 '***' 0.001 '**' 0.01 '*' 0.05 '.' 0.1 ' ' 1
Residual standard error: 0.2873 on 65 degrees of freedom
Multiple R-Squared: 0.9272,  Adjusted R-squared: 0.9205
F-statistic: 138 on 6 and 65 DF,  p-value: < 2.2e-16
```

Since all the predictor variables have statistically significant t-values, there is no redundancy in model (6.30) and as such we shall adopt it as our full model. Notice how similar the estimated regression coefficients are in models (6.29) and (6.30). Note that there is no real need to redo the diagnostic plots for model (6.30) since it is so similar to model (6.29).

Alternatively, we could consider a partial F-test to compare models (6.29) and (6.30). The R output for such a test is given below:

```
Analysis of Variance Table

Model 1: log(Price) ~ log(ParkerPoints) + log(CoatesPoints) +
    FirstGrowth + CultWine + Pomerol + VintageSuperstar
Model 2: log(Price) ~ log(ParkerPoints) + log(CoatesPoints) +
    P95andAbove  +  FirstGrowth  +  CultWine  +  Pomerol  +
    VintageSuperstar
  Res.Df     RSS  Df  Sum of Sq       F   Pr(>F)
1     65  5.3643
2     64  5.3195   1     0.0448  0.5389   0.4656
```

The p-value from the partial F-test is the same as the t-test p-value from model (6.29). This is due to the fact that only one predictor has been removed from (6.29) to obtain (6.30).

Part (b)

Based on model (6.30) we find that

1. A 1% increase in Parker points is predicted to increase price by 12.8%
2. A 1% increase in Coates points is predicted to increase price by 1.6%

Part (c)

If we consider either the full model (6.29), which includes 95andAbove, or the final model (6.30), which does not, then the predictor variable ParkerPoints has the largest estimated effect on price, since it has the largest regression coefficient. This effect is also the most statistically significant, since the corresponding t-value is the largest in magnitude (or alternatively, the corresponding p-value is the smallest).

Table 6.4 Unusually highly priced wines

Wine	Standardized residuals
Tertre-Roteboeuf	2.43
Le Pin	2.55

Table 6.5 Unusually lowly priced wines

Wine	Standardized residuals
La Fleur-Petrus	−2.73

Part (d)

The claim that "in terms of commercial impact his (Coates') influence is zero" is not supported by the regression model developed in (a). In particular, Clive Coates ratings have a statistically significant impact on price, even after adjusting for the influence of Robert Parker.

Part (e)

Based on the regression model in (a), there is no evidence of a statistically significant extra price premium paid for Bordeaux wines from the 2000 vintage that score 95 and above from Robert Parker since the coefficient of 95andAbove in the regression model is not statistically significant.

Part (f)

(i) Wines which are unusually highly priced are those with standardized residuals greater than + 2. These are given in Table 6.4.
(ii) Wines which are unusually lowly priced are those with standardized residuals less than −2. The only such wine is given in Table 6.5.

6.6 Pitfalls of Observational Studies Due to Omitted Variables

In this section we consider some of the pitfalls of regression analysis based on data from observational studies. An observational study is one in which outcomes are observed and no attempt is made to control or influence the variables of interest. As such there may be systematic differences that are not included in the regression model, which we shall discover, raises the issue of omitted variables.

6.6.1 Spurious Correlation Due to Omitted Variables

We begin by describing a well-known weakness of regression modeling based on observational data, namely that the observed association between two variables may be because both are related to a third variable that has been omitted from the regression model. This phenomenon is commonly referred to as "spurious correlation."

The term spurious correlation dates back to at least Pearson (1897). According to Stigler (2005, p. S89):

> ... Pearson studied measurements of a large collection of skulls from the Paris Catacombs, with the goal of understanding the interrelationships among the measurements. For each skull, his assistant measured the length and the breadth, and computed ... the correlation coefficient between these measures ... The correlation ... turned out to be significantly greater than zero ... But ... the discovery was deflated by his noticing that if the skulls were divided into male and female, the correlation disappeared. Pearson recognized the general nature of this phenomenon and brought it to the attention of the world. When two measurements are correlated, this may be because they are both related to a third factor that has been omitted from the analysis. In Pearson's case, skull length and skull breadth were essentially uncorrelated if the factor "sex" were incorporated in the analysis.

Neyman (1952, pp. 143–154) provides an example based on fictitious data which dramatically illustrates spurious correlation. According to Kronmal (1993, p. 379), a fictitious friend of Neyman was interested in empirically examining the theory that storks bring babies and collected data on the number of women, babies born and storks in each of 50 counties. This fictitious data set was reported in Kronmal (1993, p. 383) and it can be found on the course web site in the file storks.txt.

Figure 6.51 shows a scatter plot of the number of babies against the number of storks along with the least squares fit. Fitting the following straight-line regression model to these data produces the output shown below.

$$\text{Babies} = \beta_0 + \beta_1 \text{Storks} + e \tag{6.31}$$

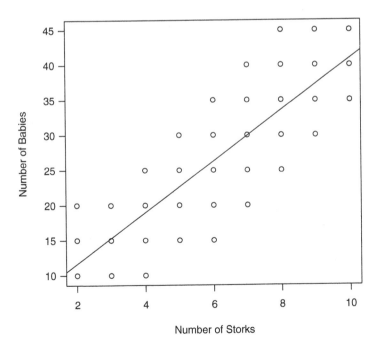

Figure 6.51 A plot of two variables from the fictitious data on storks

The regression output from R shows that there is very strong evidence of a positive linear association between the number of storks and the number of babies born (p-value < 0.0001). However, to date we have ignored the data available on the other potential predictor variable, namely, the number of women.

Regression output from R

```
Coefficients:
              Estimate   Std. Error   t value   Pr(>|t|)
(Intercept)     4.3293      2.3225      1.864     0.068  .
Storks          3.6585      0.3475     10.528    1.71e-14 ***
---

Residual standard error: 5.451 on 52 degrees of freedom
Multiple R-Squared: 0.6807,  Adjusted R-squared: 0.6745
F-statistic: 110.8 on 1 and 52 DF,  p-value: 1.707e-14
```

Figure 6.52 shows scatter plots of all three variables from the stork data set along with the least squares fits. It is apparent that there is a strong positive linear association between each of the three variables. Thus, we consider the following regression model:

$$\text{Babies} = \beta_0 + \beta_1 \text{Storks} + \beta_2 \text{Women} + e \qquad (6.32)$$

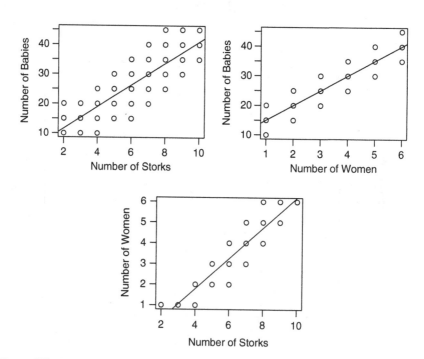

Figure 6.52 A plot of all three variables from the fictitious data on storks

Given below is the output from R for a regression model (6.32). Notice that the estimated regression coefficient for the number of storks is zero to many decimal places. Thus, correlation between the number of babies and the number of storks calculated from (6.31) is said to be **spurious** as it is due to both variables being associated with the number of women. In other words, a predictor (the number of women) exists which is related to both the other predictor (the number of storks) and the outcome variable (the number of babies), and which accounts for all of the observed association between the latter two variables. The number of women predictor variable is commonly called either an **omitted variable** or a **confounding covariate**.

Regression output from R

```
Coefficients:
               Estimate    Std. Error    t value   Pr(>|t|)
(Intercept)    1.000e+01    2.021e+00      4.948   8.56e-06   ***
Women          5.000e+00    8.272e-01      6.045   1.74e-07   ***
Storks        -6.203e-16    6.619e-01   -9.37e-16         1
---
Residual standard error: 4.201 on 51 degrees of freedom
Multiple R-Squared: 0.814,  Adjusted R-squared: 0.8067
F-statistic: 111.6 on 2 and 51 DF,  p-value: < 2.2e-16
```

6.6.2 The Mathematics of Omitted Variables

In this section we shall consider the situation in which an important predictor is omitted from a regression model. We shall denote the omitted predictor variable by v and the predictor variable included in the one-predictor regression model by x. In the fictitious stork data x corresponds to the number of storks and v corresponds to the number of women.

To make things as straightforward as possible we shall consider the situation in which Y is related to two predictors x and v as follows:

$$Y = \beta_0 + \beta_1 x + \beta_2 v + e_{Y \cdot x, v} \tag{6.33}$$

Similarly, suppose that v is related to x as follows:

$$v = \alpha_0 + \alpha_1 x + e_{v \cdot x} \tag{6.34}$$

Substituting (6.34) into (6.33) we will be able to discover what happens if omit v from the regression model. The result is as follows:

$$Y = \left(\beta_0 + \beta_2 \alpha_0\right) + \left(\beta_1 + \beta_2 \alpha_1\right) x + \left(e_{Y \cdot x, v} + \beta_2 e_{v \cdot x}\right) \tag{6.35}$$

Notice that the regression coefficient of x in (6.35) is the sum of two terms, namely, $\beta_1 + \beta_2 \alpha_1$. We next consider two distinct cases:

1. $\alpha_1 = 0$ and/or $\beta_2 = 0$: Then the omitted variable has no effect on the regression model, which includes just x as a predictor.
2. $\alpha_1 \neq 0$ and $\beta_2 \neq 0$. Then the omitted variable has an effect on the regression model, which includes just x as a predictor. For example, Y and x can be strongly linearly associated (i.e., highly correlated) even when $\beta_1 = 0$. (This is exactly the situation in the fictitious stork data.) Alternatively, Y and x can be strongly negatively associated even when $\beta_1 > 0$.

6.6.3   Omitted Variables in Observational Studies

Omitted variables are most problematic in observational studies. We next look at two real examples, which exemplify the issues.

The first example is based on a series of papers (Cochrane et al., 1978; Hinds, 1974; Jayachandran and Jarvis, 1986) that model the relationship between the prevalence of doctors and the infant mortality rate. The controversy was the subject of a 1978 *Lancet* editorial entitled "The anomaly that wouldn't go away." In the words of one of the authors of the original paper, Selwyn St Leger (2001):

> When Archie Cochrane, Fred Moore and I conceived of trying to relate mortality in developed countries to measures of health service provision little did we imagine that it would set a hare running 20 years into the future. ... The hare was not that a statistical association between health service provision and mortality was absent. Rather it was the marked positive correlation between the prevalence of doctors and infant mortality. Whatever way we looked at our data we could not make that association disappear. Moreover, we could identify no plausible mechanism that would give rise to this association.

Kronmal (1993, p. 624) reports that Sankrithi et al. (1991) found a significant negative association ($p < 0.001$) between infant mortality rate and the prevalence of doctors after adjusting for population size. Thus, this spurious correlation was due to an omitted variable.

The second example involves a series of observational studies reported in Pettiti (1998) which find evidence of beneficial effects of hormone replacement therapy (HRT) and estrogen replacement therapy (ERT) on coronary heart disease (CHD). On the other hand, Pettiti (1998), reports that "a randomized controlled trial of 2763 postmenopausal women with established coronary disease, treatment with estrogen plus progestin did not reduce the rate of CHD events". Pettiti (1998) points to the existence of omitted variables in the following discussion of the limitations of observational studies in this situation:

> Reasons to view cautiously the observational results for CHD in users of ERT and HRT have always existed. Women with healthy behaviors, such as those who follow a low-fat diet and exercise regularly, may selectively use postmenopausal hormones. These differences in behavior may not be taken into account in the analysis of observational studies because they are not measured, are poorly measured, or are unmeasurable.

In summary, the possibility of omitted variables should be considered when the temptation arises to over interpret the results of any regression analysis based on observational data. Stigler (2005) advises that we "discipline this predisposition (to accept the results of observational studies) by a heavy dose of skepticism." We finish this section by reproducing the following advice from Wasserman (2004, p. 259):

> Results from observational studies start to become believable when: (i) the results are replicated in many studies; (ii) each of the studies controlled for possible confounding variables, (iii) there is a plausible scientific explanation for the existence of a causal relationship.

6.7 Exercises

1. The multiple linear regression model can be written as

$$\mathbf{Y} = \mathbf{X}\beta + \mathbf{e}$$

where $\text{Var}(\mathbf{e}) = \sigma^2 \mathbf{I}$ and $\mathbf{I}$ is the $(n \times n)$ identity matrix so that $\text{Var}\,(\mathbf{Y} \mid \mathbf{X}) = \sigma^2\,\mathbf{I}$. The fitted or predicted values are given by

$$\hat{\mathbf{Y}} = \mathbf{X}\hat{\beta} = \mathbf{X}(\mathbf{X}'\mathbf{X})^{-1}\mathbf{X}'\mathbf{Y} = \mathbf{H}\mathbf{Y}$$

where $\mathbf{H} = \mathbf{X}(\mathbf{X}'\mathbf{X})^{-1}\mathbf{X}'$. Show that $\text{Var}\left(\hat{\mathbf{Y}} \mid \mathbf{X}\right) = \sigma^2\mathbf{H}$.

2. Chapter 5-2 of the award-winning book on baseball (Keri, 2006) makes extensive use of multiple regression. For example, since the 30 "Major League Baseball teams play eighty-one home games during the regular season and receive the largest share of their income from the ticket sales associated with these games" the author develops a least squares regression model to predict Y, yearly income (in 2005 US dollars) from ticket sales for each team from home games each year. Ticket sales data for each team for each of the years from 1997 to 2004 are used to develop the model. Thus, there are $30 \times 8 = 240$ rows of data. Twelve potential predictor variables are identified as follows: Six predictor variables measure team quality, namely:

x_1 = Number of games won in current season
x_2 = Number of games won in previous season
x_3 = Dummy variable for playoff appearance in current season
x_4 = Dummy variable for playoff appearance in previous season
x_5 = Number of winning seasons in the past 10 years
x_6 = Number of playoff appearances in the past 10 years

Three predictors measure stadium of quality, namely:

x_7 = Seating capacity
x_8 = Stadium quality rating
x_9 = Honeymoon effect

Two predictors measure market quality, namely:

x_{10} = Market size

x_{11} = Per-capita income

Finally, x_{12} = Year is included to allow for inflation. The author found that "*seven of these (predictor) variables had a statistically significant impact on attendance revenue*" (i.e., had a t-statistic significant at least at the 10% level). Describe in detail two major concerns that potentially threaten the validity of the model.

3. The analyst was so impressed with your answers to Exercise 5 in Section 3.4 that your advice has been sought regarding the next stage in the data analysis, namely an analysis of the effects of different aspects of a car on its suggested retail price. Data are available for all 234 cars on the following variables:

Y = Suggested Retail Price; x_1 = Engine size; x_2 = Cylinders;

x_3 = Horse power; x_4 = Highway mpg; x_5 = Weight x_6 = Wheel Base;

and x_7 = Hybrid, a dummy variable which is 1 for so-called hybrid cars. The first model considered for these data was

$$Y = \beta_0 + \beta_1 x_1 + \beta_2 x_2 + \beta_3 x_3 + \beta_4 x_4 + \beta_5 x_5 + \beta_6 x_6 + \beta_7 x_7 + e \qquad (6.36)$$

Output from model (6.36) and associated plots (Figures 6.53 and 6.54) appear on the following pages.

(a) Decide whether (6.36) is a valid model. Give reasons to support your answer.
(b) The plot of residuals against fitted values produces a curved pattern. Describe what, if anything can be learned about model (6.36) from this plot.
(c) Identify any bad leverage points for model (6.36).

The multivariate version of the Box-Cox method was used to transform the predictors, while a log transformation was used for the response variable to improve interpretability. This resulted in the following model

$$\log(Y) = \beta_0 + \beta_1 x_1^{0.25} + \beta_2 \log(x_2) + \beta_3 \log(x_3) \qquad (6.37)$$
$$+ \beta_4 \left(\frac{1}{x_4} \right) + \beta_5 x_5 + \beta_6 \log(x_6) + \beta_7 x_7 + e$$

Output from model (6.37) and associated plots (Figures 6.55, 6.56 and 6.57) appear on the following pages. In that output a "t" at the start of a variable name means that the variable has been transformed according to model (6.37).

(d) Decide whether (6.37) is a valid model.
(e) To obtain a final model, the analyst wants to simply remove the two insignificant predictors $(1/x_4)$ (i.e., tHighwayMPG) and $\log(x_6)$ (i.e., tWheel-Base) from (6.37). Perform a partial F-test to see if this is a sensible strategy.

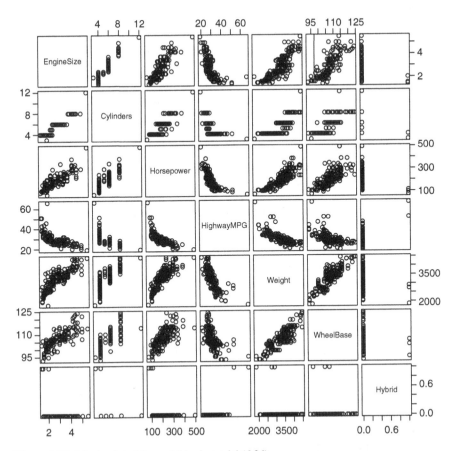

Figure 6.53 Matrix plot of the variables in model (6.36)

(f) The analyst's boss has complained about model (6.37) saying that it fails to take account of the manufacturer of the vehicle (e.g., BMW vs Toyota). Describe how model (6.37) could be expanded in order to estimate the effect of manufacturer on suggested retail price.

Output from R output from model (6.36)

```
Call:
lm(formula   = SuggestedRetailPrice      ~ EngineSize    + Cylinders +
Horsepower   + HighwayMPG      + Weight     + WheelBase    + Hybrid)
Coefficients:
                Estimate    Std. Error    t value    Pr(>|t|)
(Intercept)    -68965.793    16180.381     -4.262    2.97e-05  ***
EngineSize      -6957.457     1600.137     -4.348    2.08e-05  ***
Cylinders        3564.755      969.633      3.676    0.000296  ***
Horsepower        179.702       16.411     10.950    < 2e-16   ***
```

HighwayMPG	637.939	202.724	3.147	0.001873	**
Weight	11.911	2.658	4.481	1.18e-05	***
WheelBase	47.607	178.070	0.267	0.789444	
Hybrid	431.759	6092.087	0.071	0.943562	

```
Residual standard error: 7533 on 226 degrees of freedom
Multiple R-squared:0.7819, Adjusted R-squared: 0.7751
F-statistic: 115.7 on 7 and 226 DF, p-value: < 2.2e-16
```

box.cox Transformations to Multinormality

	Est.Power	Std.Err.	Wald(Power=0)	Wald(Power=1)
EngineSize	0.2551	0.1305	1.9551	-5.7096
Cylinders	-0.0025	0.1746	-0.0144	-5.7430
Horsepower	-0.0170	0.1183	-0.1439	-8.5976
HighwayMPG	-1.3752	0.1966	-6.9941	-12.0801
Weight	1.0692	0.2262	4.7259	0.3057
WheelBase	0.0677	0.6685	0.1012	-1.3946

	LRT	df	p.value
LR test, all lambda equal 0	78.4568	6	7.438494e-15
LR test, all lambda equal 1	305.1733	6	0.000000e+00

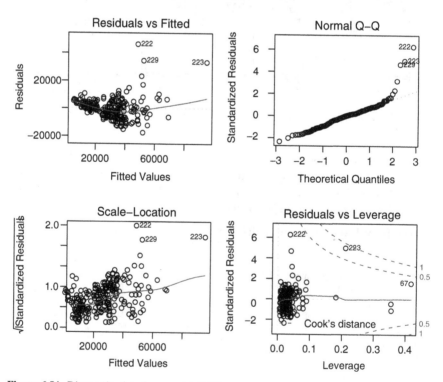

Figure 6.54 Diagnostic plots from model (6.36)

Output from R for model (6.37)

```
Call:
lm(formula = tSuggestedRetailPrice ~ tEngineSize + tCylinders +
tHorsepower + tHighwayMPG + Weight + tWheelBase + Hybrid)
Coefficients:
                Estimate    Std. Error   t value   Pr(>|t|)
(Intercept)     6.119e+00   7.492e-01      8.168   2.22e-14  ***
tEngineSize    -2.247e+00   3.352e-01     -6.703   1.61e-10  ***
tCylinders      3.950e-01   1.165e-01      3.391   0.000823  ***
tHorsepower     8.951e-01   8.542e-02     10.478   < 2e-16   ***
tHighwayMPG    -2.133e+00   4.403e+00     -0.484   0.628601
Weight          5.608e-04   6.071e-05      9.237   < 2e-16   ***
tWheelBase     -1.894e+01   4.872e+01     -0.389   0.697801
Hybrid          1.330e+00   1.866e-01      7.130   1.34e-11  ***
```

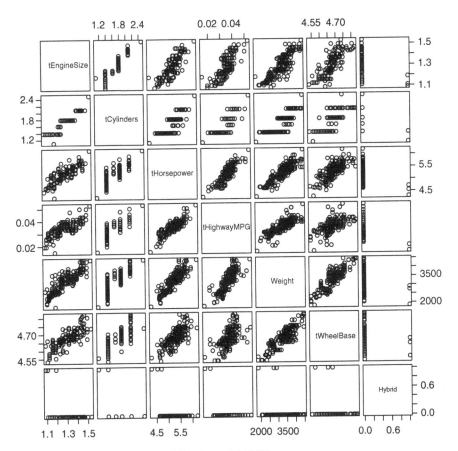

Figure 6.55 Matrix plot of the variables in model (6.37)

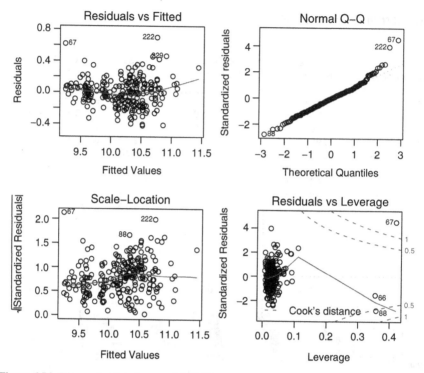

Figure 6.56 Diagnostic plots from model (6.37)

Residual standard error: 0.1724 on 226 degrees of freedom
Multiple R-Squared: 0.872, Adjusted R-squared: 0.868
F-statistic: 219.9 on 7 and 226 DF, p-value: < 2.2e-16

Output from R for model (6.37)

```
vif(m2)
tEngineSize    tCylinders   tHorsepower   tHighwayMPG    Weight
      8.67          7.17          5.96          4.59      8.20
tWheelBase      Hybrid
      4.78        1.22


Call:
lm(formula = tSuggestedRetailPrice ~ tEngineSize
+ tCylinders + tHorsepower + Weight + Hybrid)
Coefficients:
              Estimate   Std. Error   t value   Pr(>|t|)
(Intercept)  5.422e+00    3.291e-01    16.474     <2e-16   ***
tEngineSize -1.591e+00    3.157e-01    -5.041   9.45e-07   ***
tCylinders   2.375e-01    1.186e-01     2.003     0.0463   *
```

```
tHorsepower   9.049e-01   8.305e-02   10.896   <2e-16   ***
Weight        5.029e-04   5.203e-05    9.666   <2e-16   ***
Hybrid        6.340e-01   1.080e-01    5.87   1.53e-08   ***
---
Residual standard error: 0.1781 on 228 degrees of freedom
Multiple R-squared: 0.862,  Adjusted R-squared: 0.859
F-statistic: 284.9 on 5 and 228 DF,  p-value: < 2.2e-16
```

4. A book on robust statistical methods published in June 2006 considers regression models for a data set taken from Jalali-Heravi and Knouz (2002). The aim of this modeling is to predict a physical property of chemical compounds called the Krafft point based on four potential predictor variables using a data set of size $n = 32$. According to Maronna, Martin and Yohai (2006, p. 380)

The Krafft point is an important physical characteristic of the compounds called surfactants, establishing the minimum temperature at which a surfactant can be used.

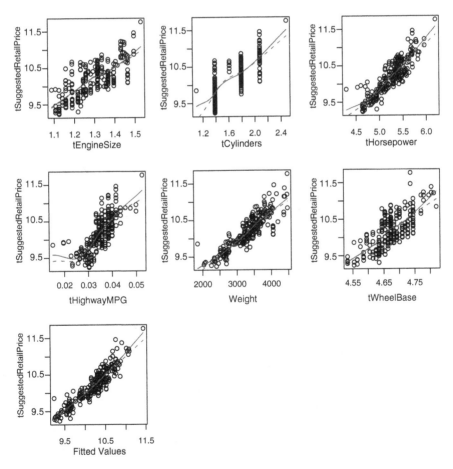

Figure 6.57 Marginal model plots from model (6.37)

The authors of the original paper sought to find a regression model to predict:
Y = Krafft Point (KPOINT)
from
x_1 = Randic Index (RA)
x_2 = Heat of formation (HEAT)
x_3 = Reciprocal of volume of the tail of the molecule (VTINV)
x_4 = Reciprocal of Dipole Moment (DIPINV)

The first model considered by Jalali-Heravi and Knouz (2002) was

$$Y = \beta_0 + \beta_1 x_1 + \beta_2 x_2 + \beta_3 x_3 + \beta_4 x_4 + e \qquad (6.38)$$

Output from model (6.38) and associated plots (Figures 6.58, 6.59 and 6.60) appear on the following pages.

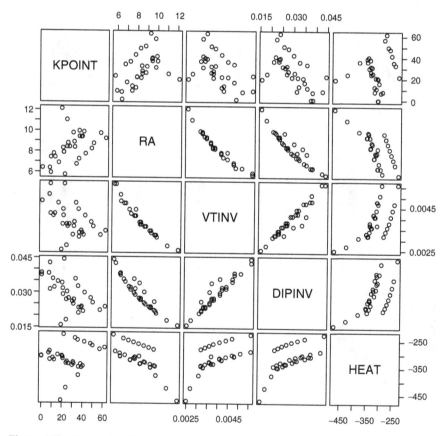

Figure 6.58   Matrix plot of the variables in model (6.38)

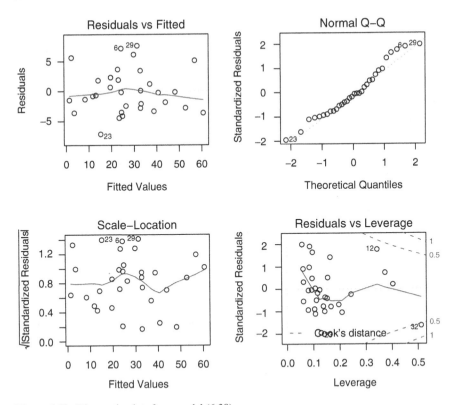

Figure 6.59   Diagnostic plots from model (6.38)

(a) Decide whether (6.38) is a valid model.
(b) The plots of standardized residuals against RA and VTINV produce curved
    patterns. Describe what, if anything can be learned about model (6.38) from
    these plots. Give a reason to support your answer.
(c) Jalali-Heravi and Knouz (2002) give "four criteria of correlation coefficient
    (r), standard deviation (s), F value for the statistical significance of the
    model and the ratio of the number of observations to the number of descrip-
    tors in the equation" for choosing between competing regression models.
    Provide a detailed critique of this suggestion.

Output from R for model (6.38)

```
Call:
lm(formula = KPOINT ~ RA + VTINV + DIPINV + HEAT)
Coefficients:
              Estimate   Std. Error   t value   Pr(>|t|)
(Intercept)   7.031e+01   3.368e+01    2.088     0.046369    *
RA            1.047e+01   2.418e+00    4.331     0.000184    ***
VTINV         9.038e+03   4.409e+03    2.050     0.050217    .
```

```
DIPINV        -1.826e+03    3.765e+02    -4.850  4.56e-05   ***
HEAT           3.550e-01    2.176e-02    16.312  1.66e-15   ***

Residual standard error: 3.919 on 27 degrees of freedom
Multiple R-Squared: 0.9446,  Adjusted R-squared: 0.9363
F-statistic: 115 on 4 and 27 DF,  p-value: < 2.2e-16
vif(m1)
        RA      VTINV     DIPINV       HEAT
25.792770 22.834190 13.621363  2.389645
```

5. An avid fan of the PGA tour with limited background in statistics has sought your help in answering one of the age-old questions in golf, namely, *what is the relative importance of each different aspect of the game on average prize money in professional golf?*

   The following data on the top 196 tour players in 2006 can be found on the book web site in the file pgatour2006.csv:

   Y, PrizeMoney = average prize money per tournament
   x_1, Driving Accuracy is the percent of time a player is able to hit the fairway with his tee shot.
   x_2, GIR, Greens in Regulation is the percent of time a player was able to hit the green in regulation. A green is considered hit in regulation if any part of the

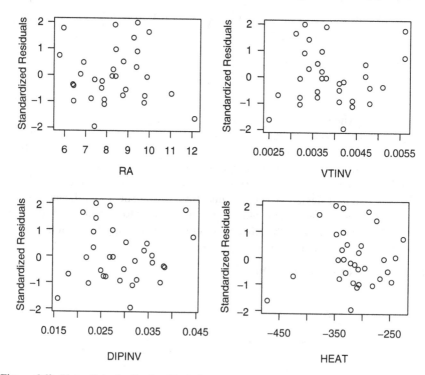

Figure 6.60 Plots of standardized residuals from model (6.38)

ball is touching the putting surface and the number of strokes taken is two or less than par.

x_3, Putting Average measures putting performance on those holes where the green is hit in regulation (GIR). By using greens hit in regulation the effects of chipping close and one putting are eliminated.

x_4, Birdie Conversion% is the percent of time a player makes birdie or better after hitting the green in regulation.

x_5, SandSaves% is the percent of time a player was able to get "up and down" once in a greenside sand bunker.

x_6, Scrambling% is the percent of time that a player misses the green in regulation, but still makes par or better.

x_7, PuttsPerRound is the average total number of putts per round.(http://www.pgatour.com/r/stats/; accessed March 13, 2007)

(a) A statistician from Australia has recommended to the analyst that they not transform any of the predictor variables but that they transform Y using the log transformation. Do you agree with this recommendation? Give reasons to support your answer.

(b) Develop a valid full regression model containing all seven potential predictor variables listed above. Ensure that you provide justification for your choice of full model, which includes scatter plots of the data, plots of standardized residuals, and any other relevant diagnostic plots.

(c) Identify any points that should be investigated. Give one or more reasons to support each point chosen.

(d) Describe any weaknesses in your model.

(e) The golf fan wants to remove all predictors with insignificant t-values from the full model in a single step. Explain why you would not recommend this approach.

In the next chapter, we will consider variable selection techniques in order to remove any redundancy from this regression model.

Chapter 7
Variable Selection

In this chapter we consider methods for choosing the "best" model from a class of multiple regression models using what are called variable selection methods. Interestingly, while there is little agreement on how to define "best," there is general agreement in the statistics literature on the consequences of variable selection on subsequent inferential procedures, (i.e., tests and confidence intervals).

We begin by introducing some terminology. The full model is the following multiple regression model containing all m potential predictor variables:

$$Y = \beta_0 + \beta_1 x_1 + \beta_2 x_2 + \ldots + \beta_m x_m + e \tag{7.1}$$

Throughout this chapter we shall assume that the full model is a valid regression model.

Variable selection methods aim to choose the subset of the predictors that is "best" in a given sense. In general, the more predictor variables included in a valid model the lower the bias of the predictions, but the higher the variance. Including too many predictors in a regression model is commonly called over-fitting while the opposite is called under-fitting. Hesterberg, Choi, Meier and Fraley (2008) make the following important point about the different goals of variable selection and prediction accuracy.

> If the main interest is in finding an interpretable model or in identifying the 'true' underlying model as closely as possible, prediction accuracy is of secondary importance (to variable selection). ... On the other hand, if prediction is the focus of interest, it is usually acceptable for the selected model to contain some extra variables, as long as the coefficients of those variables are small.

The two key aspects of variable selection methods are:

1. Evaluating each potential subset of p predictor variables
2. Deciding on the collection of potential subsets

We begin by considering the first aspect.

S.J. Sheather, *A Modern Approach to Regression with R*,
DOI: 10.1007/978-0-387-09608-7_7, © Springer Science+Business Media LLC 2009

7.1   Evaluating Potential Subsets of Predictor Variables

We shall discuss four criteria for evaluating subsets of predictor variables.

7.1.1   Criterion 1: R^2-Adjusted

Recall from Chapter 5 that R^2, the coefficient of determination of the regression model, is defined as the proportion of the total sample variability in the Y's explained by the regression model, that is,

$$R^2 = \frac{\text{SSreg}}{\text{SST}} = 1 - \frac{\text{RSS}}{\text{SST}}$$

Adding irrelevant predictor variables to the regression equation often increases R^2. To compensate for this one can define an adjusted coefficient of determination, R^2_{adj}

$$R^2_{adj} = 1 - \frac{\text{RSS}/(n-p-1)}{\text{SST}/(n-1)},$$

where p is the number of predictors in the current model. In fact, it can be shown that adding predictor variables to the current model only leads to an increase in R^2_{adj} if the corresponding partial F-test statistic exceeds 1.

The usual practice is to choose the subset of the predictors that has the *highest* value of R^2_{adj}. It can be shown that this is equivalent to choosing the subset of the predictors with the *lowest* value of $S^2 = \dfrac{\text{RSS}}{n-p-1}$, where p is the number of predictors in the subset.

We shall see that *choosing the subset of the predictors that has the highest value of R^2_{adj} tends towards over-fitting.* For example, suppose that the maximum value is $R^2_{adj} = 0.692$ for a subset of $p = 10$ predictors, $R^2_{adj} = 0.691$ for a subset of $p = 9$ predictors and $R^2_{adj} = 0.541$ for a subset of $p = 8$ predictors. Even though R^2_{adj} increases when we go from 9 to 10 predictors there is very little improvement in fit and so the nine-predictor subset is generally preferred.

The other three criteria are based on likelihood theory when both the predictors and the response are normally distributed. As such we briefly review this theory next.

Maximum Likelihood applied to Multiple Linear Regression

Suppose that $y_i, x_{1i}, \ldots x_{pi}$ are the observed values of normal random variables. Then, $y_i \mid x_{1i}, \ldots x_{pi} \sim N\left(\beta_0 + \beta_1 x_{1i} + \ldots + \beta_p x_{pi}, \sigma^2\right)$. Thus, the conditional density of y_i given $x_{1i}, \ldots x_{pi}$ is given by

$$f(y_i \mid x_{1i}, \ldots x_{pi}) = \frac{1}{\sigma\sqrt{2\pi}} \exp\left(-\frac{\left(y_i - \left\{\beta_0 + \beta_1 x_{1i} + \ldots + \beta_p x_{pi}\right\}\right)^2}{2\sigma^2}\right).$$

Assuming the n observations are independent, then given Y the likelihood function is the function of the unknown parameters $\beta_0, \beta_1, \ldots, \beta_p, \sigma^2$ given by

$$L(\beta_0, \beta_1, \ldots, \beta_p, \sigma^2 \mid Y)$$

$$= \prod_{i=1}^{n} f(y_i \mid x_i)$$

$$= \prod_{i=1}^{n} \frac{1}{\sigma\sqrt{2\pi}} \exp\left(-\frac{\left(y_i - \left\{\beta_0 + \beta_1 x_{1i} + \ldots + \beta_p x_{pi}\right\}\right)^2}{2\sigma^2}\right)$$

$$= \left(\frac{1}{\sigma\sqrt{2\pi}}\right)^n \exp\left(-\frac{1}{2\sigma^2}\sum_{i=1}^{n}\left(y_i - \left\{\beta_0 + \beta_1 x_{1i} + \ldots + \beta_p x_{pi}\right\}\right)^2\right)$$

The log-likelihood function is given by

$$\log L\left((\beta_0, \beta_1, \ldots, \beta_p, \sigma^2 \mid Y)\right)$$

$$= -\frac{n}{2}\log(2\pi) - \frac{n}{2}\log(\sigma^2) - \frac{1}{2\sigma^2}\sum_{i=1}^{n}\left(y_i - \left\{\beta_0 + \beta_1 x_{1i} + \ldots + \beta_p x_{pi}\right\}\right)^2$$

Notice that $\beta_0, \beta_1, \ldots, \beta_p$ are only included in the third term of this last equation. Thus, the maximum likelihood estimates of $\beta_0, \beta_1, \ldots, \beta_p$ can be obtained by minimizing this third term, that is, by minimizing the residual sum of squares. Thus, the maximum likelihood estimates of $\beta_0, \beta_1, \ldots, \beta_p$ are the same as the least squares estimates. With $\beta_0, \beta_1, \ldots, \beta_p$ equal to their least squares estimates, the last equation becomes

$$\log\left(L(\hat{\beta}_0, \hat{\beta}_1, \ldots, \hat{\beta}_p, \sigma^2 \mid Y)\right) = -\frac{n}{2}\log(2\pi) - \frac{n}{2}\log(\sigma^2) - \frac{1}{2\sigma^2}\text{RSS}$$

where

$$\text{RSS} = \sum_{i=1}^{n}\left(y_i - \left\{\hat{\beta}_0 + \hat{\beta}_1 x_{1i} + \ldots + \hat{\beta}_p x_{pi}\right\}\right)^2$$

Differentiating the last log-likelihood equation with respect to σ^2 and setting the result to zero gives the maximum likelihood estimate of σ^2 as

$$\hat{\sigma}^2_{\text{MLE}} = \frac{\text{RSS}}{n}.$$

Notice that this estimate differs slightly from our usual estimate of σ^2, namely,

$$S^2 = \frac{1}{(n-p-1)}\text{RSS}.$$

Substituting the result for the maximum likelihood estimate of σ^2 back into the expression for the log-likelihood we find that the likelihood associated with the maximum likelihood estimates is given by

$$\log\left(L(\hat{\beta}_0,\hat{\beta}_1,\ldots,\hat{\beta}_p,\hat{\sigma}^2\mid Y)\right) = -\frac{n}{2}\log(2\pi) - \frac{n}{2}\log(\frac{\text{RSS}}{n}) - \frac{n}{2}$$

7.1.2 Criterion 2: AIC, Akaike's Information Criterion

Akaike's information criterion (AIC) can be motivated in two ways. The most popular motivation seems to be based on balancing goodness of fit and a penalty for model complexity. AIC is defined such that *the smaller the value of AIC the better the model.* A measure of goodness of fit such that the smaller the better is minus one times the likelihood associated with the fitted model, while a measure of complexity is K, the number of estimated parameters in the fitted model. AIC is defined to be

$$\text{AIC} = 2\left[-\log\left(L(\hat{\beta}_0,\hat{\beta}_1,\ldots,\hat{\beta}_p,\hat{\sigma}^2\mid Y)\right)+K\right] \tag{7.2}$$

where $K = p + 2$, since $\beta_0, \beta_1, \ldots, \beta_p, \sigma^2$ are estimated in the fitted model. The measure of complexity is necessary since adding irrelevant predictor variables to the regression equation can increase the log-likelihood. We shall soon discover the reason for the 2 in (7.2).

Akaike's original motivation for AIC is based on the Kullback-Leibler (K-L) information measure, $I(f, g)$, which is the amount of information lost when g (which depends on θ) is used to model f, defined as

$$I(f,g) = \int f(y)\log\left(\frac{f(y)}{g(y\mid\theta)}\right)dy$$

It can also be thought as a distance measure between f and g. The following material is based on Burnham and Anderson (2004).

The Kullback-Leibler (K-L) information measure can be reexpressed as

$$\begin{aligned}
I(f,g) &= \int f(y)\log(f(y))dy - \int f(y)\log(g(y\mid\theta))dy \\
&= E_f\left[\log(f(y))\right] - E_f\left[\log(g(y\mid\theta))\right] \\
&= C - E_f\left[\log(g(y\mid\theta))\right]
\end{aligned}$$

where C is an unknown constant that does not depend on the model. Thus, to compare the Kullback-Leibler (K-L) information measure across different models $g(y\mid\theta)$ the quantity $E_f\left[\log\left(g(y\mid\theta)\right]\right.$ needs to be estimated for each model. This quantity is called the relative expected K-L information.

Akaike found a relationship between K-L information and likelihood theory. In particular, he found that the maximized log-likelihood was a biased estimate of $E_f [\log (g(y|\theta)]$ with the bias approximately equal to K, the number of estimable parameters in the model $g(y|\theta)$. In symbols,

$$\hat{E}_f \left[\log(g(y|\theta))\right] = \log\left(L(\hat{\theta}|Y)\right) - K$$

Akaike multiplied this last result by -2, reportedly, for "historical reasons," and this became Akaike's information criterion:

$$
\begin{aligned}
\text{AIC} &= -2\left[\log\left(L(\hat{\theta}|Y)\right) - K\right] \\
&= -2\left[\log\left(L(\hat{\beta}_0,\hat{\beta}_1,\dots,\hat{\beta}_p,\hat{\sigma}^2 \mid Y)\right) - (p+2)\right]
\end{aligned}
$$

since $K = (p + 2)$ parameters are estimated in the regression model.
Using results found earlier for the maximized likelihood,

$$
\begin{aligned}
\text{AIC} &= -2\left[-\frac{n}{2}\log(2\pi) - \frac{n}{2}\log\left(\frac{\text{RSS}}{n}\right) - \frac{n}{2} - (p+2)\right] \\
&= n\log\left(\frac{\text{RSS}}{n}\right) + 2p + \text{other terms}
\end{aligned}
$$

where the other terms do not depend on RSS or p and thus are the same for every model under consideration. In view of this last result, R calculates AIC using

$$\text{AIC} = n\log\left(\frac{\text{RSS}}{n}\right) + 2p$$

7.1.3   Criterion 3: AIC$_C$, Corrected AIC

Hurvich and Tsai (1989) developed AIC$_C$, a bias corrected version of AIC for use when the sample size is small, or when the number of parameters estimated is a moderate to large fraction of the sample size. Burnham and Anderson (2004) recommend that AIC$_C$ be used instead of AIC unless $n/K > 40$. Furthermore they recommend that AIC$_C$ be used in practice since as n gets large AIC$_C$ converges to AIC. AIC$_C$ is given by

$$\text{AIC}_C = -2\log\left(L(\hat{\theta}|Y)\right) + 2K + \frac{2K(K+1)}{n-K+1} = \text{AIC} + \frac{2K(K+1)}{n-K+1}$$

In our case, $K = p + 2$ so that

$$\text{AIC}_C = \text{AIC} + \frac{2(p+2)(p+3)}{n-p-1}$$

When the sample size is small, or when the number of parameters estimated is a moderate to large fraction of the sample size, it is well-known that AIC has a tendency for over-fitting since the penalty for model complexity is not strong enough. As such it is not surprising in these circumstances that the bias corrected version of AIC has a larger penalty for model complexity.

7.1.4   Criterion 4: BIC, Bayesian Information Criterion

Schwarz (1978) proposed the Bayesian information criterion as

$$\text{BIC} = -2\log\left(L(\hat{\beta}_0, \hat{\beta}_1, \ldots, \hat{\beta}_p, \hat{\sigma}^2 \mid Y)\right) + K\log(n) \tag{7.3}$$

where $K = p + 2$, the number of parameters estimated in the model. BIC is defined such that *the smaller the value of BIC the better the model*. Comparing (7.3) with (7.2), we see that BIC is similar to AIC except that the factor 2 in the penalty term is replaced by $\log(n)$. When $n \geq 8$, $\log(n) > 2$ and so the penalty term in BIC is greater than the penalty term in AIC. Thus, in these circumstances, BIC penalizes complex models more heavily than AIC, thus favoring simpler models than AIC.

The following discussion is based on Burnham and Anderson (2004). BIC is a misnomer in the sense that it is not related to information theory. Define ΔBIC_i as the difference between BIC for the *ith* model and the minimum BIC value. Then, under the assumption that all R models under consideration have equal prior probability, it can be shown that the posterior probability of model i is given by

$$p_i = P(\text{model}_i \mid \text{data}) = \frac{\exp\left(-\Delta\text{BIC}_i/2\right)}{\sum_{r=1}^{R}\exp\left(-\Delta\text{BIC}_r/2\right)}$$

In practice, BIC is generally used in a frequentist sense, thus ignoring the concepts of prior and posterior probabilities.

7.1.5   Comparison of AIC, AIC$_C$ and BIC

There has been much written about the relative merits of AIC, AIC$_C$ and BIC. Two examples of this material are given next.

Simonoff (2003, p. 46) concludes the following:

AIC and AIC$_C$ have the desirable property that they are efficient model selection criteria. What this means is that as the sample gets larger, the error obtained in making predictions using the model chosen using these criteria becomes indistinguishable from the error obtained using the best possible model among all candidate models. That is, in this large-sample predictive sense, it is as if the best approximation was known to the data analyst. Other criteria, such as the Bayesian Information Criterion, BIC ... do not have this property.

Hastie, Tibshirani and Freedman (2001, p. 208) put forward the following different point of view:

> For model selection purposes, there is no clear choice between AIC and BIC. BIC is asymptotically consistent as a selection criterion. What this means is that given a family of models, including the true model, the probability that BIC will select the correct model approaches one as the sample size $N \rightarrow \infty$. This is not the case for AIC, which tends to choose models which are too complex as $N \rightarrow \infty$. On the other hand, for finite samples, BIC often chooses models that are too simple, because of the heavy penalty on complexity.

A popular data analysis strategy which we shall adopt is to calculate R^2_{adj}, AIC, AIC_C and BIC and compare the models which minimize AIC, AIC_C and BIC with the model that maximizes R^2_{adj}.

7.2 Deciding on the Collection of Potential Subsets of Predictor Variables

There are two distinctly different approaches to choosing the potential subsets of predictor variables, namely,

1. All possible subsets
2. Stepwise methods

We shall begin by discussing the first approach.

7.2.1 All Possible Subsets

This approach is based on considering all 2^m possible regression equations and identifying the subset of the predictors of a given size that maximizes a measure of fit or minimizes an information criterion based on a monotone function of the residual sum of squares. Furnival and Wilson (1974, p. 499) developed a "simple leap and bound technique for finding the best subsets without examining all possible subsets."

With a fixed number of terms in the regression model, all four criteria for evaluating a subset of predictor variables (R^2_{adj}, AIC, AIC_C and BIC) agree that the best choice is the set of predictors with the smallest value of the residual sum of squares. Thus, for example, if a subset with a fixed number of terms maximizes R^2_{adj} (i.e., minimizes RSS) among all subsets of size p, then this subset will also minimize AIC, AIC_C and BIC among all subsets of fixed size p. Note however, when the comparison is across models with different numbers of predictors the four methods (R^2_{adj}, AIC, AIC_C and BIC) can give quite different results.

Example: Bridge construction (cont.)

Recall from Chapter 6 that our aim is to model

$Y = \text{Time} = \text{design time in person-days}$

based on the following potential predictor variables

x_1 = DArea = Deck area of bridge (000 sq ft)
x_2 = CCost = Construction cost ($000)
x_3 = Dwgs = Number of structural drawings
x_4 = Length = Length of bridge (ft)
x_5 = Spans = Number of spans

Recall further that we found that the following full model

$$\log(Y) = \beta_0 + \beta_1 \log(x_1) + \beta_2 \log(x_2) + \beta_3 \log(x_3) + \beta_4 \log(x_4)$$
$$+ \beta_5 \log(x_5) + e \tag{7.4}$$

is a valid model for the data. Given below again is the output from R associated with fitting model (7.4).

Regression output from R

```
Call:
lm(formula = log(Time) ~ log(DArea) + log(CCost) + log(Dwgs) +
log(Length) + log(Spans))
Coefficients:
             Estimate   Std. Error   t value   Pr(>|t|)
(Intercept)   2.28590     0.61926      3.691    0.00068   ***
log(DArea)   -0.04564     0.12675     -0.360    0.72071
log(CCost)    0.19609     0.14445      1.358    0.18243
log(Dwgs)     0.85879     0.22362      3.840    0.00044   ***
log(Length)  -0.03844     0.15487     -0.248    0.80530
log(Spans)    0.23119     0.14068      1.643    0.10835
---
Signif. codes: 0 '***' 0.001 '**' 0.01 '*' 0.05 '.' 0.1 ' ' 1

Residual standard error: 0.3139 on 39 degrees of freedom
Multiple R-Squared: 0.7762, Adjusted R-squared: 0.7475
F-statistic: 27.05 on 5 and 39 DF, p-value: 1.043e-11
```

Notice that while the overall F-test for model (7.4) is highly statistically significant, only one of the estimated regression coefficients is statistically significant (i.e., log(Dwgs) with a p-value < 0.001). Thus, we wish to choose a subset of the predictors using variable selection.

We begin our discussion of variable selection in this example by identifying the subset of the predictors of a given size that maximizes adjusted R-squared (i.e., minimizes RSS). Figure 7.1 shows plots of adjusted R-squared against the number of predictors in the model for the optimal subsets of predictors. For example, the optimal subset of predictors of size 2 consists of the predictors log(Dwgs) and log(Spans). In addition, the model with the three predictors log(CCost), log(Dwgs) and log(Spans) maximizes adjusted R-squared.

Table 7.1 gives the values of R^2_{adj}, AIC, AIC_C and BIC for the best subset of each size. Highlighted in bold are the minimum values of AIC, AIC_C and BIC along with the maximum value of R^2_{adj}. Notice from Table 7.1 that AIC judges the predictor subset of size 3 to be "best" while AIC_C and BIC judge the subset of size 2 to be

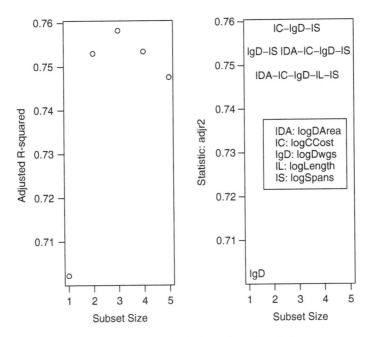

Figure 7.1 Plots of R^2_{adj} against subset size for the best subset of each size

Table 7.1 Values of R^2_{adj}, AIC, AIC_C and BIC for the best subset of each size

Subset size	Predictors	R^2_{adj}	AIC	AIC_C	BIC
1	log(Dwgs)	0.702	−94.90	−94.31	−91.28
2	log(Dwgs), log(Spans)	0.753	−102.37	**−101.37**	**−96.95**
3	log(Dwgs), log(Spans), log(CCost)	**0.758**	**−102.41**	−100.87	−95.19
4	log(Dwgs), log(Spans), log(CCost), log(DArea)	0.753	−100.64	−98.43	−91.61
5	log(Dwgs), log(Spans), log(CCost), log(DArea), log(Length)	0.748	−98.71	−95.68	−87.87

"best." While the maximum value of R^2_{adj} corresponds to the predictor subset of size 3, using the argument described earlier we could choose the subset of size 2 to be "best" in terms of R^2_{adj}.

Regression output from R

```
Call:
lm(formula = log(Time) ~ log(Dwgs) + log(Spans))

Coefficients:
```

	Estimate	Std. Error	t value	Pr(>|t|)	
(Intercept)	2.66173	0.26871	9.905	1.49e-12	***
log(Dwgs)	1.04163	0.15420	6.755	3.26e-08	***
log(Spans)	0.28530	0.09095	3.137	0.00312	*

Residual standard error: 0.3105 on 42 degrees of freedom
Multiple R-Squared: 0.7642,   Adjusted R-squared: 0.753
F-statistic: 68.08 on 2 and 42 DF, p-value: 6.632e-14

Call:
lm(formula = log(Time) ~ log(Dwgs) + log(Spans) + log(CCost))

Coefficients:

	Estimate	Std. Error	t value	Pr(>|t|)	
(Intercept)	2.3317	0.3577	6.519	7.9e-08	***
log(Dwgs)	0.8356	0.2135	3.914	0.000336	***
log(Spans)	0.1963	0.1107	1.773	0.083710	.
log(CCost)	0.1483	0.1075	1.380	0.175212	

Residual standard error: 0.3072 on 41 degrees of freedom
Multiple R-Squared: 0.7747,   Adjusted R-squared: 0.7582
F-statistic: 46.99 on 3 and 41 DF, p-value: 2.484e-13

Given above is the output from R associated with fitting the best models with 2 and 3 predictor variables. Notice that both predictor variables are judged to be statistically significant in the two-variable model, while just one variable is judged to be statistically significant in the three-variable model. Later in this chapter we shall see that the p-values obtained after variable selection are much smaller than their true values. In view of this, it seems that the three-variable model over-fits the data and as such *the two-variable model is to be preferred.*

7.2.2 Stepwise Subsets

This approach is based on examining just a sequential subset of the 2^m possible regression models. Arguably, the two most popular variations on this approach are *backward elimination* and *forward selection*.

Backward elimination starts with all potential predictor variables in the regression model. Then, at each step, it deletes the predictor variable such that the resulting model has the lowest value of an information criterion. (This amounts to deleting the predictor with the **largest** p-value each time.) This process is continued until all variables have been deleted from the model or the information criterion increases.

Forward selection starts with no potential predictor variables in the regression equation. Then, at each step, it adds the predictor such that the resulting model has the lowest value of an information criterion. (This amounts to adding the predictor with the **smallest** p-value each time.) This process is continued until all variables have been added to the model or the information criterion increases.

Backward elimination and forward selection consider at most $m + (m-1) + (m-2) + ... + 1 = m(m+1)/2$ of the 2^m possible predictor subsets. Thus, backward elimination and forward selection do not necessarily find the model that minimizes the information criteria across all 2^m possible predictor subsets. In addition, there is no guarantee that backward elimination and forward selection will produce the same final model. However, in practice they produce the same model in many different situations.

Example: Bridge construction (cont.)

We wish to perform variable selection using backward elimination and forward selection based on AIC and BIC. Given below is the output from R associated with backward elimination based on AIC.

Output from R: Backward Elimination based on AIC

```
Start: AIC= -98.71
log(Time) ~ log(DArea) + log(CCost) + log(Dwgs) + log(Length) + log
(Spans)
                    Df    Sum of Sq      RSS         AIC
- log(Length)        1        0.006     3.850     -100.640
- log(DArea)         1        0.013     3.856     -100.562
<none>                                  3.844      -98.711
- log(CCost)         1        0.182     4.025      -98.634
- log(Spans)         1        0.266     4.110      -97.698
- log(Dwgs)          1        1.454     5.297      -86.277

Step: AIC= -100.64
log(Time) ~ log(DArea) + log(CCost) + log(Dwgs) + log(Spans)
                    Df    Sum of Sq      RSS         AIC
- log(DArea)         1        0.020     3.869     -102.412
<none>                                  3.850     -100.640
- log(CCost)         1        0.181     4.030     -100.577
- log(Spans)         1        0.315     4.165      -99.101
- log(Dwgs)          1        1.449     5.299      -88.260

Step: AIC= -102.41
log(Time) ~ log(CCost) + log(Dwgs) + log(Spans)
                    Df    Sum of Sq      RSS         AIC
<none>                                  3.869     -102.412
- log(CCost)         1        0.180     4.049     -102.370
- log(Spans)         1        0.297     4.166     -101.089
- log(Dwgs)          1        1.445     5.315      -90.128
```

Thus, backward elimination based on AIC chooses the model with the three predictors log(CCost), log(Dwgs) and log(Spans). It can be shown that backward elimination based on BIC chooses the model with the two predictors log(Dwgs) and log(Spans).

Forward selection based on AIC (shown below) arrives at the same model as backward elimination based on AIC. It can be shown that forward selection based on BIC arrives at the same model as backward elimination based on BIC. We are again faced with a choice between the two-predictor and three-predictor models discussed earlier.

Output from R: Forward selection based on AIC

```
Start: AIC= -41.35
log(Time)  ~  1
                Df   Sum of Sq       RSS        AIC
+ log(Dwgs)      1      12.176     4.998    -94.898
+ log(CCost)     1      11.615     5.559    -90.104
+ log(DArea)     1      10.294     6.880    -80.514
+ log(Length)    1      10.012     7.162    -78.704
+ log(Spans)     1       8.726     8.448    -71.274
<none>                            17.174    -41.347
Step: AIC= -94.9
log(Time)  ~  log(Dwgs)
                Df   Sum of Sq       RSS        AIC
+ log(Spans)     1       0.949     4.049   -102.370
+ log(CCost)     1       0.832     4.166   -101.089
+ log(Length)    1       0.669     4.328    -99.366
+ log(DArea)     1       0.476     4.522    -97.399
<none>                             4.998    -94.898
Step: AIC= -102.37
log(Time)  ~  log(Dwgs) + log(Spans)
                Df   Sum of Sq       RSS        AIC
+ log(CCost)     1       0.180     3.869   -102.412
<none>                             4.049   -102.370
+ log(DArea)     1       0.019     4.030   -100.577
+ log(Length)    1       0.017     4.032   -100.559
Step: AIC= -102.41
log(Time)  ~  log(Dwgs) + log(Spans) + log(CCost)
                Df   Sum of Sq       RSS        AIC
<none>                             3.869   -102.412
+ log(DArea)     1       0.020     3.850   -100.640
+ log(Length)    1       0.013     3.856   -100.562
```

7.2.3 Inference After Variable Selection

An important caution associated with variable selection (or model selection as it is also referred to) is that the selection process changes the properties of the estimators as well as the standard inferential procedures such as tests and confidence intervals. The regression coefficients obtained after variable selection are biased. In addition, the p-values obtained after variable selection from F- and t-statistics are generally much smaller than their true values. These issues are well summarized in the following quote from Leeb and Potscher (2005, page 22):

> The aim of this paper is to point to some intricate aspects of data-driven model selection that do not seem to have been widely appreciated in the literature or that seem to be viewed too optimistically. In particular, we demonstrate innate difficulties of data-driven model selection. Despite occasional claims to the contrary, no model selection procedure—implemented on a machine or not—is immune to these difficulties. The main points we want to make and that will be elaborated upon subsequently can be summarized as follows:

1. Regardless of sample size, the model selection step typically has a dramatic effect on the sampling properties of the estimators that can not be ignored. In particular, the sampling properties of post-model-selection estimators are typically significantly different from the nominal distributions that arise if a fixed model is supposed.
2. As a consequence, naive use of inference procedures that do not take into account the model selection step (e.g., using standard t-intervals as if the selected model had been given prior to the statistical analysis) can be highly misleading.

7.3   Assessing the Predictive Ability of Regression Models

Given that the model selection process changes the properties of the standard inferential procedures, a standard approach to assessing the predictive ability of different regression models is to evaluate their performance on a new data set (i.e., one not used in the development of the models). In practice, this is often achieved by randomly splitting the data into:

1. A training data set
2. A test data set

The training data set is used to develop a number of regression models, while the test data set is used to evaluate the performance of these models. We illustrate these steps using the following example.

Example: Prostate cancer

Hastie, Tibshirani and Friedman (2001) analyze data taken from Stamey et al. (1989). According to Hastie, Tibshirani and Friedman:

> The goal is to predict the log-cancer volume (lacavol) from a number of measurements including log prostate weight (lweight), age, log of benign prostatic hyperplasia (lpbh), seminal vesicle invasion (svi), log of capsular penetration (lcp), Gleason score (gleason), and percent of Gleason scores 4 or 5 (pgg45).

Hastie, Tibshirani and Friedman (2001, p. 48) "randomly split the dataset into a training set of size 67 and a test set of size 30." These data sets can be found on the book web site in the files prostateTraining.txt and postateTest.txt. We first consider the training set.

7.3.1   Stage 1: Model Building Using the Training Data Set

We begin by plotting the training data. Figure 7.2 contains a scatter plot matrix of response variable and the eight predictor variables.

Looking at Figure 7.2, we see that the relationship between the response variable (lpsa) and each of the predictor variables appears to be linear. There is also no evidence of nonlinearity amongst the eight predictor variables. Thus we shall consider

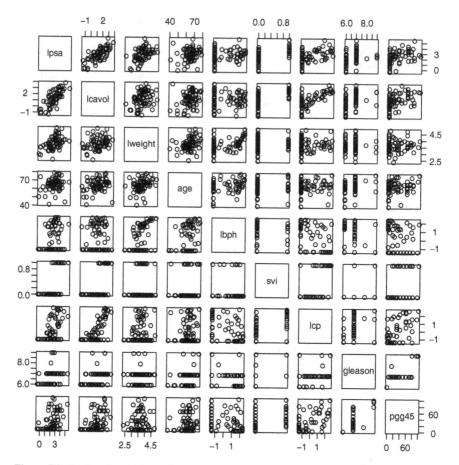

Figure 7.2 Scatter plot matrix of the response variable and each of the predictors

the following full model with all eight potential predictor variables for the training data set:

$$
\begin{aligned}
\text{lpsa} = \beta_0 &+ \beta_1\text{lcavol} + \beta_2\text{lweight} + \beta_3\text{age} + \beta_4\text{lbph} + \beta_5\text{svi} \\
&+ \beta_6\text{lcp} + \beta_7\text{gleason} + \beta_8\text{pgg45} + e
\end{aligned}
\tag{7.5}
$$

Figure 7.3 contains scatter plots of the standardized residuals against each predictor and the fitted values for model (7.5). Each of the plots in Figure 7.3 shows a random pattern. Thus, model (7.5) appears to be a valid model for the data.

Figure 7.4 contains a plot of lpsa against the fitted values. The straight-line fit to this plot provides a reasonable fit. This provides further evidence that model (7.5) is a valid model for the data.

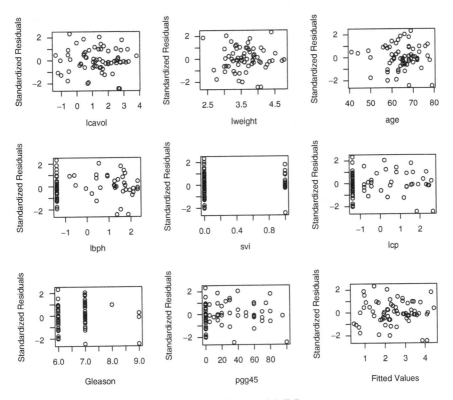

Figure 7.3 Plots of the standardized residuals from model (7.5)

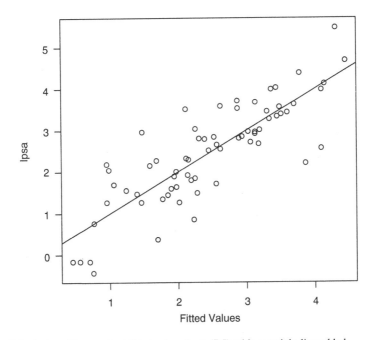

Figure 7.4 A plot of lpsa against fitted values from (7.5) with a straight line added

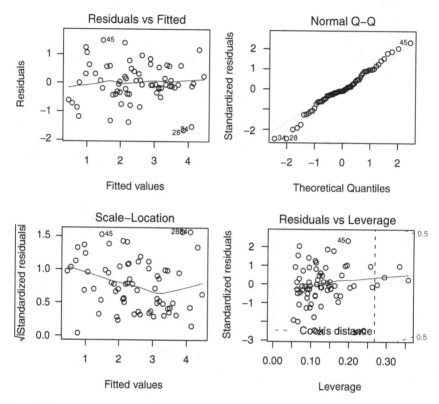

Figure 7.5 Diagnostic plots from R for model (7.5)

Figure 7.5 shows the diagnostic plots provided by R for model (7.5). Apart from a hint of decreasing error variance, these plots further confirm that model (7.5) is a valid model for the data.

The dashed vertical line in the bottom right-hand plot of Figure 7.5 is the usual cut-off for declaring a point of high leverage (i.e., $2 \times (p+1)/n = 18/67 = 0.269$). Thus, there are no bad leverage points.

Figure 7.6 contains the recommended marginal model plots for model (7.5). The nonparametric estimates of each pair-wise relationship are marked as solid curves, while the smooths of the fitted values are marked as dashed curves. The two curves in each plot match quite well thus providing further evidence that (7.5) is a valid model.

Below is the output from R associated with fitting model (7.5).

Regression output from R

```
Call:
lm(formula = lpsa ~ lcavol + lweight + age + lbph + svi + lcp +
gleason + pgg45)
Coefficients:
```

	Estimate	Std. Error	t value	Pr(>|t|)	
(Intercept)	0.429170	1.553588	0.276	0.78334	
lcavol	0.576543	0.107438	5.366	1.47e-06	***
lweight	0.614020	0.223216	2.751	0.00792	**
age	-0.019001	0.013612	-1.396	0.16806	
lbph	0.144848	0.070457	2.056	0.04431	*
svi	0.737209	0.298555	2.469	0.01651	*
lcp	-0.206324	0.110516	-1.867	0.06697	.
gleason	-0.029503	0.201136	-0.147	0.88389	
pgg45	0.009465	0.005447	1.738	0.08755	.

```
Signif.codes: 0 '***' 0.001 '**' 0.01 '*' 0.05 '.' 0.1 ' ' 1
Residual standard error: 0.7123 on 58 degrees of   freedom
Multiple R-Squared: 0.6944, Adjusted R-squared: 0.6522
F-statistic: 16.47 on 8 and 58 DF, p-value: 2.042e-12
```

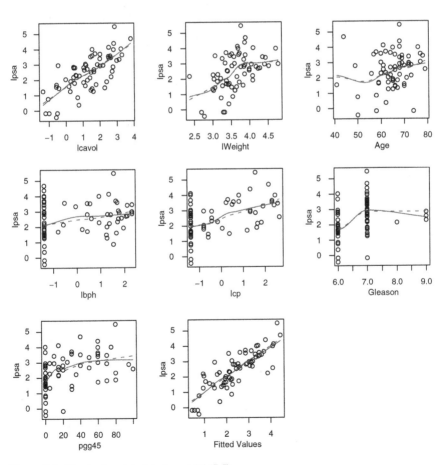

Figure 7.6 Marginal model plots for model (7.5)

Notice that the overall F-test for model (7.5) is highly statistically significant and four of the estimated regression coefficients are statistically significant (i.e., lcavol, lweight, lbph and svi).

Finally, we show in Figure 7.7 the added-variable plots associated with model (7.5). Case 45 appears to be highly influential in the added-variable plot for lweight, and, as such, it should be investigated. We shall return to this issue later. For now we shall continue under the assumption that (7.5) is a valid model.

The variance inflation factors for the training data set are as follows:

```
lcavol  lweight      age     lbph       svi      lcp  gleason      pgg45
2.318496 1.472295 1.356604 1.383429 2.045313 3.117451 2.644480 3.313288
```

None of these exceed 5 and so multicollinearity is not a serious issue.

We next consider variable selection in this example by identifying the subset of the predictors of a given size that maximizes adjusted R-squared (i.e., minimizes RSS). Figure 7.8 shows plots of adjusted R-squared against the number of predictors in the model for the optimal subsets of predictors. Table 7.2 gives the values of R^2_{adj}, AIC, AIC_{C} and BIC for the best subset of each size. Highlighted in bold are the minimum values of AIC, AIC_{C} and BIC along with the maximum value of R^2_{adj}.

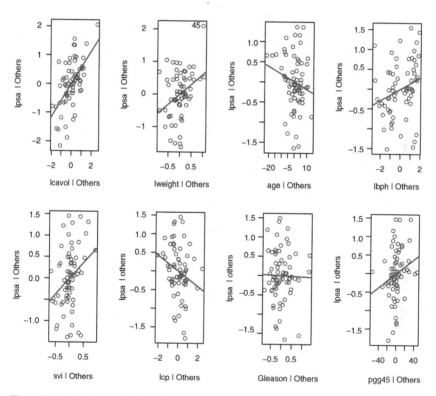

Figure 7.7 Added-variable plots for model (7.5)

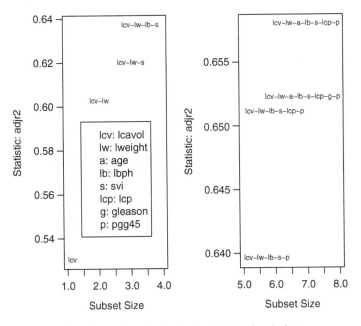

Figure 7.8 Plots of R^2_{adj} against subset size for the best subset of each size

Table 7.2 Values of R^2_{adj}, AIC, AIC_C and BIC for the best subset of each size

Subset size	Predictors	R^2_{adj}	AIC	AIC_C	BIC
1	lcavol	0.530	−23.374	−22.99	−18.96
2	lcavol, lweight	0.603	−33.617	−32.97	**−27.00**
3	lcavol, lweight, svi	0.620	−35.683	−34.70	−26.86
4	lcavol, lweight, svi, lbph	0.637	−37.825	**−36.43**	−26.80
5	lcavol, lweight, svi, lbph, pgg45	0.640	−37.365	−35.47	−24.14
6	lcavol, lweight, svi, lbph, pgg45, lcp	0.651	−38.64	−36.16	−23.21
7	lcavol, lweight, svi, lbph, pgg45, lcp, age	**0.658**	**−39.10**	−35.94	−21.47
8	lcavol, lweight, svi, lbph, pgg45, lcp, age, gleason	0.652	−37.13	−33.20	−17.29

Notice from Table 7.2 that AIC judges the predictor subset of size 7 to be "best" while AIC_C judges the subset of size 4 to be "best" and BIC judge the subset of size 2 to be "best." While the maximum value of corresponds to the predictor subset of size 7, using the argument described earlier in this chapter, one could choose the subset of size 4 to be "best" in terms of R^2_{adj}.

Given below is the output from R associated with fitting the best models with two-, four- and seven-predictor variables to the training data.

Regression output from R

```
Call:
lm(formula = lpsa ~ lcavol + lweight)
Coefficients:
            Estimate    Std. Error   t value   Pr(>|t|)
(Intercept) -1.04944     0.72904      -1.439   0.154885
lcavol       0.62761     0.07906       7.938   4.14e-11   ***
lweight      0.73838     0.20613       3.582   0.000658   ***
---
Residual standard error: 0.7613 on 64 degrees of freedom
Multiple R-Squared: 0.6148, Adjusted R-squared: 0.6027
F-statistic: 51.06 on 2 and 64 DF, p-value: 5.54e-14

Call:
lm(formula = lpsa ~ lcavol + lweight + svi + lbph)

Coefficients:
            Estimate    Std. Error   t value   Pr(>|t|)
(Intercept) -0.32592     0.77998      -0.418   0.6775
lcavol       0.50552     0.09256       5.461   8.85e-07   ***
lweight      0.53883     0.22071       2.441   0.0175     *
svi          0.67185     0.27323       2.459   0.0167     *
lbph         0.14001     0.07041       1.988   0.0512     .
---
Residual standard error: 0.7275 on 62 degrees of freedom
Multiple R-Squared: 0.6592, Adjusted R-squared: 0.6372
F-statistic: 29.98 on 4 and 62 DF, p-value: 6.911e-14

Call:
lm(formula=lpsa~lcavol+lweight+svi+lbph+pgg45+lcp +age)

Coefficients:
            Estimate    Std. Error   t value   Pr(>|t|)
(Intercept)  0.259062    1.025170     0.253    0.8014
lcavol       0.573930    0.105069     5.462    9.88e-07   ***
lweight      0.619209    0.218560     2.833    0.0063     **
svi          0.741781    0.294451     2.519    0.0145     *
lbph         0.144426    0.069812     2.069    0.0430     *
pgg45        0.008945    0.004099     2.182    0.0331     *
lcp         -0.205417    0.109424    -1.877    0.0654     .
age         -0.019480    0.013105    -1.486    0.1425
---
Residual standard error: 0.7064 on 59 degrees of freedom
Multiple R-Squared: 0.6943, Adjusted R-squared: 0.658
F-statistic: 19.14 on 7 and 59 DF, p-value: 4.496e-13
```

Notice that both predictor variables are judged to be "statistically significant" in the two-variable model, three variables are judged to be "statistically significant" in the four-variable model and five variables are judged to be "statistically significant" in the seven-variable model. However, the p-values obtained after variable selection are much smaller than their true values. In view of this, it seems that the four- and seven-variable models over-fit the data and as such *the two-variable model seems to be preferred.*

7.3.2 Stage 2: Model Comparison Using the Test Data Set

We can now use the test data to compare the two-, four- and seven-variable models we identified above.

Given below is the output from R associated with fitting the best models with two-, four and seven-predictor variables to the 30 cases in the test data.

Regression output from R

```
Call:
lm(formula = lpsa ~ lcavol + lweight)

Coefficients:
             Estimate   Std. Error   t value   Pr(>|t|)
(Intercept)   0.7354      0.9572       0.768     0.449
lcavol        0.7478      0.1294       5.778   3.81e-06   ***
lweight       0.1968      0.2473       0.796     0.433
---
Residual standard error: 0.721 on 27 degrees of freedom
Multiple R-Squared: 0.5542, Adjusted R-squared: 0.5212
F-statistic: 16.78 on 2 and 27 DF, p-value: 1.833e-05


Call:
lm(formula = lpsa ~ lcavol + lweight + svi + lbph)

Coefficients:
             Estimate   Std. Error   t value   Pr(>|t|)
(Intercept)   0.52957     0.93066      0.569     0.5744
lcavol        0.59555     0.12655      4.706   7.98e-05   ***
lweight       0.26215     0.24492      1.070     0.2947
svi           0.95051     0.32214      2.951     0.0068   **
lbph         -0.05337     0.09237     -0.578     0.5686
---
Residual standard error: 0.6445 on 25 degrees of freedom
Multiple R-Squared: 0.6703, Adjusted R-squared: 0.6175
F-statistic: 12.7 on 4 and 25 DF, p-value: 8.894e-06

Call:
lm(formula=lpsa~lcavol+lweight+svi+lbph+pgg45+lcp+age)

Coefficients:
             Estimate   Std. Error   t value   Pr(>|t|)
(Intercept)   0.873329    1.490194     0.586     0.56381
lcavol        0.481237    0.165881     2.901     0.00828   **
lweight       0.313601    0.257112     1.220     0.23549
svi           0.619278    0.423109     1.464     0.15744
lbph         -0.090696    0.121368    -0.747     0.46281
pgg45         0.001316    0.006370     0.207     0.83819
lcp           0.180850    0.166970     1.083     0.29048
age          -0.004958    0.022220    -0.223     0.82550
---
Residual standard error: 0.6589 on 22 degrees of  freedom
Multiple R-Squared: 0.6967, Adjusted R-squared:  0.6001
F-statistic: 7.218 on 7 and 22 DF, p-value: 0.0001546
```

Notice that in the test data just one-predictor variable is judged to be "statistically significant" in the two-variable model, two variables are judged to be "statistically significant" in the four-variable model and just one variable is judged to be "statistically significant" in the seven-variable model. Thus, based on the test data none of these models is very convincing.

7.3.2.1 What Has Happened?

Put briefly, this situation is due to

- Case 45 in the training set accounts for most of the statistical significance of the predictor variable lweight
- Splitting the data into a training set and a test set by randomly assigning cases does not always work well in small data sets.

We discuss each of these issues in turn.

7.3.2.2 Case 45 in the Training Set

We reconsider variable selection in this example by identifying the subset of the predictors of a given size that maximizes adjusted R-squared (i.e., minimizes RSS) for the training data set with and without case 45. Figure 7.9 shows plots of adjusted R-squared (for models with up to 5 predictors) against the number of predictors in the model for the optimal subsets of predictors for the training data set with and without case 45. Notice how the optimal two-, three- and five-variable models change with the omission of just case 45. Thus, case 45 has a dramatic effect on variable selection. It goes without saying that case 45 in the training set should be thoroughly investigated.

7.3.2.3 Splitting the Data into a Training Set and a Test Set

Snee (1977, p. 421) demonstrated the advantages of splitting the data into a training set and a test set such that "the two sets cover approximately the same region and have the same statistical properties." Random splits, especially in small samples do not always have these desirable properties. In addition, Snee (1977) described the DUPLEX algorithm for data splitting which has the desired properties. For details on the algorithm see Montgomery, Peck and Vining (2001, pp. 536–537).

Figure 7.10 provides an illustration of the difference between the training and test data sets. It shows a scatter plot of lpsa against lweight with different symbols used for the training and test data sets. The least squares regression line for each data set is also marked on Figure 7.10. While case 45 in the training data set does not stand out in Figure 7.10, case 9 in the test data set stands out due to its very high value of lweight.

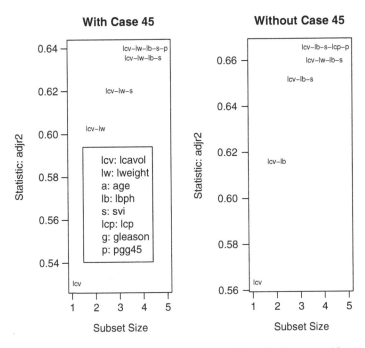

Figure 7.9 Plots of R^2_{adj} for the best subset of sizes 1 to 5 with and without case 45

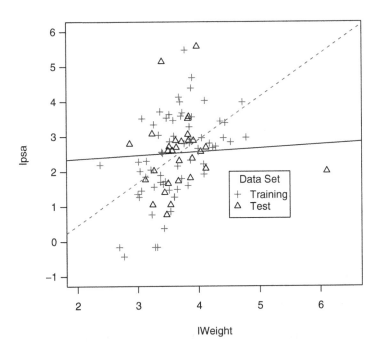

Figure 7.10 Plot of lpsa against lweight for both the training and test data sets

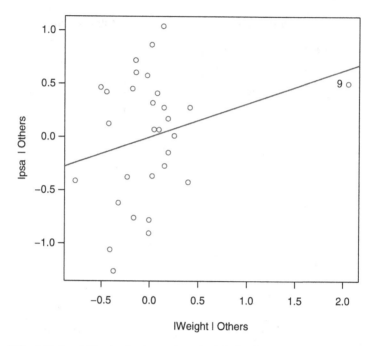

Figure 7.11 Added-variable plot for the predictor lweight for the test data

   To further illustrate the dramatic effect due to case 9 in the test data set, Figure 7.11 shows an added-variable plot for the predictor lweight based on the full model for the test data.

   In summary, case 45 in the training data and case 9 in the test data need to be thoroughly investigated before any further statistical analyses are performed. This example once again illustrates the importance of carefully examining any regression fit in order to determine outliers and influential points. If cases 9 and 45 are found to be valid data points and not associated with special cases, then a possible way forward is to use variable selection techniques based on robust regression – see Maronna, Martin and Yohai (2006, Chapter 5) for further details.

7.4   Recent Developments in Variable Selection – LASSO

In this section we briefly discuss LASSO, least absolute shrinkage and selection operator (Tibshirani, 1996), which we shall discover is a method that effectively performs variable selection and regression coefficient estimation simultaneously. There has been much interest in LASSO as evidenced by the fact that according to the Web of Science, Tibshirani's 1996 LASSO paper has been cited more than 400 times as of June, 2008.

The LASSO estimates of the regression coefficients from the full model (7.1) are obtained from the following constrained version of least squares:

$$\min \sum_{i=1}^{n} \left(y_i - \left\{ \beta_0 + \beta_1 x_{1i} + \ldots + \beta_p x_{pi} \right\} \right)^2 \text{ subject to } \sum_{j=1}^{p} \left| \beta_j \right| \leq s \qquad (7.6)$$

for some number $s \geq 0$. Using a Lagrange multiplier argument, it can be shown that (7.6) is equivalent to minimizing the residual sum of squares plus a penalty term on the absolute value of the regression coefficients, that is,

$$\min \sum_{i=1}^{n} \left(y_i - \left\{ \beta_0 + \beta_1 x_{1i} + \ldots + \beta_p x_{pi} \right\} \right)^2 + \lambda \sum_{j=1}^{p} \left| \beta_j \right| \qquad (7.7)$$

for some number $\lambda \geq 0$. When the value of s in (7.6) is very large (or equivalently in $\lambda = 0$ (7.7)), the constraint in (7.6) (or equivalently the penalty term in (7.7)) has no effect and the solution is just the set of least squares estimates for model (7.1). Alternatively, for small values of s (or equivalently large values of λ) some of the resulting estimated regression coefficients are exactly zero, effectively omitting predictor variables from the fitted model. Thus, LASSO performs variable selection and regression coefficient estimation simultaneously.

Zhou, Hastie and Tibshirani (2007) develop versions of AIC and BIC for LASSO that can be used to find an "optimal" value or λ or equivalently s. They suggest using BIC to find the "optimal" LASSO model when sparsity of the model is of primary concern.

LARS, least angle regression (Efron et al., 2004) provides a clever and hence very efficient way of computing the complete Lasso sequence of solutions as s is varied from 0 to infinity. In fact, Zhou, Hastie and Tibshirani (2007) show that it is possible to find the optimal lasso fit with the computational effort equivalent to obtaining a single least squares fit. Thus, the LASSO has the potential to revolutionize variable selection. A more detailed discussion of LASSO is beyond the scope of this book.

Finally, Figure 7.12 contains a flow chart which summarizes the steps in developing a multiple linear regression model.

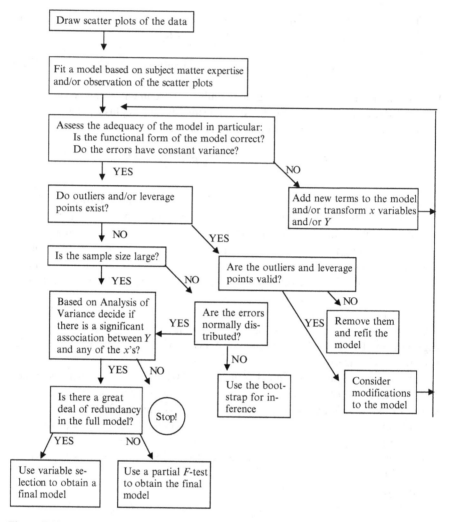

Figure 7.12 Flow chart for multiple linear regression

7.5   Exercises

1. The generated data set in this question is taken from Mantel (1970). The data are given in Table 7.3 and can be found on the book web site in the file Mantel.txt. Interest centers on using variable selection to choose a subset of the predictors to model Y. The data were generated such that the full model

$$Y = \beta_0 + \beta_1 X_1 + \beta_2 X_2 + \beta_3 X_3 + e \qquad (7.8)$$

is a valid model for the data.

Output from R associated with different variable selection procedures based on model (7.8) appears below.

Table 7.3 Mantel's generated data

Case	Y	X1	X2	X3
1	5	1	1004	6
2	6	200	806	7.3
3	8	-50	1058	11
4	9	909	100	13
5	11	506	505	13.1

(a) Identify the optimal model or models based on R^2_{adj}, AIC and BIC from the approach based on all possible subsets.
(b) Identify the optimal model or models based on AIC and BIC from the approach based on forward selection.
(c) Carefully explain why different models are chosen in (a) and (b).
(d) Decide which model you would recommend. Give detailed reasons to support your choice.

Output from R: Correlations between the predictors in model (7.8)

```
                X1                X2                X3
X1       1.0000000        -0.9999887         0.6858141
X2      -0.9999887         1.0000000        -0.6826107
X3       0.6858141        -0.6826107         1.0000000
```

Approach 1: All Possible Subsets

Figure 7.13 shows a plot of adjusted R-squared against the number of predictors in the model for the optimal subsets of predictors.
Table 7.4 gives the values of R^2_{adj}, AIC and BIC for the best subset of each size.

Approach 2: Stepwise Subsets

Forward Selection Based on AIC

```
Start: AIC= 9.59
Y ~ 1
            Df     Sum of Sq         RSS          AIC
+ X3         1       20.6879      2.1121      -0.3087
+ X1         1        8.6112     14.1888       9.2151
+ X2         1        8.5064     14.2936       9.2519
<none>                           22.8000       9.5866
Step: AIC= -0.31
Y ~ X3
            Df     Sum of Sq         RSS          AIC
<none>                            2.11211     -0.30875
+ X2         1        0.06633      2.04578      1.53172
+ X1         1        0.06452      2.04759      1.53613
```

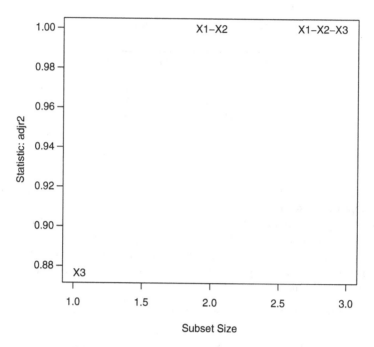

Figure 7.13 Plots of R^2_{adj} for the best subset of each size

Table 7.4 Values of R^2_{adj}, AIC and BIC for the best subset of each size

Subset size	Predictors	R^2_{adj}	AIC	BIC
1	X3	0.8765	−0.3087	−1.0899
2	X1, X2	1.0000	−316.2008	−317.3725
3	X1, X2, X3	1.0000	−314.7671	−316.3294

Forward Selection Based on BIC*

```
Start: AIC= 9.2
Y ~ 1
               Df    Sum of Sq        RSS         AIC
+ X3            1      20.6879     2.1121     -1.0899
+ X1            1       8.6112    14.1888      8.4339
+ X2            1       8.5064    14.2936      8.4707
<none>                           22.8000      9.1961
Step: AIC= -1.09
Y ~ X3
               Df    Sum of Sq        RSS         AIC
<none>                            2.11211    -1.08987
+ X2            1       0.06633    2.04578     0.36003
+ X1            1       0.06452    2.04759     0.36444
```

---

* The R command step which was used here labels the output as AIC even when the BIC penalty term is used.

Output from R

```
Call:
lm(formula = Y ~ X3)

Coefficients:
             Estimate   Std. Error   t value   Pr(>|t|)
(Intercept)    0.7975      1.3452      0.593     0.5950
X3             0.6947      0.1282      5.421     0.0123   *
---

Residual standard error: 0.8391 on 3 degrees of freedom
Multiple R-Squared: 0.9074, Adjusted R-squared: 0.8765
F-statistic: 29.38 on 1 and 3 DF, p-value: 0.01232

Call:
lm(formula = Y ~ X1 + X2)

Coefficients:
              Estimate    Std. Error     t value    Pr(>|t|)
(Intercept) -1.000e+03   4.294e-12   -2.329e+14    <2e-16   ***
X1           1.000e+00   4.250e-15    2.353e+14    <2e-16   ***
X2           1.000e+00   4.266e-15    2.344e+14    <2e-16   ***
---

Residual standard error:1.607e-14 on 2 degrees of freedom
Multiple R-Squared: 1, Adjusted R-squared: 1
F-statistic: 4.415e+28 on 2 and 2 DF, p-value: < 2.2e-16

Call:
lm(formula = Y ~ X1 + X2 + X3)

Coefficients:
              Estimate    Std. Error     t value    Pr(>|t|)
(Intercept) -1.000e+03   1.501e-11   -6.660e+13   9.56e-15   ***
X1           1.000e+00   1.501e-14    6.661e+13   9.56e-15   ***
X2           1.000e+00   1.501e-14    6.664e+13   9.55e-15   ***
X3           4.108e-15   1.186e-14    3.460e-01   0.788
---

Residual standard error: 2.147e-14 on 1 degrees of freedom
Multiple R-Squared:          1,   Adjusted R-squared:          1
F-statistic: 1.648e+28 on 3 and 1 DF, p-value: 5.726e-15
```

2. The real data set in this question first appeared in Hald (1952). The data are given in Table 7.5 and can be found on the book web site in the file Haldcement. txt. Interest centers on using variable selection to choose a subset of the predictors to model Y. Throughout this question we shall assume that the full model below is a valid model for the data

$$Y = \beta_0 + \beta_1 X_1 + \beta_2 X_2 + \beta_3 X_3 + \beta_4 X_4 + e \qquad (7.9)$$

Output from R associated with different variable selection procedures based on model (7.9) appears on the following pages:

Table 7.5 Hald's real data

Y	x_1	x_2	x_3	x_4
78.5	7	26	6	60
74.3	1	29	15	52
104.3	11	56	8	20
87.6	11	31	8	47
95.9	7	52	6	33
109.2	11	55	9	22
102.7	3	71	17	6
72.5	1	31	22	44
93.1	2	54	18	22
115.9	21	47	4	26
83.8	1	40	23	34
113.3	11	66	9	12
109.4	10	68	8	12

(a) Identify the optimal model or models based on R^2_{adj}, AIC, AIC_C, BIC from the approach based on all possible subsets.
(b) Identify the optimal model or models based on AIC and BIC from the approach based on forward selection.
(c) Identify the optimal model or models based on AIC and BIC from the approach based on backward elimination.
(d) Carefully explain why the models chosen in (a), (b) & (c) are not all the same.
(e) Recommend a final model. Give detailed reasons to support your choice.

Output from R: Correlations between the predictors in model (7.9)

```
            x1            x2            x3            x4
x1   1.0000000    0.2285795   -0.82413376   -0.24544511
x2   0.2285795    1.0000000   -0.13924238   -0.97295500
x3  -0.8241338   -0.1392424    1.00000000    0.02953700
x4  -0.2454451   -0.9729550    0.02953700    1.00000000
```

Approach 1: All Possible Subsets

Figure 7.14 shows a plot of adjusted R-squared against the number of predictors in the model for the optimal subsets of predictors. Table 7.6 shows the best subset of each size.

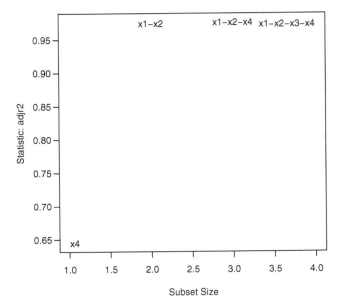

Figure 7.14 Plots of R^2_{adj} for the best subset of each size

Table 7.6 Values of R^2_{adj}, AIC, AIC_C and BIC for the best subset of each size

Subset size	Predictors	R^2_{adj}	AIC	AIC_C	BIC
1	X4	0.6450	58.8516	61.5183	59.9815
2	X1, X2	0.9744	25.4200	30.4200	27.1148
3	X1, X2, X4	0.9764	24.9739	33.5453	27.2337
4	X1, X2, X3, X4	0.9736	26.9443	40.9443	29.7690

Approach 2: Stepwise Subsets

Backward elimination based on AIC

```
Start: AIC= 26.94
Y ~ x1 + x2 + x3 + x4
           DF    Sum of Sq        RSS        AIC
- x3        1       0.109      47.973     24.974
- x4        1       0.247      48.111     25.011
- x2        1       2.972      50.836     25.728
<none>                         47.864     26.944
- x1        1      25.951      73.815     30.576

Step: AIC= 24.97
Y ~ x1 + x2 + x4
           Df    Sum of Sq        RSS        AIC
<none>                           47.97      24.97
- x4        1        9.93        57.90      25.42
- x2        1       26.79        74.76      28.74
- x1        1      820.91       868.88      60.63
```

Backward elimination based on BIC

```
Start: AIC= 29.77
Y ~ x1 + x2 + x3 + x4
              Df     Sum of Sq        RSS          AIC
- x3          1         0.109      47.973       27.234
- x4          1         0.247      48.111       27.271
- x2          1         2.972      50.836       27.987
<none>                             47.864       29.769
- x1          1        25.951      73.815       32.836

Step: AIC= 27.23
Y ~ x1 + x2 + x4
              Df     Sum of Sq        RSS          AIC
- x4          1          9.93       57.90        27.11
<none>                              47.97        27.23
- x2          1         26.79       74.76        30.44
- x1          1        820.91      868.88        62.32

Step: AIC= 27.11
Y ~ x1 + x2
              Df     Sum of Sq        RSS          AIC
<none>                              57.90        27.11
- x1          1        848.43      906.34        60.31
- x2          1       1207.78     1265.69        64.65
```

Forward selection based on AIC

```
Start: AIC= 71.44
Y ~ 1
              Df     Sum of Sq        RSS          AIC
+ x4          1       1831.90      883.87        58.85
+ x2          1       1809.43      906.34        59.18
+ x1          1       1450.08     1265.69        63.52
+ x3          1        776.36     1939.40        69.07
<none>                            2715.76        71.44

Step: AIC= 58.85
Y ~ x4
              Df     Sum of Sq        RSS          AIC
+ x1          1        809.10       74.76        28.74
+ x3          1        708.13      175.74        39.85
<none>                             883.87        58.85
+ x2          1         14.99      868.88        60.63

Step: AIC= 28.74
Y ~ x4 + x1
              Df     Sum of Sq        RSS          AIC
+ x2          1        26.789      47.973       24.974
+ x3          1        23.926      50.836       25.728
<none>                             74.762       28.742
```

```
Step: AIC= 24.97
Y ~ x4 + x1 + x2
                Df      Sum of Sq           RSS        AIC
<none>                                   47.973     24.974
+ x3            1          0.109          47.864     26.944
```

Forward selection based on BIC

```
Start: AIC= 72.01
Y ~ 1
                Df      Sum of Sq           RSS        AIC
+ x4            1        1831.90         883.87      59.98
+ x2            1        1809.43         906.34      60.31
+ x1            1        1450.08        1265.69      64.65
+ x3            1         776.36        1939.40      70.20
<none>                                  2715.76      72.01

Step: AIC= 59.98
Y ~ x4
                Df      Sum of Sq           RSS        AIC
+ x1            1         809.10          74.76      30.44
+ x3            1         708.13         175.74      41.55
<none>                                   883.87      59.98
+ x2            1          14.99         868.88      62.32

Step: AIC= 30.44
Y ~ x4 + x1
                Df      Sum of Sq           RSS        AIC
+ x2            1         26.789         47.973      27.234
+ x3            1         23.926         50.836      27.987
<none>                                   74.762      30.437

Step: AIC= 27.23
Y ~ x4 + x1 + x2
                Df      Sum of Sq           RSS        AIC
<none>                                   47.973     27.234
+ x3            1          0.109          47.864     29.769
```

Output from R

```
Call:
lm(formula = Y ~ x4)

Coefficients:
                Estimate    Std. Error     t value    Pr(>|t|)
(Intercept)     117.5679        5.2622      22.342     1.62e-10    ***
x4               -0.7382        0.1546      -4.775     0.000576    ***
---
Residual standard error: 8.964 on 11 degrees of freedom
Multiple R-Squared: 0.6745, Adjusted R-squared: 0.645
F-statistic: 22.8 on 1 and 11 DF, p-value: 0.0005762
```

```
Call:
lm(formula = Y ~ x1 + x2)

Coefficients:
              Estimate   Std. Error   t value   Pr(>|t|)
(Intercept)   52.57735      2.28617     23.00   5.46e-10   ***
x1             1.46831      0.12130     12.11   2.69e-07   ***
x2             0.66225      0.04585     14.44   5.03e-08   ***
---
Residual standard error: 2.406 on 10 degrees of freedom
Multiple R-Squared: 0.9787, Adjusted R-squared: 0.9744
F-statistic: 229.5 on 2 and 10 DF, p-value: 4.407e-09

vif(om2)
        x1              x2
1.055129        1.055129

Call:
lm(formula = Y ~ x1 + x2 + x4)

Coefficients:
              Estimate   Std. Error   t value   Pr(>|t|)
(Intercept)   71.6483      14.1424       5.066   0.000675   ***
x1             1.4519       0.1170      12.410   5.78e-07   ***
x2             0.4161       0.1856       2.242   0.051687   .
x4            -0.2365       0.1733      -1.365   0.205395
---
Residual standard error: 2.309 on 9 degrees of freedom
Multiple R-Squared: 0.9823, Adjusted R-squared: 0.9764
F-statistic: 166.8 on 3 and 9 DF, p-value: 3.323e-08

vif(om3)
        x1              x2              x4
1.066330        18.780309       18.940077


Call:
lm(formula = Y ~ x1 + x2 + x3 + x4)
Coefficients:
              Estimate   Std. Error   t value   Pr(>|t|)
(Intercept)   62.4054      70.0710       0.891   0.3991
x1             1.5511       0.7448       2.083   0.0708   .
x2             0.5102       0.7238       0.705   0.5009
x3             0.1019       0.7547       0.135   0.8959
x4            -0.1441       0.7091      -0.203   0.8441
---
Residual standard error: 2.446 on 8 degrees of freedom
Multiple R-Squared: 0.9824, Adjusted R-squared: 0.9736
F-statistic: 111.5 on 4 and 8 DF, p-value: 4.756e-07

vif(om4)
        x1              x2              x3              x4
38.49621        254.42317       46.86839        282.51286
```

3. This is a continuation of Exercise 5 in Chapter 6. The golf fan was so impressed with your answers to part 1 that your advice has been sought re the next stage in the data analysis, namely using model selection to remove the redundancy in full the model developed in part 1.

$$\log(Y) = \beta_0 + \beta_1 x_1 + \beta_2 x_2 + \beta_3 x_3 + \beta_4 x_4 + \beta_5 x_5 + \beta_6 x_6 + \beta_7 x_7 + e \quad (7.10)$$

where
$Y = PrizeMoney$; $x_1 = Driving\ Accuracy$; $x_2 = GIR$; $x_3 = PuttingAverage$; $x_4 = BirdieConversion$; $x_5 = SandSaves$; $x_6 = Scrambling$; and $x_7 = PuttsPerRound$.

Interest centers on using variable selection to choose a subset of the predictors to model the transformed version of Y. Throughout this question we shall assume that model (7.10) is a valid model for the data.

(a) Identify the optimal model or models based on R_{adj}^2, AIC, AIC_C, BIC from the approach based on all possible subsets.
(b) Identify the optimal model or models based on AIC and BIC from the approach based on backward selection.
(c) Identify the optimal model or models based on AIC and BIC from the approach based on forward selection.
(d) Carefully explain why the models chosen in (a) & (c) are not the same while those in (a) and (b) are the same.
(e) Recommend a final model. Give detailed reasons to support your choice.
(f) Interpret the regression coefficients in the final model. Is it necessary to be cautious about taking these results to literally?

Chapter 8
Logistic Regression

Thus far in this book we have been concerned with developing models where the response variable is numeric and ideally follows a normal distribution. In this chapter, we consider the situation in which the response variable is based on a series of "yes"/"no" responses, such as whether a particular restaurant is recommended by being included in a prestigious guide. Ideally such responses follow a binomial distribution in which case the appropriate model is a logistic regression model.

8.1 Logistic Regression Based on a Single Predictor

We begin this chapter by considering the case of predicting a binomial random variable Y based on a single predictor variable x via logistic regression. Before considering logistic regression we briefly review some facts about the binomial distribution.

The binomial distribution

A binomial process is one that possesses the following properties:

1. There are m identical trials
2. Each trial results in one of two outcomes, either a "success," S or a "failure," F
3. θ, the probability of "success" is the same for all trials
4. Trials are independent

The trials of a binomial process are called Bernoulli trials.

Let Y = number of successes in m trials of a binomial process. Then Y is said to have a binomial distribution with parameters m and θ. The short-hand notation for this is as follows:

$$Y \sim \text{Bin}(m, \theta)$$

The probability that Y takes the integer value j ($j = 0, 1, \ldots, m$) is given by

$$P(Y = j) = \binom{m}{j} \theta^j (1-\theta)^{m-j} = \frac{m!}{j!(m-j)!} \theta^j (1-\theta)^{m-j} \quad j = 1, \ldots, m$$

S.J. Sheather, *A Modern Approach to Regression with R*,
DOI: 10.1007/978-0-387-09608-7_8, © Springer Science+Business Media LLC 2009

The mean and variance of Y are given by

$$E(Y) = m\theta, Var(Y) = m\theta(1-\theta)$$

In the logistic regression setting, we wish to model θ and hence Y on the basis of predictors $x_1, x_2, ..., x_p$.

We shall begin by considering the case of a single predictor variable x. In this case

$$(Y \mid x_i) \sim \text{Bin}(m_i, \theta(x_i)) \quad i = 1, ..., n$$

The sample proportion of "successes" at each i is given by y_i / m_i. Notice that

$$\text{E}(y_i/m_i \mid x_i) = \theta(x_i) \text{ and } \text{Var}(y_i/m_i \mid x_i) = \theta(x_i)(1-\theta(x_i))/m_i$$

We shall consider the sample proportion of "successes," y_i / m_i as the response since:

1. y_i/m_i is an unbiased estimate of $\theta(x_i)$
2. y_i/m_i varies between 0 and 1

Notice that the variance of the response y_i/m_i, depends on $\theta(x_i)$ and as such it is not constant. In addition, this variance is also therefore unknown. Thus, least squares regression is an inappropriate technique for analyzing Binomial responses.

Example: Michelin and Zagat guides to New York City restaurants

In November 2005, Michelin published its first ever guide to hotels and restaurants in New York City (Anonymous, 2005). According to the guide, inclusion in the guide is based on Michelin's "meticulous and highly confidential evaluation process (in which) Michelin inspectors – American and European – conducted anonymous visits to New York City restaurants and hotels. ... Inside the premier edition of the *Michelin Guide New York City* you'll find a selection of restaurants by level of comfort; those with the best cuisine have been awarded our renowned Michelin stars. ... From the best casual, neighborhood eateries to the city's most impressive gourmet restaurants, the *Michelin Guide New York City* provides trusted advice for an unbeatable experience, every time."

On the other hand, the *Zagat Survey 2006: New York City Restaurants* (Gathje and Diuguid, 2005) is purely based on views submitted by customers using mail-in or online surveys.

We shall restrict our comparison of the two restaurant guides to the 164 French restaurants that are included in the *Zagat Survey 2006: New York City Restaurants*. We want to be able to model θ, the probability that a French restaurant is included in the *2006 Michelin Guide New York City*, based on customer views from the *Zagat Survey 2006: New York City Restaurants*. We begin looking at the effect of x, customer ratings of food on θ. Table 8.1 classifies the 164 French restaurants included in the *Zagat Survey 2006: New York City Restaurants* according to whether they were included in the *Michelin Guide New York City* for each value of

Table 8.1 French restaurants in the Michelin guide broken down by food ratings

Food rating, x_i	InMichelin, y_i	NotInMichelin, $m_i\text{-}y_i$	m_i	y_i/m_i
15	0	1	1	0.00
16	0	1	1	0.00
17	0	8	8	0.00
18	2	13	15	0.13
19	5	13	18	0.28
20	8	25	33	0.24
21	15	11	26	0.58
22	4	8	12	0.33
23	12	6	18	0.67
24	6	1	7	0.86
25	11	1	12	0.92
26	1	1	2	0.50
27	6	1	7	0.86
28	4	0	4	1.00

the food rating. For example, $m_i = 33$ French restaurants in the *Zagat Survey 2006: New York City Restaurants* received a food rating of $x_i = 20$ (out of 30). Of these 33, $y_i = 8$ were included in the *Michelin Guide New York City* and $m_i - y_i = 25$ were not. In this case, the observed proportion of "successes" at $x = 20$ is given by $y_i/m_i = 8/33 = 0.24$. The data in Table 8.1 can be found on the book web site in the file MichelinFood.txt. Figure 8.1 contains a plot of the sample proportions of "success" against Zagat food ratings.

It is clear from Figure 8.1 that the shape of the underlying function, $\theta(x)$ is not a straight line. Instead it appears S-shaped, with very low values of the x-variable resulting in zero probability of "success" and very high values of the x-variable resulting in a probability of "success" equal to one.

8.1.1   The Logistic Function and Odds

A popular choice for the S-shaped function evident in Figure 8.1 is the logistic function, that is,

$$\theta(x) = \frac{\exp(\beta_0 + \beta_1 x)}{1 + \exp(\beta_0 + \beta_1 x)} = \frac{1}{1 + \exp(-\{\beta_0 + \beta_1 x\})}$$

Solving this last equation for $\beta_0 + \beta_1 x$ gives

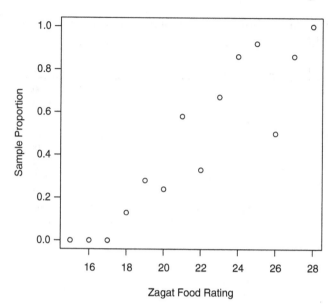

Figure 8.1 Plot of the sample proportion of "successes" against food ratings

$$\beta_0 + \beta_1 x = \log\left(\frac{\theta(x)}{1-\theta(x)}\right)$$

Thus, if the chosen function is correct, a plot of $\log\left(\frac{\theta(x)}{1-\theta(x)}\right)$ against x will produce a straight line. The quantity $\log\left(\frac{\theta(x)}{1-\theta(x)}\right)$ is called a *logit*.

The quantity $\frac{\theta(x)}{1-\theta(x)}$ is known as *odds*. The concept of odds has two forms, namely, the *odds in favor* of "success" and the *odds against* "success." The odds in favor of "success" are defined as the ratio of the probability that "success" will occur, to the probability that "success" will not occur. In symbols, let $\theta = P(\text{success})$ then,

$$\text{Odds in favor of success} = \frac{P(\text{success})}{1-P(\text{success})} = \frac{\theta}{1-\theta}.$$

Thus, *the odds in logistic regression are in the form of odds in favor of a "success."*

The odds against "success" are defined as the ratio of the probability that "success" will not occur, to the probability that "success" will occur. In symbols,

$$\text{Odds against success} = \frac{1-P(\text{success})}{P(\text{success})} = \frac{1-\theta}{\theta}.$$

Bookmakers quote odds as odds against "success" (i.e., winning). A horse quoted at the fixed odds of 20 to 1 (often written as the ratio 20/1) is expected to lose 20 and win just 1 out of every 21 races.

Let x denote the Zagat food rating for a given French restaurant and $\theta(x)$ denote the probability that this restaurant is included in the Michelin guide. Then our logistic regression model for the response, $\theta(x)$ based on the predictor variable x is given by

$$\theta(x) = \frac{1}{1 + \exp(-\{\beta_0 + \beta_1 x\})} \qquad (8.1)$$

Given below is the output from R for model (8.1).

Logistic regression output from R

```
Call:
glm(formula = cbind(InMichelin, NotInMichelin) ~ Food, family =
binomial)
Coefficients:
              Estimate    Std. Error   z value    Pr(>|z|)
(Intercept)  -10.84154      1.86236    -5.821     5.84e-09   ***
Food           0.50124      0.08768     5.717     1.08e-08   ***
---
(Dispersion parameter for binomial family taken to be 1)

    Null deviance: 61.427 on 13 degrees of freedom
Residual deviance: 11.368 on 12 degrees of freedom
AIC: 41.491
```

The fitted model is

$$\hat{\theta}(x) = \frac{1}{1 + \exp(-\{\hat{\beta}_0 + \hat{\beta}_1 x\})} = \frac{1}{1 + \exp(-\{-10.842 + 0.501x\})}$$

Figure 8.2 shows a plot of the of the sample proportions of "success" (i.e., inclusion in the Michelin guide) against x, Zagat food rating. The fitted logistic regression model is marked on this plot as a smooth curve.

Rearranging the fitted model equation gives the log(odds) or logit

$$\log\left(\frac{\hat{\theta}(x)}{1 - \hat{\theta}(x)}\right) = \hat{\beta}_0 + \hat{\beta}_1 x = -10.842 + 0.501x$$

Notice that the log(odds) or logit is a linear function of x. The estimated odds for being included in the Michelin guide are given by

$$\left(\frac{\hat{\theta}(x)}{1 - \hat{\theta}(x)}\right) = \exp(\hat{\beta}_0 + \hat{\beta}_1 x) = \exp(-10.842 + 0.501x)$$

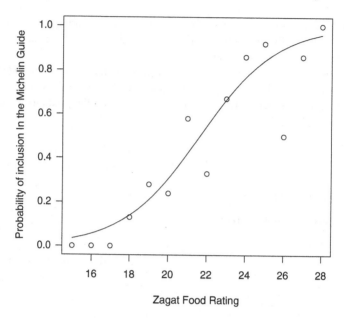

Figure 8.2   Logistic regression fit to the data in Figure 8.1

For example, if x, Zagat food rating is increased by

- One unit then the odds for being included in the Michelin guide increases by $\exp(0.501) = 1.7$
- Five units then the odds for being included in the Michelin guide increases by $\exp(5 \times 0.501) = 12.2$

Table 8.2 gives the estimated probabilities and odds obtained from the logistic model (8.1). Taking the ratio of successive entries in the last column of Table 8.2 (i.e., $0.060/0.036 = 0.098/0.060 = \dots = 24.364/14.759 = 1.7$) reproduces the result that increasing x (Zagat food rating) by one unit increases the odds of being included in the Michelin guide by 1.7.

Notice from Table 8.2 that the odds are greater than 1 when the probability is greater than 0.5. In these circumstances the probability of "success" is greater than the probability of "failure."

8.1.2   Likelihood for Logistic Regression with a Single Predictor

We next look at how likelihood can be used to estimate the parameters in logistic regression.

Table 8.2 Estimated probabilities and odds obtained from the logistic model

x, Zagat food rating	$\hat{\theta}(x)$, estimated probability of inclusion in the Michelin guide	$\hat{\theta}(x)/(1-\hat{\theta}(x))$ estimated odds
15	0.035	0.036
16	0.056	0.060
17	0.089	0.098
18	0.140	0.162
19	0.211	0.268
20	0.306	0.442
21	0.422	0.729
22	0.546	1.204
23	0.665	1.988
24	0.766	3.281
25	0.844	5.416
26	0.899	8.941
27	0.937	14.759
28	0.961	24.364

Let y_i= number of successes in m_i trials of a binomial process where $i = 1,\ldots, n$. Then

$$y_i \mid x_i \sim \text{Bin}(m_i, \theta(x_i))$$

So that

$$P(Y_i = y_i \mid x_i) = \binom{m_i}{y_i}\theta(x_i)^{y_i}\left(1 - \theta(x_i)\right)^{m_i - y_i}$$

Assume further that

$$\theta(x_i) = \frac{1}{1 + \exp(-\{\beta_0 + \beta_1 x_i\})}$$

So that

$$\log\left(\frac{\theta(x_i)}{1 - \theta(x_i)}\right) = \beta_0 + \beta_1 x_i$$

Assuming the n observations are independent, then the likelihood function is the function of the unknown probability of success $\theta(x_i)$ given by

$$L = \prod_{i=1}^{n} P(Y_i = y_i \mid x_i) = \prod_{i=1}^{n} \binom{m_i}{y_i}\theta(x_i)^{y_i}\left(1 - \theta(x_i)\right)^{m_i - y_i}$$

The log-likelihood function is given by

$$
\begin{aligned}
\log(L) &= \sum_{i=1}^{n} \left[\log \binom{m_i}{y_i} + \log\left(\theta(x_i)^{y_i}\right) + \log\left((1-\theta(x_i))^{m_i-y_i}\right) \right] \\
&= \sum_{i=1}^{n} \left[y_i \log\left(\theta(x_i)\right) + (m_i - y_i)\log\left(1-\theta(x_i)\right) + \log\binom{m_i}{y_i} \right] \\
&= \sum_{i=1}^{n} \left[y_i \log\left(\frac{\theta(x_i)}{1-\theta(x_i)}\right) + m_i \log\left(1-\theta(x_i)\right) + \log\binom{m_i}{y_i} \right] \\
&= \sum_{i=1}^{n} \left[y_i\left(\beta_0 + \beta_1 x_i\right) - m_i \log\left(1+\exp(\beta_0 + \beta_1 x_i)\right) + \log\binom{m_i}{y_i} \right]
\end{aligned}
$$

since

$$
\log\left(1-\theta(x_i)\right) = \log\left(1 - \frac{\exp(\beta_0 + \beta_1 x)}{1+\exp(\beta_0 + \beta_1 x)}\right) = \log\left(\frac{1}{1+\exp(\beta_0 + \beta_1 x)}\right)
$$

The parameters β_0 and β_1 can be estimated by maximizing the log-likelihood. This has to be done using an iterative method such as Newton-Raphson or iteratively reweighted least squares.

The standard approach to testing

$$
H_0 : \beta_1 = 0
$$

is to use what is called a Wald test statistic

$$
Z = \frac{\hat{\beta}_1}{\text{estimated se}\left(\hat{\beta}_1\right)}
$$

where the estimated standard error is calculated based on the iteratively reweighted least squares approximation to the maximum likelihood estimate. The Wald test statistic is then compared to a standard normal distribution to test for statistical significance. Confidence intervals based on the Wald statistic are of the form

$$
\hat{\beta}_1 \pm z_{1-\alpha/2}\,\text{estimated se}\left(\hat{\beta}_1\right)
$$

8.1.3   *Explanation of Deviance*

In logistic regression the concept of the residual sum of squares is replaced by a concept known as the *deviance*. In the case of logistic regression the deviance is defined to be

$$G^2 = 2 \sum_{i=1}^{n} \left[y_i \log \left(\frac{y_i}{\hat{y}_i} \right) + (m_i - y_i) \log \left(\frac{m_i - y_i}{m_i - \hat{y}_i} \right) \right]$$

where

$$\hat{y}_i = m_i \hat{\theta}(x_i).$$

The degrees of freedom (df) associated with the deviance are given by

$$df = n - (\text{number of } \beta's \text{ estimated})$$

The deviance associated with a given logistic regression model (M) is based on comparing the maximized log-likelihood under (M) with the maximized log-likelihood under (S), the so-called saturated model that has a parameter for each observation. In fact, the deviance is given by twice the difference between these maximized log-likelihoods.

The saturated model, (S) estimates $\theta(x_i)$ by the observed proportion of "successes" at x_i, i.e., by y_i/m_i. In symbols, $\hat{\theta}_S(x_i) = y_i/m_i$. In the current example, these estimates can be found in Table 8.1. Let $\hat{\theta}_M(x_i)$ denote the estimate of $\theta(x_i)$ obtained from the logistic regression model. In the current example, these estimates can be found in Table 8.2. Let $\hat{y}_1$ denote the predicted value of y_i obtained from the

logistic regression model then $\hat{y}_i = m_i \hat{\theta}_M(x_i)$ or $\hat{\theta}_M(x_i) = \frac{\hat{y}_i}{m_i}$.

Recall that the log-likelihood function is given by

$$\log(L) = \sum_{i=1}^{n} \left[y_i \log(\theta(x_i)) + (m_i - y_i) \log(1 - \theta(x_i)) + \log\binom{m_i}{y_i} \right]$$

Thus, the deviance is given by

$$G^2 = 2\left[\log(L_S) - \log(L_M)\right]$$

$$= 2\sum_{i=1}^{n}\left[y_i \log\left(\frac{y_i}{m_i}\right) + (m_i - y_i)\log\left(1 - \frac{y_i}{m_i}\right)\right]$$

$$-2\sum_{i=1}^{n}\left[y_i \log\left(\frac{\hat{y}_i}{m_i}\right) + (m_i - y_i)\log\left(1 - \frac{\hat{y}_i}{m_i}\right)\right]$$

$$= 2\sum_{i=1}^{n}\left[y_i \log\left(\frac{y_i}{\hat{y}_i}\right) + (m_i - y_i)\log\left(\frac{m_i - y_i}{m_i - \hat{y}_i}\right)\right]$$

When each m_i, the number of trials at x_i, is *large enough* the deviance can be used to as a goodness-of-fit test for the logistic regression model as we explain next.

We wish to test

H_0: logistic regression model (8.1) is appropriate

against

H_A: logistic model is inappropriate so a saturated model is needed

Under the null hypothesis and when each m_i is *large enough*, the deviance G^2 is approximately distributed as χ^2_{n-p-1}, where n = the number of binomial samples, p = the of predictors in the model (i.e., $p + 1$ = number of parameters estimated). In this case, $n = 14$, $p = 1$, and so we have 12 df. In R, the deviance associated with model (8.1) is referred to as the `Residual deviance` while the null deviance is based on model (8.1) with β_1 set to zero.

Logistic regression output from R

```
    Null deviance: 61.427 on 13 degrees of freedom
Residual deviance: 11.368 on 12 degrees of freedom
```

So that the *p*-value is

$$P(G^2 > 11.368) = 0.498$$

Thus, we are unable to reject H_0. In other words, the deviance goodness-of-fit test finds that the logistic regression model (8.1) is an adequate fit overall for the Michelin guide data.

8.1.4 Using Differences in Deviance Values to Compare Models

The difference in deviance can be used to compare nested models. For example, we can compare the null and residual deviances to test

$$H_0 : \theta(x) = \frac{1}{1 + \exp(-\beta_0)} \quad (\text{i.e.}, \beta_1 = 0)$$

against

$$H_A : \theta(x) = \frac{1}{1 + \exp(-\{\beta_0 + \beta_1 x\})} (\text{i.e.}, \beta_1 \neq 0)$$

The difference in these two deviances is given by

$$G_{H_0}^2 - G_{H_A}^2 = 61.427 - 11.368 = 50.059$$

This difference is to be compared to χ^2 a distribution with $\mathrm{df}_{H_0} - \mathrm{df}_{H_A} = 13 - 12 = 1$ degree of freedom. The resulting p-value is given by

$$P(G_{H_0}^2 - G_{H_A}^2 > 50.059) = 1.49\mathrm{e}\text{-}12$$

Earlier, we found that the corresponding p-value based on the Wald test equals 1.08e-08. We shall see that Wald tests and tests based on the difference in deviances can result in quite different p-values.

8.1.5 R^2 for Logistic Regression

Recall that for linear regression

$$R^2 = 1 - \frac{\mathrm{RSS}}{\mathrm{SST}}.$$

Since the deviance, $G^2 = 2\left[\log(L_S) - \log(L_M)\right]$ in logistic regression is a generalization of the residual sum of squares in linear regression, one version of R^2 for logistic regression model is given by

$$R_{\mathrm{dev}}^2 = 1 - \frac{G_{H_A}^2}{G_{H_0}^2}$$

For the single predictor logistic regression model (8.1) for the Michelin guide data,

$$R_{\mathrm{dev}}^2 = 1 - \frac{11.368}{61.427} = 0.815 \cdot$$

There are other ways to define R^2 for logistic regression. Menard (2000) provides a review and critique of these, and ultimately recommends R_{dev}^2.

Pearson goodness-of-fit statistic

An alternative measure of the goodness-of-fit of a logistic regression model is the Pearson X^2 statistic which is given by

$$X^2 = \sum_{i=1}^{n} \frac{\left(y_i/m_i - \hat{\theta}(x_i)\right)^2}{\hat{\text{Var}}(y_i/m_i)} = \sum_{i=1}^{n} \frac{\left(y_i/m_i - \hat{\theta}(x_i)\right)^2}{\hat{\theta}(x_i)\left(1 - \hat{\theta}(x_i)\right)/m_i}$$

The degrees of freedom associated with this statistic are the same as those associated with the deviance, namely,

$$\text{Degrees of freedom} = n - p - 1,$$

where n = the number of binomial samples, p = the number of predictors in the model (i.e., $p + 1$ = number of parameters estimated). In this case, $n = 14$, $p = 1$, and so we have 12 df. The Pearson X^2 statistic is also approximately distributed as X^2_{n-p-1}, when each m_i is *large enough*. In this situation, the Pearson X^2 statistic and the deviance G^2 generally produce similar values, as they do in the current example.

Logistic regression output from R

```
Pearson's X^2 = 11.999
```

We next look at diagnostic procedures for logistic regression. We begin by considering the concept of residuals in logistic regression.

8.1.6 Residuals for Logistic Regression

There are at least three types of residuals for logistic regression, namely,

- Response residuals
- Pearson residuals and standardized Pearson residuals
- Deviance residuals and standardized deviance residuals

Response residuals are defined as the response minus the fitted values, that is,

$$r_{\text{response},i} = y_i/m_i - \hat{\theta}(x_i)$$

where $\hat{\theta}(x_i)$ is the ith fitted value from the logistic regression model. However, since the variance of y_i/m_i is not constant, response residuals can be difficult to interpret in practice.

The problem of nonconstant variance of y_i/m_i is overcome by *Pearson residuals*, which are defined to be

$$r_{\text{Pearson},i} = \frac{\left(y_i/m_i - \hat{\theta}(x_i)\right)}{\sqrt{\hat{V}ar\left(y_i/m_i\right)}} = \frac{\left(y_i/m_i - \hat{\theta}(x_i)\right)}{\sqrt{\hat{\theta}(x_i)\left(1 - \hat{\theta}(x_i)\right)/m_i}}.$$

Notice that

$$\sum_i^n r_{\text{Pearson},i}^2 = \sum_{i=1}^n \frac{\left(y_i/m_i - \hat{\theta}(x_i)\right)^2}{\hat{\theta}(x_i)\left(1 - \hat{\theta}(x_i)\right)/m_i} = X^2.$$

This is commonly cited as the reason for the name Pearson residuals.

Pearson residuals do not account for the variance of $\hat{\theta}(x_i)$. This issue is overcome by *standardized Pearson residuals*, which are defined to be

$$\begin{aligned}
sr_{\text{Pearson},i} &= \frac{\left(y_i/m_i - \hat{\theta}(x_i)\right)}{\sqrt{\hat{V}ar\left(y_i/m_i - \hat{\theta}(x_i)\right)}} \\
&= \frac{\left(y_i/m_i - \hat{\theta}(x_i)\right)}{\sqrt{(1 - h_{ii})\hat{\theta}(x_i)\left(1 - \hat{\theta}(x_i)\right)/m_i}} = \frac{r_{\text{Pearson},i}}{\sqrt{(1 - h_{ii})}}
\end{aligned}$$

where h_{ii} is the ith diagonal element of the hat matrix obtained from the weighted least squares approximation to the MLE.

Deviance residuals are defined in an analogous manner to Pearson residuals with the Pearson goodness-of-fit statistic replaced by the deviance G^2, that is,

$$\sum_i^n r_{\text{Deviance},i}^2 = G^2$$

Thus, *deviance residuals* are defined by

$$r_{\text{Deviance},i} = \text{sign}\left(y_i/m_i - \hat{\theta}(x_i)\right)g_i$$

where $G^2 = \sum_{i=1}^n g_i^2$. Furthermore, *standardized deviance residuals* are defined to be

$$sr_{\text{Deviance},i} = \left. r_{\text{Deviance},i} \middle/ \sqrt{1 - h_{ii}} \right.$$

Table 8.3 gives the values of the response residuals, Pearson residuals and the deviance residuals for the Michelin guide data in Table 8.1. The Pearson residuals and deviance residuals are quite similar, since most of the m_i are somewhat larger than 1. Figure 8.3 shows plots of standardized Pearson and deviance residuals against Food Rating. Both plots produce very similar nonrandom patterns. Thus, model (8.1) is a valid model.

Table 8.3 Three types of residuals for the Michelin guide data in Table 8.1

Food rating, x_i	Response, y_i/m_i	$\hat{\theta}(x_i)$	Response residuals	Pearson residuals	Deviance residuals
15	0.000	0.025	−0.035	−0.190	−0.266
16	0.000	0.042	−0.056	−0.244	−0.340
17	0.000	0.069	−0.089	−0.886	−1.224
18	0.133	0.111	−0.006	−0.069	−0.070
19	0.176	0.175	0.067	0.693	0.670
20	0.229	0.265	−0.064	−0.798	−0.815
21	0.519	0.38	0.155	1.602	1.589
22	0.250	0.509	−0.213	−1.482	−1.485
23	0.667	0.638	0.001	0.012	0.012
24	0.857	0.75	0.091	0.567	0.599
25	0.909	0.836	0.073	0.693	0.749
26	0.500	0.896	−0.399	−1.878	−1.426
27	0.857	0.936	−0.079	−0.862	−0.748
28	1.000	0.961	0.039	0.405	0.567
				$X^2 = 11.999$	$G^2 = 11.368$

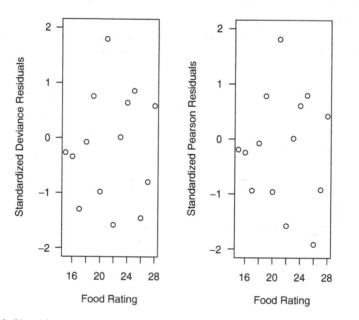

Figure 8.3 Plots of standardized residuals against Food Rating

According to Simonoff (2003, p. 133):

> The Pearson residuals are probably the most commonly used residuals, but the deviance residuals (or standardized deviance residuals) are actually preferred, since their distribution is closer to that of least squares residuals.

8.2 Binary Logistic Regression

A very important special case of logistic regression occurs when all the m_i equal 1. Such data are called binary data. As we shall see below, in this situation the goodness-of-fit measures X^2 and G^2 are problematic and plots of residuals can be difficult to interpret. To illustrate these points we shall reconsider the Michelin guide example, this time using the data in its binary form.

Example: Michelin and Zagat guides to New York City restaurants (cont.)

We again consider the 164 French restaurants included in the *Zagat Survey 2006: New York City Restaurants*. This time we shall consider each restaurant separately and classify each one according to whether they were included in the in the *Michelin Guide New York City*. As such we define the following binary response variable:

$y_i = 1$ if the restuarant is included in the Michelin guide
$y_i = 0$ if the restuarant is NOT included in the Michelin guide

We shall consider the following potential predictor variables:

x_1= Food = customer rating of the food (out of 30)
x_2= Décor = customer rating of the decor (out of 30)
x_3= Service = customer rating of the service (out of 30)
x_4= Price = the price (in $US) of dinner (including one drink and a tip)

The data can be found on the book web site in the file MichelinNY.csv. The first six rows of the data are given in Table 8.4.

Table 8.4 Partial listing of the Michelin Guide data with a binary response

InMichelin, y_i	Restaurant name	Food	Decor	Service	Price
0	14 Wall Street	19	20	19	50
0	212	17	17	16	43
0	26 Seats	23	17	21	35
1	44	19	23	16	52
0	A	23	12	19	24
0	A.O.C.	18	17	17	36

Let $\theta(x_1)$ denote the probability that a French restaurant with Zagat food rating x_1 is included in the Michelin guide. We shall first consider the logistic regression model with the single predictor x_1 given by (8.1). In this case the response variable, y_i is binary (i.e., takes values 0 or 1) and so each m_i equals 1.

Figure 8.4 shows a plot of y_i against x_1, food rating. The points in this figure have been jittered in both the vertical and horizontal directions to avoid over plotting. It is evident from Figure 8.4 that the proportion of y_i equalling one increase as Food Rating increases.

Figure 8.5 shows separate box plots of Food Rating for French restaurants included in the Michelin Guide and those that are not. It is clear from Figure 8.5 that the distribution of food ratings for French restaurants included in the Michelin Guide has a larger mean than the distribution of food ratings for French restaurants not included in the Michelin Guide. On the other hand the variability in food ratings is similar in the two groups. Later we see that comparisons of means and variances of predictor variables across the two values of the binary outcome variable is an important step in model building.

Given below is the output from R for model (8.1) using the binary data in Table 8.4.

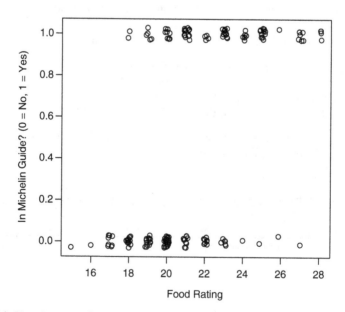

Figure 8.4 Plot of y_i versus food rating

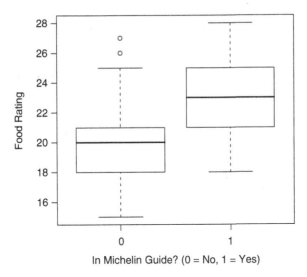

Figure 8.5 Box plots of Food Ratings

Logistic regression output from R

```
Call:
glm(formula = y ~ Food, family = binomial(), data = MichelinNY)
Coefficients:
             Estimate   Std. Error   z value   Pr(>|z|)
(Intercept)  -10.84154     1.86234    -5.821   5.83e-09 ***
Food           0.50124     0.08767     5.717   1.08e-08 ***
---
(Dispersion parameter for binomial family taken to be 1)

    Null deviance: 225.79 on 163 degrees of freedom
Residual deviance: 175.73 on 162 degrees of freedom
AIC: 179.73

Number of Fisher Scoring iterations: 4
```

For comparison purposes, given below is the output from R for model (8.1) using the cross-tabulated data in Table 8.1.

```
Call:
glm(formula = cbind(InMichelin, NotInMichelin) ~ Food, family =
binomial)

Coefficients:
             Estimate   Std. Error   z value   Pr(>|z|)
(Intercept)  -10.84154     1.86236    -5.821   5.84e-09 ***
Food           0.50124     0.08768     5.717   1.08e-08 ***
---

(Dispersion parameter for binomial family taken to be 1)
```

```
     Null deviance: 61.427 on 13 degrees of freedom
 Residual deviance: 11.368 on 12 degrees of freedom
 AIC: 41.491
```

Notice that while the *model coefficients* (and standard errors etc.,) *are the same*, the *deviance and AIC values differ* in the two sets of output. Why? We consider this issue next.

8.2.1 Deviance for the Case of Binary Data

For binary data all the m_i are equal to one. Thus, the saturated model, S estimates $\theta(x_i)$ by the observed proportion of "successes" at x_i, i.e., by y_i. In symbols $\hat{\theta}_S(x_i) = y_i$. Let $\hat{\theta}_M(x_i)$ denote the estimate of $\theta(x_i)$ obtained from the logistic regression model. Let $\hat{y}_i$ denote the predicted value of y_i obtained from the logistic regression model then $\hat{y}_i = \hat{\theta}_M(x_i)$. Since $m_i = 1$ the log-likelihood function is given by

$$\log(L) = \sum_{i=1}^{n}\left[y_i \log\left(\theta(x_i)\right) + \left(1 - y_i\right)\log\left(1 - \theta(x_i)\right) + \log\left(\tbinom{1}{y_i}\right)\right]$$

Thus, the deviance is given by

$$
\begin{aligned}
G^2 &= 2\left[\log\left(L_S\right) - \log\left(L_M\right)\right]\\
&= 2\sum_{i=1}^{n}\left[y_i \log\left(y_i\right) + \left(1 - y_i\right)\log\left(1 - y_i\right)\right]\\
&\quad -2\sum_{i=1}^{n}\left[y_i \log\left(\hat{y}_i\right) + \left(1 - y_i\right)\log\left(1 - \hat{y}_i\right)\right]\\
&= -2\sum_{i=1}^{n}\left[y_i \log\left(\hat{y}_i\right) + \left(1 - y_i\right)\log\left(1 - \hat{y}_i\right)\right]
\end{aligned}
$$

since using L'Hopital's rule with $f(y) = -\log(y)$ and $g(y) = 1/y$

$$\lim_{y\to 0} f(y) = \infty, \lim_{y\to 0} g(y) = \infty$$

so that

$$\lim_{y\to 0}\left(-y\log(y)\right) = \lim_{y\to 0}\frac{f(y)}{g(y)} = \lim_{y\to 0}\frac{f'(y)}{g'(y)} = \lim_{y\to 0}\frac{-y^{-1}}{-y^{-2}} = \lim_{y\to 0} y = 0$$

Notice that the two terms in $\log(L_S)$ above are zero for each i, thus the deviance only depends on $\log(L_M)$. As such *the deviance does not provide an assessment of the* *goodness-of-fit of model (M) when all the m_i are equal to one.* Furthermore, the distribution of the deviance is not χ^2, even in any approximate sense.

However, *even when all the m_i are equal to one, the distribution of the difference in deviances is approximately χ^2.*

8.2.2 Residuals for Binary Data

Figure 8.6 shows plots of standardized Pearson residuals and standardized deviance residuals against the predictor variable, Food Rating for model (8.1) based on the binary data in Table 8.4.

Both plots in Figure 8.6 produce very similar highly nonrandom patterns. In each plot the standardized residuals fall on two smooth curves, the one for which all the standardized residuals are positive corresponds to the cases for which y_i equals one, while the one for which all the standardized residuals are negative corresponds to the cases for which y_i equals zero. Such a phenomenon can exist irrespective of whether the fitted model is valid or not. *In summary, residual plots are problematic when the data are binary.* Thus, we need to find another method other than residual plots to check the validity of logistic regression models based on binary data.

In the current example with just one predictor we can aggregate the binary data in Table 8.4 across values of the food rating to produce the data in Table 8.1. Most of the values of m_i are somewhat greater than 1 and so in this situation, residual plots are interpretable in the usual manner. Unfortunately, however, aggregating binary data does not work well when there are a number of predictor variables.

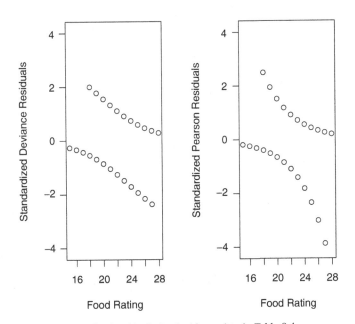

Figure 8.6 Plots of standardized residuals for the binary data in Table 8.4

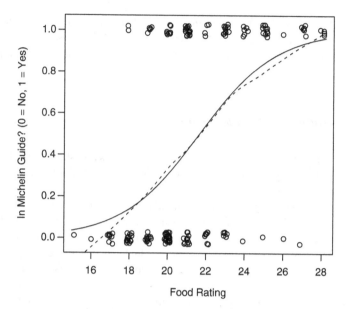

Figure 8.7 Plot of y_i versus food rating with the logistic and loess fits added

Figure 8.7 shows a plot of y_i against x_1, Food Rating. The points in this figure have been jittered in both the vertical and horizontal directions to avoid over-plotting. Figure 8.7 also includes the logistic fit for model (8.1) and as a solid curve and the loess fit (with $\alpha = 2/3$). The two fits agree reasonably (except possibly at the bottom) indicating that model (8.1) is an adequate model for the data.

We shall return to model checking plots with nonparametric fits later. In the meantime, we shall discuss transforming predictor variables.

8.2.3 Transforming Predictors in Logistic Regression for Binary Data

In this section we consider the circumstances under which the logistic regression model is appropriate for binary data and when it is necessary to transform predictor variables. The material in this section is based on Kay and Little (1987) and Cook and Weisberg (1999b, pp. 499–501).

Suppose that Y is a binary random variable (i.e., takes values 0 and 1) and that X is a single predictor variable. Then

$$
\begin{aligned}
\theta(x) &= E(Y \mid X = x) \\
&= 1 \times P(Y = 1 \mid X = x) + 0 \times P(Y = 0 \mid X = x) \\
&= P(Y = 1 \mid X = x)
\end{aligned}
$$

First suppose that X is a discrete random variable (e.g., a dummy variable), then

$$\frac{\theta(x)}{1-\theta(x)} = \frac{P(Y=1\,|\,X=x)}{P(Y=0\,|\,X=x)}$$

$$= \frac{P(Y=1 \cap X=x)}{P(Y=0 \cap X=x)}$$

$$= \frac{P(X=x\,|\,Y=1)P(Y=1)}{P(X=x\,|\,Y=0)P(Y=0)}$$

Taking logs of both sides of this last equation gives

$$\log\left(\frac{\theta(x)}{1-\theta(x)}\right) = \log\left(\frac{P(Y=1)}{P(Y=0)}\right) + \log\left(\frac{P(X=x\,|\,Y=1)}{P(X=x\,|\,Y=0)}\right)$$

when X is a discrete random variable. Similarly when X is a continuous random variable, it can be shown that

$$\log\left(\frac{\theta(x)}{1-\theta(x)}\right) = \log\left(\frac{P(Y=1)}{P(Y=0)}\right) + \log\left(\frac{f(x\,|\,Y=1)}{f(x\,|\,Y=0)}\right)$$

where $f(x|Y = j)$, $j = 0,1$, is the conditional density function of the predictor given the value of the response.

Thus, the log odds equal the sum of two terms, the first of which does not depend on X and thus can be ignored when discussing transformations of X. We next look at the second term for a specific density.

Suppose that $f(x|Y = j)$, $j = 0,1$, is a normal density, with mean μ_j and variance σ_j^2, $j = 0,1$. Then

$$f(x\,|\,y=j) = \frac{1}{\sigma_j\sqrt{2\pi}}\exp\left\{-\frac{(x-\mu_j)^2}{2\sigma_j^2}\right\}, \quad j=0,1$$

So that,

$$\log\left(\frac{f(x\,|\,Y=1)}{f(x\,|\,Y=0)}\right)$$

$$= \log\left(\frac{\sigma_0}{\sigma_1}\right) + \left[\frac{-(x-\mu_1)^2}{2\sigma_1^2} + \frac{(x-\mu_0)^2}{2\sigma_0^2}\right]$$

$$= \log\left(\frac{\sigma_0}{\sigma_1}\right) + \left(\frac{\mu_0^2}{2\sigma_0^2} - \frac{\mu_1^2}{2\sigma_1^2}\right) + \left(\frac{\mu_1}{\sigma_1^2} - \frac{\mu_0}{\sigma_0^2}\right)x + \frac{1}{2}\left(\frac{1}{\sigma_0^2} - \frac{1}{\sigma_1^2}\right)x^2$$

Thus,

$$\log\left(\frac{\theta(x)}{1-\theta(x)}\right) = \log\left(\frac{P(Y=1)}{P(Y=0)}\right) + \log\left(\frac{f(x|Y=1)}{f(x|Y=0)}\right)$$

$$= \beta_0 + \beta_1 x + \beta_2 x^2$$

where

$$\beta_0 = \log\left(\frac{P(Y=1)}{P(Y=0)}\right) + \log\left(\frac{\sigma_0}{\sigma_1}\right) + \left(\frac{\mu_0^2}{2\sigma_0^2} - \frac{\mu_1^2}{2\sigma_1^2}\right),$$

$$\beta_1 = \left(\frac{\mu_1}{\sigma_1^2} - \frac{\mu_0}{\sigma_0^2}\right), \beta_2 = \frac{1}{2}\left(\frac{1}{\sigma_0^2} - \frac{1}{\sigma_1^2}\right)$$

Thus, *when the predictor variable X is normally distributed with a different vari-*
ance for the two values of Y, the log odds are a quadratic function of x.
   When $\sigma_1^2 = \sigma_0^2 = \sigma^2$, the log odds simplifies to

$$\log\left(\frac{\theta(x)}{1-\theta(x)}\right) = \beta_0 + \beta_1 x$$

where

$$\beta_1 = \left(\frac{\mu_1 - \mu_0}{\sigma^2}\right)$$

Thus, *when the predictor variable X is normally distributed with the same vari-*
ance for the two values of Y, the log odds are a linear function of x, with the slope,
β_1 *equal to the difference in the mean of X across the two groups divided by the*
common variance of X in each group.
   The last result can be extended to the case where we have *p predictor variables*
which have multivariate normal conditional distributions. If the variance–covari-
ance matrix of the predictors differs across the two groups then the log odds are a
function of x_i, x_i^2 *and* $x_i x_j$ *(i, j = 1,..., p; i ≠ j)*
   If the densities $f(x|Y=j)$, $j = 0,1$ are skewed the log odds can depend on both x
and $\log(x)$. It does, for example, for the gamma distribution. Cook and Weisberg
(1999b, p. 501) give the following advice:

> When conducting a binary regression with a skewed predictor, it is often easiest to assess the
> need for x and $\log(x)$ by including them both in the model so that their relative contributions
> can be assessed directly.

Alternatively, if the skewed predictor can be transformed to have a normal distribution conditional on Y, then just the transformed version of X should be included
in the logistic regression model.

Next, suppose that the conditional distribution of X is Poisson with mean λ_j. Then

$$P(X = x \mid y = j) = \frac{e^{-\lambda_j} \lambda_j^x}{x!}, \quad j = 0,1$$

So that,

$$\log\left(\frac{P(X = x \mid Y = 1)}{P(X = x \mid Y = 0)}\right) = x\log\left(\frac{\lambda_1}{\lambda_0}\right) + (\lambda_0 - \lambda_1)$$

Thus,

$$\log\left(\frac{\theta(x)}{1 - \theta(x)}\right) = \log\left(\frac{P(Y = 1)}{P(Y = 0)}\right) + \log\left(\frac{P(X = x \mid Y = 1)}{P(X = x \mid Y = 0)}\right) = \beta_0 + \beta_1 x$$

where

$$\beta_0 = \log\left(\frac{P(Y = 1)}{P(Y = 0)}\right) + (\lambda_0 - \lambda_1), \beta_1 = \log\left(\frac{\lambda_1}{\lambda_0}\right)$$

Thus, *when the predictor variable X has a Poisson distribution, the log odds are a linear function of x. When X is a dummy variable, it can be shown that the log odds are also a linear function of x.*

Figure 8.8 shows separate box plots of each of the four potential predictors, namely, Food Rating, Décor Rating, Service Rating and Price for French restaurants included in the Michelin Guide and those that are not. It is evident from Figure 8.8 that while the distributions of the first three predictors are reasonably symmetric the distribution of Price is quite skewed. Thus, we shall include both Price and log(Price) as potential predictors in our logistic regression model.

Examining Figure 8.8 further, we see that for each predictor the distribution of results for French restaurants included in the Michelin Guide has a larger mean than the distribution of results for French restaurants not included in the Michelin Guide.

Let $\theta(\mathbf{x}) = \theta(x_1, x_2, x_3, x_4, \log(x_4))$ denote the probability that a French restaurant with the following predictor variables:

x_1 = Food rating, x_2 = Décor rating, x_3 = Service rating, x_4 = Price, $\log(x_4)$ = log(Price). We next consider the following logistic regression model with these four predictor variables:

$$\theta(\mathbf{x}) = \frac{1}{1 + \exp\left(-\{\beta_0 + \beta_1 x_1 + \beta_2 x_2 + \beta_3 x_3 + \beta_4 x_4 + \beta_5 \log(x_4)\}\right)} \tag{8.2}$$

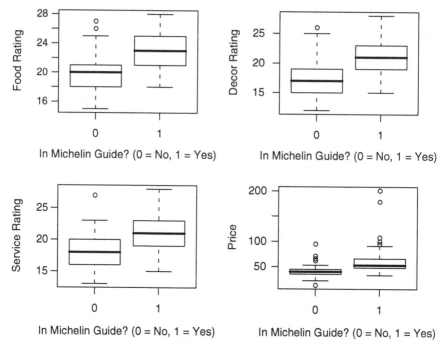

Figure 8.8   Box plots of the four predictor variables

Given that residual plots are difficult to interpret for binary data, we shall examine marginal model plots instead.

8.2.4   Marginal Model Plots for Binary Data

Consider the situation when there are just two predictors x_1 and x_2. We wish to visually assess whether

$$\theta(\mathbf{x}) = \frac{1}{1+\exp\left(-\left\{\beta_0 + \beta_1 x_1 + \beta_2 x_2\right\}\right)} \tag{M1}$$

models $\theta(\mathbf{x}) = E(Y \mid \mathbf{X} = \mathbf{x}) = P(Y = 1 \mid \mathbf{X} = \mathbf{x})$ adequately. Again we wish to compare the fit from (M1) with a fit from a nonparametric regression model (F1) where

$$\theta(\mathbf{x}) = f(x_1, x_2) \tag{F1}$$

Under model (F1), we can estimate $E_{F_1}(Y \mid x_1)$ by adding a nonparametric fit to the plot of Y against x_1. We want to check that the estimate of $E_{F_1}(Y \mid x_1)$ is close to the estimate of $E_{M_1}(Y \mid x_1)$.

Under model (M1), Cook and Weisberg (1997) utilized the following result:

$$E_{M_1}(Y \mid x_1) = E\left[E_{M_1}(Y \mid x) \mid x_1\right] \tag{8.3}$$

The result follows from the well-known general result re conditional expectations.

Under model (M1), we can estimate

$$E_{M_1}(Y \mid x) = \theta(\mathbf{x}) = \frac{1}{1 + \exp\left(-\{\beta_0 + \beta_1 x_1 + \beta_2 x_2\}\right)}$$

by the fitted values

$$\hat{Y} = \hat{\theta}(\mathbf{x}) = \frac{1}{1 + \exp\left(-\{\hat{\beta}_0 + \hat{\beta}_1 x_1 + \hat{\beta}_2 x_2\}\right)}.$$

Utilizing (8.3) we can therefore estimate

$$E_{M_1}(Y \mid x_1) = E\left[E_{M_1}(Y \mid x) \mid x_1\right]$$

by estimating $E\left[E_{M_1}(Y \mid x) \mid x_1\right]$ with an estimate of $E\left[\hat{Y} \mid x_1\right]$.

In summary, we wish to compare estimates under models (F1) and (M1) by comparing nonparametric estimates of $E(Y \mid x_1)$ and $E\left[\hat{Y} \mid x_1\right]$. If the two nonparametric estimates agree then we conclude that x_1 is modelled correctly by model (M1). If **not** then we conclude that x_1 is **not** modelled correctly by model (M1).

The left-hand plot in Figure 8.9 is a plot of Y and against x_1, Food Rating with the loess estimate of $E(Y \mid x_1)$ included. The right-hand plot in Figure 8.9 is a plot of $\hat{Y}$ from model (8.2) against x_1, Food Rating with the loess estimate of $E\left[\hat{Y} \mid x_1\right]$ included.

In general, it is difficult to compare curves in different plots. Thus, following Cook and Weisberg (1997) we shall from this point on include both nonparametric curves on the plot of Y against x_1. The plot of Y against x_1 with the loess fit for $\hat{Y}$ against x_1 and the loess fit for Y against x_1 both marked on it is called a **marginal model plot** for Y and x_1.

Figure 8.10 contains **marginal model plots** for Y and each predictor in model (8.2). The solid curve is the loess estimate of $E(Y|\text{Predictor})$ while the dashed curve is the loess estimate of $E[\hat{Y}|\text{Predictor}]$ where the fitted values are from model (8.2). The bottom right-hand plot uses these fitted values, that is,

$$\hat{\beta}_0 + \hat{\beta}_1 x_1 + \hat{\beta}_2 x_2 + \hat{\beta}_3 x_3 + \hat{\beta}_4 x_4 + \hat{\beta}_5 \log(x_4)$$

as the horizontal axis.

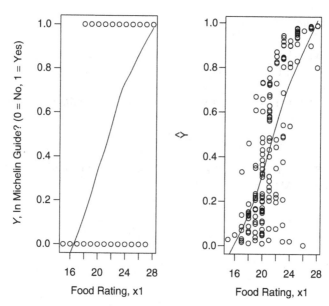

Figure 8.9 Plots of Y and against $\hat{Y}$ x_1, Food Rating

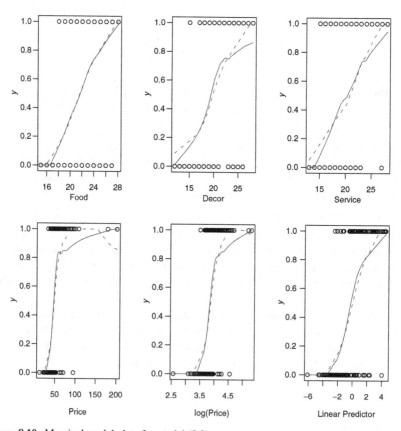

Figure 8.10 Marginal model plots for model (8.2)

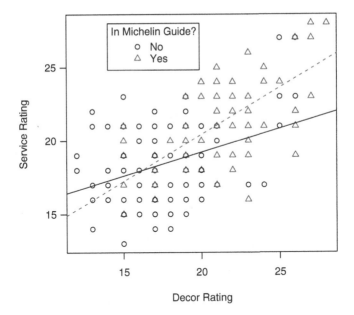

Figure 8.11 Plots of Décor and Service ratings with different slopes for each value of y

There is reasonable agreement between the two fits in each of the marginal model plots in Figure 8.10 except for the plots involving Décor and Service and to a lesser extent Price. At this point, one possible approach is to consider adding extra predictor terms involving Décor and Service to model (8.2).

Recall that when we have *p predictor variables which have multivariate normal conditional distributions, if the variance–covariance matrix of the predictors differs across the two groups then the log odds are a function of x_i, x_i^2 and $x_i x_j$ (i,j = 1,..., p; i ≠ j).* A quadratic term in x_i is needed as a predictor if the variance of x_i differs across the two values of y. The product term $x_i x_j$ is needed as a predictor if the covariance of x_i and x_j differs across the two values of y (i.e., if the regression of x_i on x_j (or vice versa) has a different slope for the two values of y.)

Next we investigate the covariances between the predictors Décor and Service. Figure 8.11 contains a plot of Décor and Service with different estimated slopes for each value of y. It is evident from Figure 8.11 that the slopes in this plot differ. In view of this we shall expand model (8.2) to include a two-way interaction terms between x_2 = Décor rating and x_3 = Service rating. Thus we shall consider the following model:

$$\theta(\mathbf{x}) = \frac{1}{1 + \exp\left(-\{\beta'\mathbf{x}\}\right)} \tag{8.4}$$

where $\mathbf{x}' = \left(x_1, x_2, x_3, x_4, \log(x_4), x_2 x_3\right)'$ and $\beta' = \left(\beta_1, \beta_2, \beta_3, \beta_4, \beta_5, \beta_6\right)'$.

Figure 8.12 contains **marginal model plots** for Y and the first five predictors in model (8.4). The solid curve is the loess estimate of E(Y|predictor) while the

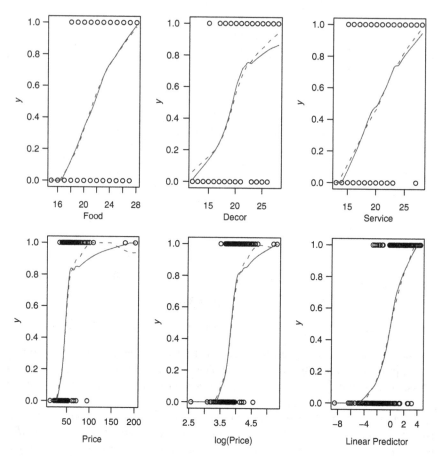

Figure 8.12 Marginal model plots for model (8.4)

dashed curve is the loess estimate of $E[\hat{Y}|\text{predictor}]$. The bottom right-hand plot uses $\hat{\beta}'\mathbf{x}$ as the horizontal axis.

Comparing the plots in Figure 8.12 with those in Figure 8.10, we see that there is better agreement between the two sets of fits in Figure. 8.12, especially for the variables, Décor and Service. There is still somewhat of an issue with the marginal model plot for Price, especially at high values.

Regression output from R

```
Analysis of Deviance Table
Model 1:  y ~ Food + Decor +  Service + Price + log(Price)
Model 2:  y ~ Food + Decor +  Service + Price + log(Price)
                    + Service:Decor
    Resid. Df    Resid. Dev     Df    Deviance    P(>|Chi|)
1        158       136.431
2        157       129.820      1       6.611        0.010
```

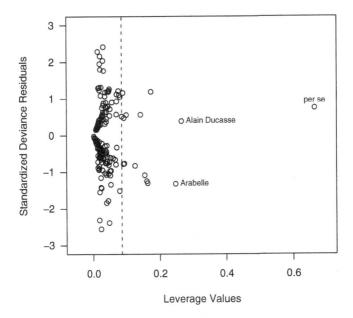

Figure 8.13 A plot of leverage against standardized deviance residuals for (8.4)

Recall that the difference in deviance can be used to compare nested models. For example, we can compare models (8.2) and (8.4) in this way. The output above from R shows that the addition of the interaction term for Décor and Service has significantly reduced the deviance (p-value = 0.010).

We next examine leverage values and standardized deviance residuals for model (8.4) (see Figure 8.13). The leverage values are obtained from the weighted least squares approximation to the maximum likelihood estimates. According to Pregibon (1981, p. 173) the average leverage is equal to $(p + 1)/n = 7/164 = 0.0427$. We shall use the usual cut-off of twice the average, which in this case equals 0.085. The three points with the largest leverage values evident in Figure 8.13 correspond to the restaurants Arabelle, Alain Ducasse and per se. The price of dinner at each of these restaurants is $71, $179 and $201, respectively. Looking back at the box plots of Price in Figure 8.8 we see that these last two values are the highest values of Price. Thus, for at least two of these points their high leverage values are mainly due to their extreme values of Price.

We next look at the output from R for model (8.4).

Output from R

```
Call:
glm(formula = y ~ Food + Decor + Service + Price + log(Price) +
Service:Decor, family = binomial(), data = MichelinNY)
```

```
Coefficients:
                Estimate   Std. Error   z value   Pr(>|z|)
(Intercept)    -70.85308     15.45786    -4.584   4.57e-06  ***
Food             0.66996      0.18276     3.666   0.000247  ***
Decor            1.29788      0.49299     2.633   0.008471  **
Service          0.91971      0.48829     1.884   0.059632   .
Price           -0.07456      0.04416    -1.688   0.091347   .
log(Price)      10.96400      3.22845     3.396   0.000684  ***
Decor:Service   -0.06551      0.02512    -2.608   0.009119  **
---
(Dispersion parameter for binomial family taken to be 1)

    Null deviance: 225.79  on 163  degrees of freedom
Residual deviance: 129.82  on 157  degrees of freedom
AIC: 143.82
Number of Fisher Scoring iterations: 6
```

Given that the variable Price is only marginally statistically significant (Wald p-value = 0.091), we shall momentarily remove it from the model. Thus, we shall consider the following model:

$$\theta(\mathbf{x}) = \frac{1}{1 + \exp\left(-\{\beta'\mathbf{x}\}\right)} \tag{8.5}$$

where $\mathbf{x}' = (x_1, x_2, x_3, \log(x_4), x_2 x_3)'$, $\beta' = (\beta_1, \beta_2, \beta_3, \beta_5, \beta_7)'$. We next test

$$H_0 : \beta_4 = 0 \text{ (i.e., model (8.5))}$$

against

$$H_A : \beta_4 \neq 0 \text{ (i.e., model (8.4))}$$

using the difference in deviance between the two models. The output from R for this test is given next.

Output from R

```
Analysis of Deviance Table
Model 1: y ~ Food + Decor + Service + log(Price)
                + Service:Decor
Model 2: y ~ Food + Decor + Service + Price + log(Price)
                + Service:Decor
  Resid. Df   Resid. Dev   Df   Deviance   P(>|Chi|)
1       158      131.229
2       157      129.820    1      1.409       0.235
```

The p-value from the difference in deviances (p-value = 0.235) is higher than the corresponding Wald p-value for the coefficient of Price (p-value = 0.091). As foreshadowed earlier, this example illustrates that Wald tests and tests based on the difference in deviances can result in quite different p-values. Additionally, in view of the leverage problems associated with the variable Price (which may lead to under estimation of the standard error of its regression coefficient), it seems that model (8.5) is to be preferred over model (8.4). The output from R for model (8.5) is given next.

Output from R

```
Call:
glm(formula = y ~ Food + Decor + Service + log(Price)
+ Service:Decor, family = binomial(), data = MichelinNY)
Coefficients:
                Estimate    Std. Error    z value    Pr(>|z|)
(Intercept)    -63.76436      14.09848     -4.523     6.10e-06    ***
Food             0.64274       0.17825      3.606     0.000311    ***
Decor            1.50597       0.47883      3.145     0.001660    **
Service          1.12633       0.47068      2.393     0.016711    *
log(Price)       7.29827       1.81062      4.031     5.56e-05    ***
Decor:Service   -0.07613       0.02448     -3.110     0.001873    **
(Dispersion parameter for binomial family taken to be 1)

 Null deviance: 225.79 on 163 degrees of freedom
Residual deviance: 131.23 on 158 degrees of freedom
AIC: 143.23
Number of Fisher Scoring iterations: 6
```

All of the regression coefficients in model (8.5) are highly significant at the 5% level. Interestingly, the coefficients of the predictors Food, Service, Décor and log(Price) are positive implying that (all other things equal) higher Food, Service and Décor ratings and higher log(Price) in the Zagat guide increases the chance of a French restaurant being included in the Michelin Guide, as one would expect. The coefficient of the interaction term between Service and Décor is negative moderating the main effects of Service and Décor.

We next check the validity of model (8.5) using marginal model plots (see Figure 8.14). These marginal model plots show reasonable agreement across the two sets of fits indicating that (8.5) is a valid model.

As a final validity check we examine leverage values and standardized deviance residuals for model (8.5) (see Figure 8.15). We shall again use the usual cut-off of 0.073, equal to twice the average leverage value. A number of points are highlighted in Figure 8.15 that are worthy of further investigation. After that removing the variable Price, the expensive restaurants Alain Ducasse and per se are no longer points of high leverage.

Table 8.5 provides a list of the points highlighted as outliers in Figure 8.14. As one would expect, the restaurants either have a low estimated probability of being included in the Michelin Guide and are actually included (i.e., $y = 1$) or have a high estimated probability of being included in the Michelin Guide and are not included (i.e., $y = 0$). The former group of "lucky" restaurants consists of
   Gavroche, Odeon, Paradou and Park Terrace Bistro

The latter group of "unlucky" restaurants consists of
   Atelier, Café du Soleil and Terrace in the Sky.

Finally, we shall examine just one of the restaurants listed in Table 8.5, namely, Atelier. Zagat's 2006 review of Atelier (Gathje and Diuguid, 2005) reads as follows:

"Dignified" dining "for adults" is the métier at the Ritz-Carlton Central Park's "plush" New French, although the food rating is in question following the departure of chef Alain Allegretti; offering a "stately environment" where the "charming" servers "have ESP", it caters to a necessarily well-heeled clientele.

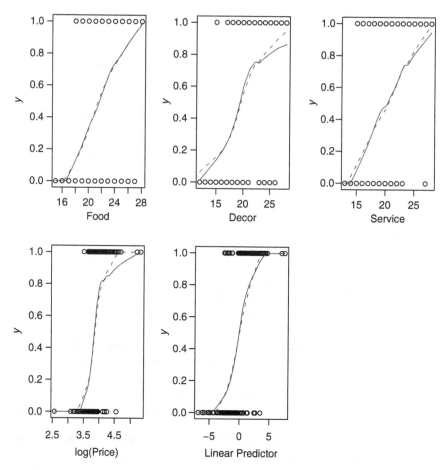

Figure 8.14 Marginal model plots for model (8.5)

One plausible explanation for the exclusion of Atelier from the Michelin Guide is that the Michelin inspectors rated Atelier after the departure of chef Alain Allegretti. Interestingly, Atelier is listed as "Closed" in the 2007 Zagat Guide.

8.3 Exercises

1. Chapter 6 of Bradbury (2007), a book on baseball, uses regression analysis to compare the success of the 30 Major League Baseball teams. For example, the author considers the relationship between x_i, market size (i.e., the population in millions of the city associated with each team) and Y_i, the number of times team i made the post-season playoffs in the $m_i=10$ seasons between 1995 and 2004.

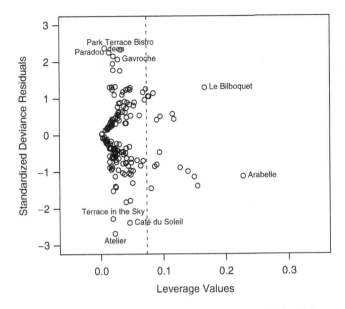

Figure 8.15 A plot of leverage against standardized deviance residuals for (8.5)

Table 8.5 "Lucky" and "unlucky" restaurants according to model (8.5)

Case	Estimated probability	y	Restaurant name	Food	Decor	Service	Price
14	0.971	0	Atelier	27	25	27	95
37	0.934	0	Café du Soleil	23	23	17	44
69	0.125	1	Gavroche	19	15	17	42
133	0.103	1	Odeon	18	17	17	42
135	0.081	1	Paradou	19	17	18	38
138	0.072	1	Park Terrace Bistro	21	20	20	33
160	0.922	0	Terrace in the Sky	23	25	21	62

The author found that "it is hard to find much correlation between market size and … success in making the playoffs. The relationship … is quite weak." The data is plotted in Figure 8.16 and it can be found on the book web site in the file playoffs.txt. The output below provides the analysis implied by the author's comments.

(a) Describe in detail two major concerns that potentially threaten the validity of the analysis implied by the author's comments.

(b) Using an analysis which is appropriate for the data, show that there is very strong evidence of a relationship between Y and x.

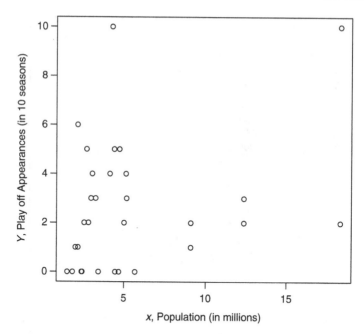

Figure 8.16 A plot of Y_i against x_i

R output for Question 1:

```
Call:
lm(formula = PlayoffAppearances ~ Population)
Coefficients:
              Estimate   Std. Error    t value     Pr(>|t|)
(Intercept)    1.7547       0.7566       2.319       0.0279 *
Population     0.1684       0.1083       1.555       0.1311
---
Signif. codes: 0 ' ' 0.001 ' ' 0.01 ' ' 0.05 '.' 0.1 ' ' 1
Residual standard error: 2.619 on 28 degrees of freedom
Multiple R-squared: 0.07952, Adjusted R-squared: 0.04664
F-statistic: 2.419 on 1 and 28 DF, p-value: 0.1311
```

2. This question is based on one of the data sets discussed in an unpublished manuscript by Powell, T. and Sheather, S. (2008) entitled "*A Theory of Extreme
   Competition*". According to Powell and Sheather:

   This paper develops a model of competitive performance when populations compete ....
   We present a theoretical framework ... and empirical tests in chess and ... national pageants. The findings show that performance in these domains is substantially predictable
   from a few observable features of population and economic geography.

   In this question we shall consider data from the Miss America pageant, which
   was founded in Atlantic City in 1921, and 81 pageants have been conducted
   through 2008. In particular we will develop a logistic regression model for the

proportion of top ten finalists for each US state for the years 2000 to 2008. According to Powell and Sheather:

Eligibility for the Miss America pageant is limited to never-married female U.S. citizens between the ages of 17 and 24. To measure population size, we obtained data for this demographic segment for each U.S. state and the District of Columbia from the 2000 U.S. Census. As a measure of participation inducements, we obtained data on the number of qualifying pageants conducted in each state, on the assumption that qualifying pageants reflect state-level infrastructure and resource commitments. As a geographic measure, we used the latitude and longitude of each state capital and Washington DC, on the assumption that state locations convey information about the regional cultural geography of beauty pageants (in particular, beauty pageants are widely believed to receive greater cultural support south of the Mason-Dixon line). To measure search efficacy, we obtained data on the total land and water area (in square miles) for each state and the District of Columbia, on the assumption that search is more difficult over larger geographic areas.

They consider the following outcome variable and potential predictor variables:

Y = Number of times each US state (and the District of Columbia) has produced a top ten finalist for the years 2000–2008

x_1 = log(population size)

x_2 = Log(average number of contestants in each state's final qualifying pageant each year between 2002 and 2007)

x_3 = Log(geographic area of each state and the District of Columbia)

x_4 = Latitude of each state capitol and

x_5 = Longitude of each state capitol, and

The data can be found on the course web site in the file. MissAmericato2008.txt.

(a) Develop a logistic regression model that predicts y from x_1, x_2, x_3, x_4 and x_5 such that each of the predictors is significant at least at the 5% level. Use marginal model plots to check the validity of the full model and the final model (if it is different from the full model).

(b) Identify any leverage points in the final model developed in (a). Decide if they are "bad" leverage points.

(c) Interpret the regression coefficients of the final model developed in (a).

3. Data on 102 male and 100 female athletes were collected at the Australian Institute of Sport. The data are available on the book web site in the file ais.txt. Develop a logistic regression model for gender ($y = 1$ corresponds to female) or ($y = 0$ corresponds to male) based on the following predictors (which is a subset of those available):

RCC, read cell count
WCC, white cell count
BMI, body mass index

(Hint: Use marginal model plots to aid model development.)

4. A number of authors have analyzed the following data on heart disease. Of key interest is the development of a model to determine whether a particular patient has heart disease (i.e., Heart Disease = 1), based on the following predictors:

x_1 = Systolic blood pressure
x_2 = A measure of cholesterol
x_3 = A dummy variable (= 1 for patients with a family history)
x_4 = A measure of obesity and
x_5 = Age.

We first consider the following logistic regression model with these five predictor variables:

$$\theta(\mathbf{x}) = \frac{1}{1 + \exp\left(-\left\{\beta_0 + \beta_1 x_1 + \beta_2 x_2 + \beta_3 x_3 + \beta_4 x_4 + \beta_5 x_5\right\}\right)} \tag{8.6}$$

where

$$\theta(x) = E(Y \mid X = x) = P(Y = 1 \mid X = x)$$

Output for model (8.6) is given below along with associated plots (Figures 8.17 and 8.18). The data (HeartDiseare, CSV) can be found on the book web site.

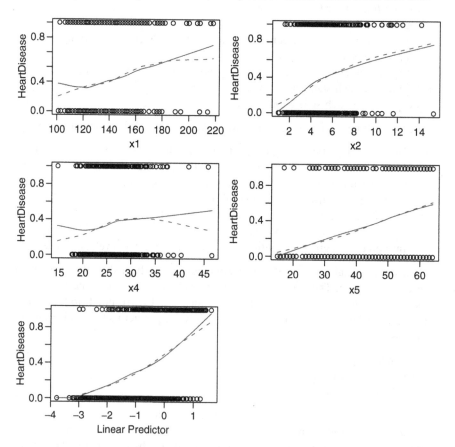

Figure 8.17   Marginal model plots for model (8.6)

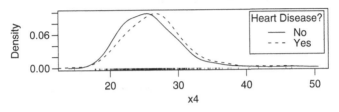

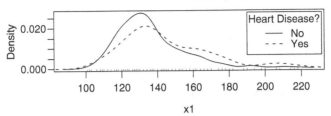

Figure 8.18 Kernel density estimates of x_1 and x_4

(a) Is model (8.6) a valid model for the data? Give reasons to support your answer.

(b) What extra predictor term or terms would you recommend be added to model (8.6) in order to improve it. Please give reasons to support each extra term.

(c) Following your advice in (b), extra predictor terms were added to model (8.6) to form model (8.7). We shall denote these extra predictors as $f_1(x_1)$ and $f_2(x_4)$ (so as not to give away the answer to (b)). Marginal model plots from model (8.7) are shown in Figure 8.19. Is model (8.7) a valid model for the data? Give reasons to support your answer.

(d) Interpret the estimated coefficient of x_3 in model (8.7).

Output from R for model (8.6)

```
Call:
glm(formula = HeartDisease ~ x1 + x2 + x3 + x4 + x5, family =
binomial(), data = HeartDisease)
Coefficients:
              Estimate    Std. Error    z value    Pr(>|z|)
(Intercept) -4.313426      0.943928     -4.570     4.89e-06    ***
x1           0.006435      0.005503      1.169     0.24223
x2           0.186163      0.056325      3.305     0.00095     ***
x3           0.903863      0.221009      4.090     4.32e-05    ***
x4          -0.035640      0.028833     -1.236     0.21643
x5           0.052780      0.009512      5.549     2.88e-08    ***
(Dispersion parameter for binomial family taken to be 1)

    Null deviance: 596.11 on 461 degrees of freedom
Residual deviance: 493.62 on 456 degrees of freedom
AIC: 505.62
Number of Fisher Scoring iterations: 4
```

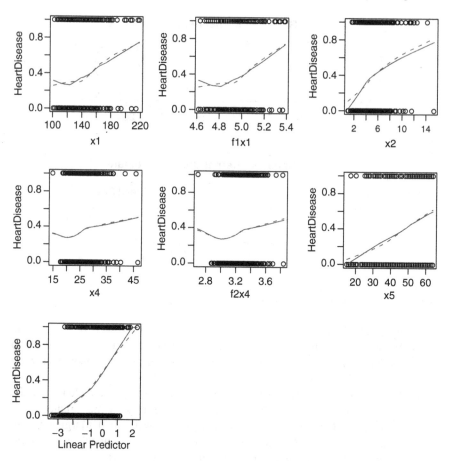

Figure 8.19   Marginal model plots for model (8.7)

Output from R for model (8.7)

```
Call:
glm(formula = HeartDisease ~ x1 + f1x1 + x2 + x3 + x4 + f2x4 +
x5, family = binomial(), data = HeartDisease)
Coefficients:
             Estimate    Std. Error   z value    Pr(>|z|)
(Intercept)  75.204768    33.830217     2.223    0.026215   *
x1            0.096894     0.052664     1.840    0.065792   .
f1x1        -13.426632     7.778559    -1.726    0.084328   .
x2            0.201285     0.057220     3.518    0.000435   ***
x3            0.941056     0.224274     4.196    2.72e-05   ***
x4            0.384608     0.208016     1.849    0.064467   .
f2x4        -11.443233     5.706058    -2.005    0.044915   *
x5            0.056111     0.009675     5.800    6.64e-09   ***
(Dispersion parameter for binomial family taken to be 1)
```

```
 Null deviance: 596.11 on 461 degrees of freedom
Residual deviance: 486.74 on 454 degrees of freedom
AIC: 502.74
Number of Fisher Scoring iterations: 4
```

5. This difficult realistic problem is based on a case study from Shmueli, Patel and Bruce (2007, pp. 262–264). The aim of the case is to develop a logistic regression model which will improve the cost effectiveness of the direct marketing campaign of a national veterans' organization. The response rate to recent marketing campaigns was such that 5.1% of those contacted made a donation to the organization. Weighted sampling of recent campaigns was used to produce a data set with 3,120 records consisting of 50% donors and 50% nondonors. The data are available after free registration at the author's book web site http:// www.dataminingbook.com/. Randomly split the data file into a training file (FundTrain.csv) and a test file (FundTest.csv) both with 1,560 records. The outcome variable is

TARGET_B which = 1 for donors and 0 otherwise.

The following predictor variables are available

HOMEOWNER = 1 for homeowners and 0 otherwise

NUMCHLD = number of children

INCOME = household income rating on a seven-point scale

GENDER = 1 for male and 0 for female

WEALTH = wealth rating on a ten-point scale (0 to 9)

(Each wealth rating has a different meaning in each state.)

HV = Average Home Value in potential donor's neighborhood

(in hundreds of dollars)

ICmed = Median Family Income in potential donor's neighborhood

(in hundreds of dollars)

ICavg = Average Family Income in potential donor's neighborhood

(in hundreds of dollars)

IC15 = % earning less than $15 K in potential donor's neighborhood

NUMPROM = Lifetime number of promotions received to date

RAMNTALL = Dollar amount of lifetime gifts to date

LASTGIFT = Dollar amount of most recent gift

TOTALMONTHS = Number of months from last donation to the last time the case was updated

TIMELAG = Number of months between first and second gift

AVGGIFT = Average dollar amount of gifts to date.

ZIP = Code for potential donor's zip code (2 = 20000 – 39999,

3 = 40000 - 59999, 4 = 60000 - 79999 & 5 = 8000 - 99999)

PART 1: Using the training data

(a) Fit a logistic regression model using each of the predictor variables except ZIP. At this stage do not transform any of the predictors.

(b) Use marginal model plots to show that the model in part (a) is not a valid model.

(c) Decide which predictor variables may benefit from being transformed and find a reasonable transformation for each of these variables.

(d) Since the wealth ratings have a different meaning within each state, create one or more predictors which represents the interaction between ZIP and WEALTH. Investigate the relationship between TARGET_B and these predictor(s).

(e) Fit a logistic regression model to the training data utilizing what you discovered in (c) and (d).

(f) Use marginal model plots to decide whether the model in part (e) is a valid model or not.

(g) Consider adding further interaction terms to your model in (e). Establish a final model for TARGET_B.

PART 2: Using the test data

(a) Use the logistic regression model you have developed in part 1 to predict whether a person will make a donation or not.

(b) Compare your predictions in part (a) with the actual results in TARGET_B. Quantify how well your model worked.

6. Dr. Hans Riedwyl, a statistician at the University of Berne was asked by local authorities to analyze data on Swiss Bank notes. In particular, the statistician was asked to develop a model to predict whether a particular banknote is counterfeit ($y = 0$) or genuine ($y = 1$) based on the following physical measurements (in millimeters) of 100 genuine and 100 counterfeit Swiss Bank notes:

Length = length of the banknote
Left = length of the left edge of the banknote
Right = length of the right edge of the banknote
Top = distance from the image to the top edge
Bottom = distance from the image to the bottom edge
Diagonal = length of the diagonal

The data were originally reported in Flury and Riedwyl (1988) and they can be found in alr3 library and on the book web site in the file banknote.txt. Figure 8.20 contains a plot of Bottom and Diagonal with different symbols for the two values of y.

(a) Fit a logistic regression model using just the last two predictor variables listed above (i.e., Bottom and Diagonal). R will give warnings including "fitted probabilities numerically 0 or 1 occurred".

(b) Compare the predicted values of y from the model in (a) with the actual values of y and show that they coincide. This is a consequence of the fact that the residual deviance is zero to many decimal places. Looking at Figure 8.20 we see

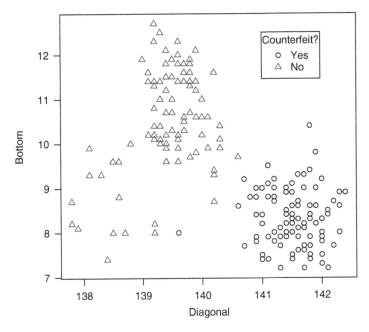

Figure 8.20 A plot of two of the predictors of counterfeit Swiss Bank notes

that the two predictors completely separate the counterfeit ($y = 0$) and genuine ($y = 1$) banknotes – thus producing a perfect logistic fit with zero residual deviance. A number of authors, including Atkinson and Riani (2000, p. 251), comment that for perfect logistic fits, the estimates of the β's approach infinity and the z-values approach zero.

Chapter 9
Serially Correlated Errors

In many situations data are collected over time. It is common for such data sets to exhibit serial correlation, that is, results from the current time period are correlated with results from earlier time periods. Thus, these data sets violate the assumption that the errors are independent, an important assumption necessary for the validity of least-squares-based regression methods. We begin by discussing the concept of autocorrelation, the correlation between a variable at different time points. We then show how generalized least squares (GLS) can be used to fit models with autocorrelated errors. Finally, we demonstrate the benefits of transforming GLS models into least squares (LS) models when it comes to examining model diagnostics.

9.1 Autocorrelation

Throughout this section and the next we shall consider the following example, which we first discussed in Chapter 3.

Estimating the price elasticity of a food product (cont.)

Recall that we want to understand the effect of price on sales and in particular to develop a model to estimate the percentage effect on sales of a 1% increase in price. This example is based on a case from Carlson (1997, p. 37). In Chapter 3, we considered weekly sales (in thousands of units) of Brand 1 at a major US supermarket chain over a year as a function of the price each week. In particular, we considered a model of the form

$$\log(\text{Sales}_t) = \beta_0 + \beta_1 \log(\text{Price}_t) + e \qquad (9.1)$$

where Sales_t denotes sales of brand 1 in week t and Price_t denotes the price of brand 1 in week t. We found a nonrandom pattern (somewhat similar to a roller coaster) in the plot of standardized residuals from model (9.1). Thus, we were not satisfied with model (9.1).

Two other potential predictor variables are available, namely,

S.J. Sheather, *A Modern Approach to Regression with R*,
DOI: 10.1007/978-0-387-09608-7_9, © Springer Science + Business Media LLC 2009

Week = week of the year

Promotion$_t$ = A dummy variable which indicates whether a
                promotion occurred for brand 1 in week t with
                0 = No promotion and
                1 = Price reduction advertised in the newspaper and in an in-store
                display

The data can be found on the course web site in the file confood2.txt. Table 9.1
gives the first four rows of the data.

Figure 9.1 contains a plot of log(Sales$_t$) against log(Price$_t$). We see from Figure
9.1 that log(Sales$_t$) and log(Price$_t$) appear to be linearly related, with promotions
having a dramatic effect on log(Sales$_t$). However, Figure 9.1 ignores the fact that
the data are collected over time.

Figure 9.2 contains a plot of log(Sales) against Week (a so-called time series plot).
It is clear from Figure 9.2 that weeks with above average values of log(Sales) are generally followed by above average values of log(Sales) and that weeks with below average
values of log(Sales) are generally followed by below average values of log(Sales).
Another way of expressing this is to say that log(Sales) in week t are positively

Table 9.1 An incomplete listing of the sales data (confood2.txt)

Week, t	Promotion$_t$	Price$_t$	Sales$_t$	SalesLag1$_t$
1	0	0.67	611	NA
2	0	0.66	673	611
3	0	0.67	710	673
4	0	0.66	478	710

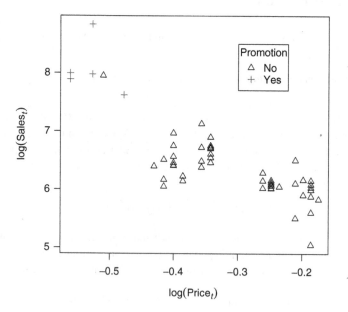

Figure 9.1 A scatter plot of log(Sales$_t$) against log(Price$_t$)

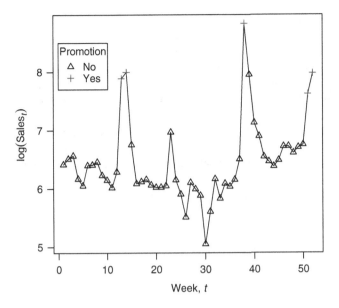

Figure 9.2 A time series plot of log(Sales$_t$)

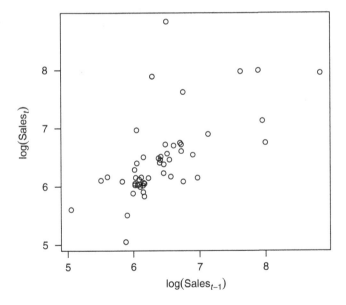

Figure 9.3 Plot of log(Sales) in week t against log(Sales) in week $t-1$

correlated with log(Sales) in week $t-1$. The latter quantity (i.e., log(Sales) in week $t-1$) is commonly referred to as log(Sales) lagged by 1 week or log(SalesLag1).

Figure 9.3 contains a plot of log(Sales) in week t against log(Sales) in week $t-1$, (i.e., of log(Sales) against log(SalesLag1)). We see from Figure 9.3 that there is a positive correlation between log(Sales) in week t and log(Sales) in week $t - 1$. Such a correlation is commonly referred to as *lag 1 autocorrelation*.

A natural question to ask at this stage is whether there is also a positive correlation between log(Sales) in week t and log(Sales) in weeks $t - 2$, $t - 3$, ..., i.e. between $Y_t = \log(\text{Sales})_t$ and Y_{t-2}, Y_{t-3}, etc. We could ascertain this by looking at scatter plots of Y_t and Y_{t-2}, Y_t and Y_{t-3}, etc., as in Figure 9.3. However, it is both cumbersome and time consuming to produce so many scatter plots.

Instead of producing lots of scatter plots like Figure 9.3, it is common statistical practice to look at values of the correlation between Y and the various values of lagged Y for different periods. Such values are called *autocorrelations*. The autocorrelation of lag l is the correlation between Y and values of Y lagged by l periods, i.e., between Y_t and Y_{t-l}, i.e.,

$$\text{Autocorrelation}(l) = \frac{\sum_{t=l+1}^{n} (y_t - \bar{y})(y_{t-l} - \bar{y})}{\sum_{t=1}^{n} (y_t - \bar{y})^2}$$

Figure 9.4 contains a plot of the first 17 autocorrelations of log(Sales). The dashed lines correspond to $-2/\sqrt{n}$ and $+2/\sqrt{n}$, since autocorrelations are declared to be statistically significantly different from zero if they are less than $-2/\sqrt{n}$ or greater than $+2/\sqrt{n}$ (i.e., if they are more than two standard errors away from zero).

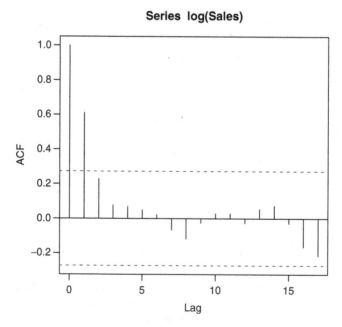

Figure 9.4 Autocorrelation function for log(Sales)

We see from Figure 9.4 that just the lag 1 autocorrelation function exceeds the normal two standard error cut-off value. Thus, last week's value of log(Sales) significantly affects this week's value of log(Sales).

Ignoring the autocorrelation effect

In order to demonstrate the effect of ignoring autocorrelation, we shall first fit a model without including it. Thus, we shall consider the model

$$\log(\text{Sales}_t) = \beta_0 + \beta_1 \log(\text{Price}_t) + \beta_2 t + \beta_3 \text{Promotion}_t + e \qquad (9.2)$$

We begin somewhat naively by assuming the errors are independent. Figure 9.5 contains diagnostic plots of the standardized residuals from least squares for model (9.2).

The top right plot in Figure 9.5 is highly nonrandom with positive (negative) standardized residuals generally followed by positive (negative) standardized residuals. Thus, there is positive autocorrelation present in the standardized residuals. To investigate this further, we next examine a plot of the autocorrelation function of the standardized residuals from model (9.2) (see Figure 9.6).

We see from Figure 9.6 that the lag 1 autocorrelation is highly statistically significant for the standardized residuals. Thus, there is strong evidence that the errors in model (9.2) are correlated over time thus violating the assumption of independence of the errors. We shall return to this example in the next section at which point we will allow for the autocorrelation that is apparent.

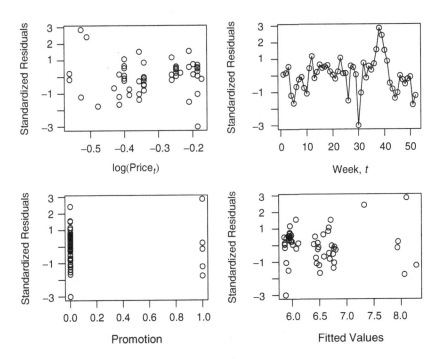

Figure 9.5 Plots of standardized residuals from LS fit of model (9.2)

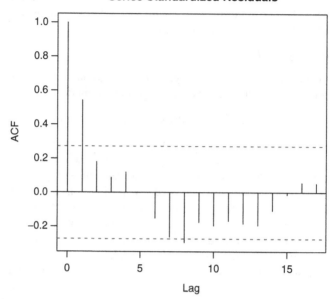

Figure 9.6 Autocorrelation function of the standardized residuals from model (9.2)

9.2   Using Generalized Least Squares When the Errors Are AR(1)

We next examine methods based on generalized least squares which allow the errors to be autocorrelated (or serially correlated, as this is often called).

We shall begin by considering the simplest situation, namely, when Y_t can be predicted from a single predictor, X_t and the errors e_t follow an autoregressive process of order 1 (AR(1)), that is,

$$Y_t = \beta_0 + \beta_1 x_t + e_t, \text{ where } e_t = \rho e_{t-1} + \upsilon_t \text{ and } \upsilon_t \text{ are iid } N(0, \sigma_\upsilon^2)$$

The errors have the following properties:

$$\mathrm{E}(e_t) = \mathrm{E}(\rho e_{t-1} + \upsilon_t) = \rho \mathrm{E}(e_{t-1}) + \mathrm{E}(\upsilon_t) = 0$$

and

$$\sigma_e^2 = \mathrm{Var}(e_t) = \mathrm{E}(e_t^2)$$
$$= \mathrm{E}\left[(\rho e_{t-1} + \upsilon_t)^2\right]$$
$$= \rho^2 \mathrm{E}(e_{t-1}^2) + \mathrm{E}(\upsilon_t^2) + 2\rho \mathrm{E}(e_{t-1}) \mathrm{E}(\upsilon_t)$$
$$= \rho^2 \sigma_e^2 + \sigma_\upsilon^2$$

since v_t is independent of e_{t-1}. Rearranging this last equation gives

$$\sigma_e^2 = \frac{\sigma_v^2}{1-\rho^2}$$

Thus, the first-order autocorrelation among the errors, e_t is given by

$$\text{Corr}(e_t, e_{t-1}) = \frac{\text{Cov}(e_t, e_{t-1})}{\sqrt{\text{Var}(e_t)\text{Var}(e_{t-1})}} = \frac{E(e_t e_{t-1})}{\sqrt{\sigma_e^2 \sigma_e^2}} = \rho$$

since

$$E(e_t e_{t-1}) = E\left[(\rho e_{t-1} + v_t)e_{t-1}\right] = \rho E(e_{t-1}^2) + E(v_t)E(e_{t-1}) = \rho \sigma_e^2$$

In a similar way, we can show that

$$\text{Corr}(e_t, e_{t-l}) = \rho^l \quad l = 1, 2, \ldots$$

When $\rho < 1$, these correlations get smaller as l increases.

Hill, Griffiths and Judge (2001, p. 264) show that the least squares estimate of β_1 has the following properties:

$$E\left(\hat{\beta}_{1LS}\right) = \beta_1$$

and

$$\text{Var}\left(\hat{\beta}_{1LS}\right) = \frac{\sigma_e^2}{SXX}\left(1 + \frac{1}{SXX}\sum_{i \neq j}\sum(x_i - \bar{x})(x_j - \bar{x})\rho^{|i-j|}\right)$$

When the errors e_t are independent ($\rho = 0$) this reduces to

$$\text{Var}\left(\hat{\beta}_{1LS}\right) = \frac{\sigma_e^2}{SXX}$$

agreeing with what we found in Chapter 2.

Thus, *using least squares and ignoring autocorrelation when it exists* will result in consistent estimates of β_1 but incorrect estimates of the variance of $\hat{\beta}_{1LS}$ *invalidating resulting confidence intervals and hypothesis tests.*

9.2.1   *Generalized Least Squares Estimation*

Define the ($n \times 1$) vector, **Y** and the $n \times (p + 1)$ matrix, **X** by

$$Y = \begin{pmatrix} y_1 \\ y_2 \\ \vdots \\ y_n \end{pmatrix} \qquad X = \begin{pmatrix} 1 & x_{11} \cdots x_{1p} \\ 1 & x_{21} \cdots x_{2p} \\ \vdots & \vdots \\ 1 & x_{n1} \cdots x_{np} \end{pmatrix}$$

Also define the $(p + 1) \times 1$ vector, β of unknown regression parameters and the $(n \times 1)$ vector, $\mathbf{e}$ of errors

$$\beta = \begin{pmatrix} \beta_0 \\ \beta_1 \\ \vdots \\ \beta_p \end{pmatrix} \qquad \mathbf{e} = \begin{pmatrix} e_1 \\ e_2 \\ \vdots \\ e_n \end{pmatrix}$$

In general matrix notation, the linear regression model is

$$Y = X\beta + \mathbf{e}$$

However, instead of assuming that the errors are independent we shall assume that

$$\mathbf{e} \sim N(0, \Sigma)$$

where Σ is a symmetric $(n \times n)$ matrix with (i, j) element equal to $\mathrm{Cov}(e_i, e_j)$.

Consider the case when the errors e_t follow an autoregressive process of order 1 (AR(1)), that is, when

$$e_t = \rho e_{t-1} + \upsilon_t \text{ and } \upsilon_t \text{ are i.i.d. } N(0, \sigma_v^2)$$

Then, it can be shown that

$$\Sigma = \sigma_e^2 \begin{pmatrix} 1 & \rho \cdots & \rho^{n-1} \\ \rho & \ddots & \vdots \\ \vdots & & \\ \rho^{n-1} & \cdots & 1 \end{pmatrix} = \frac{\sigma_v^2}{1-\rho^2} \begin{pmatrix} 1 & \rho \cdots & \rho^{n-1} \\ \rho & \ddots & \vdots \\ \vdots & & \\ \rho^{n-1} & \cdots & 1 \end{pmatrix}$$

since

$$\mathrm{Cov}(e_t, e_{t-1}) = \mathrm{E}(e_t e_{t-1}) = \rho \sigma_e^2$$

It can be shown that the log-likelihood function is given by

$$\log\left(L(\beta, \rho, \sigma_e^2 \mid Y)\right)$$

$$= -\frac{n}{2} \log(2\pi) - \frac{1}{2} \log(\det(\Sigma)) - \frac{1}{2}(Y - X\beta)' \Sigma^{-1} (Y - X\beta)$$

The maximum likelihood estimates of β, ρ, σ_e^2 can be obtained by maximizing this function. Given ρ, σ_e^2 (or estimates of these quantities), minimizing the third term in the log-likelihood gives $\hat{\beta}_{GLS}$ the generalized least squares (GLS) estimator of β. It can be shown that

$$\hat{\beta}_{GLS} = (\mathbf{X}'\boldsymbol{\Sigma}^{-1}\mathbf{X})^{-1}\mathbf{X}'\boldsymbol{\Sigma}^{-1}\mathbf{Y}$$

Comparing this with the least squares estimator of β

$$\hat{\beta}_{LS} = (\mathbf{X}'\mathbf{X})^{-1}\mathbf{X}'\mathbf{Y}$$

the important role of the inverse of the variance–covariance matrix of the errors is clearly apparent.

Estimating the price elasticity of a food product (cont.)

Given below is the output from R associated with fitting model (9.2) using maximum likelihood and assuming that the errors are AR(1).

Output from R

```
Generalized least squares fit by maximum likelihood
Model: log(Sales) ~ log(Price) + Promotion + Week
Data: confood2
     AIC          BIC        logLik
 6.537739      18.2452     2.731131
Correlation Structure: AR(1)
Formula: ~Week
Parameter estimate(s):
      Phi
0.5503593

Coefficients:
                  Value      Std.Error      t-value      p-value
(Intercept)     4.675667     0.2383703     19.615142      0.000
log(Price)     -4.327391     0.5625564     -7.692368      0.000
Promotion       0.584650     0.1671113      3.498565      0.001
Week            0.012517     0.0046692      2.680813      0.010

Residual standard error: 0.2740294
Degrees of freedom: 52 total; 48 residual

Approximate 95% confidence intervals

Coefficients:
                    lower            est.           upper
(Intercept)     4.196391300      4.67566686      5.15494243
log(Price)     -5.458486702     -4.32739122     -3.19629575
Promotion       0.248649971      0.58464986      0.92064974
Week            0.003129195      0.01251724      0.02190529
```

```
Correlation structure:
          lower          est.          upper
Phi  0.2867453    0.5503593    0.7364955

Residual standard error:
    lower           est.          upper
 0.2113312      0.2740294    0.3553291
```

Figure 9.7 shows a plot of the autocorrelation function of the generalized least squares (GLS) residuals from model (9.2) with AR(1) errors. We see from Figure 9.7 that the lag 1 autocorrelation of just under 0.6 is highly statistically significant for the GLS residuals. This is not surprising when one considers that these residuals correspond to a model where we assumed the errors to be AR(1). *The high positive autocorrelation in the GLS residuals can produce nonrandom patterns in diagnostic plots based on these residuals even when the fitted model is correct.* Instead, we will transform model (9.2) with AR(1) errors into a related model with uncorrelated errors so that we can use diagnostic plots based on least squares residuals.

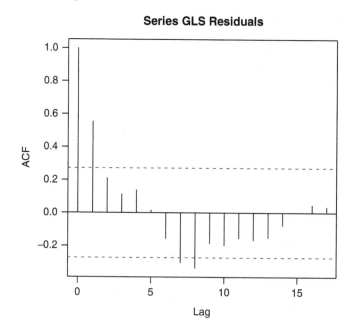

Figure 9.7   Autocorrelation function of the GLS residuals from model (9.2)

9.2.2   Transforming a Model with AR(1) Errors into a Model with iid Errors

We wish to transform the regression model

$$Y_t = \beta_0 + \beta_1 x_t + e_t = \beta_0 + \beta_1 x_t + \rho e_{t-1} + v_t$$

with AR(1) errors e_t into a related model with uncorrelated errors so that we can *use least squares for diagnostics*. Writing this last equation for Y_{t-1} gives

$$Y_{t-1} = \beta_0 + \beta_1 x_{t-1} + e_{t-1}$$

Multiplying this last equation by ρ gives

$$\rho Y_{t-1} = \rho \beta_0 + \rho \beta_1 x_{t-1} + \rho e_{t-1}$$

Subtracting the second equation from the first gives

$$Y_t - \rho Y_{t-1} = \beta_0 + \beta_1 x_t + e_t - \left(\rho \beta_0 + \rho \beta_1 x_{t-1} + \rho e_{t-1}\right)$$

Recall that

$$e_t = \rho e_{t-1} + v_t$$

So,

$$
\begin{aligned}
Y_t - \rho Y_{t-1} &= \beta_0 + \beta_1 x_t + \rho e_{t-1} + v_t - \left(\rho \beta_0 + \rho \beta_1 x_{t-1} + \rho e_{t-1}\right) \\
&= (1-\rho)\beta_0 + \beta_1 \left(x_t - \rho x_{t-1}\right) + v_t
\end{aligned}
$$

Define, what is commonly referred to as the Cochrane-Orcutt transformation (Cochrane and Orcutt, 1949),

$$Y_t^* = Y_t - \rho Y_{t-1}, x_{t2}^* = x_t - \rho x_{t-1} \text{ and } x_{t1}^* = 1 - \rho \text{ for } t = 2,...,n$$

then the last model equation can be rewritten as

$$Y_t^* = \beta_0 x_{t1}^* + \beta_1 x_{t2}^* + v_t \quad t = 2,...,n \tag{9.3}$$

Since the last equation is only valid for $t = 2,...,n$, we still need to deal with the first observation Y_1. The first observation in the regression model is given by

$$Y_1 = \beta_0 + \beta_1 x_1 + e_1$$

with error variance

$$\mathrm{Var}(e_1) = \sigma_e^2 = \frac{\sigma_v^2}{1-\rho^2}$$

Multiplying each term in the equation for Y_1 by $\sqrt{1-\rho^2}$ gives

$$\sqrt{1-\rho^2}\,Y_1 = \sqrt{1-\rho^2}\,\beta_0 + \sqrt{1-\rho^2}\,\beta_1 x_1 + \sqrt{1-\rho^2}\,e_1$$

Define what is commonly referred to as the Prais-Winsten transformation (Prais and Winsten, 1954),

$$Y_1^* = \sqrt{1-\rho^2}\,Y_1,\; x_{12}^* = \sqrt{1-\rho^2}\,x_1,\; x_{11}^* = \sqrt{1-\rho^2}\,\text{and } e_1^* = \sqrt{1-\rho^2}\,e_1$$

Then the model equation for Y_1 can be rewritten as

$$Y_1^* = \beta_0 x_{11}^* + \beta_1 x_{12}^* + e_1^* \tag{9.4}$$

where $\mathrm{Var}(e_1^*) = (1-\rho^2)\sigma_e^2 = \sigma_v^2$ matching the variance of the error term in (9.3). We shall see that Y_1^* is generally a point of high leverage when we use least squares to calculate generalized least squares estimates.

If we multiply each term in (9.3) and (9.4) by $\sqrt{1-\rho^2}$ then we find that we can equivalently define

$$Y_1^* = Y_1,\, Y_t^* = \left(Y_t - \rho Y_{t-1}\right)\sqrt{1-\rho^2}\; t = 2,\dots,n$$

In the examples in this chapter we shall use this version rather than (9.3) and (9.4).

9.2.3   A General Approach to Transforming GLS into LS

We next seek a general method for transforming a GLS model into a LS model. Consider the linear model

$$\mathbf{Y} = \mathbf{X}\beta + \mathbf{e} \tag{9.5}$$

where the errors are assumed to have mean 0 and variance–covariance matrix Σ. Earlier we found that the generalized least squares estimator of Σ is given by

$$\hat{\beta}_{\mathrm{GLS}} = (\mathbf{X}'\Sigma^{-1}\mathbf{X})^{-1}\mathbf{X}'\Sigma^{-1}\mathbf{Y}$$

where Σ is a symmetric (n×n) matrix with (i, j) element equal to $\mathrm{Cov}(e_i, e_j)$.

Since Σ is a symmetric positive-definite matrix it can be written as

$$\Sigma = SS'$$

where S is a lower triangular matrix[1] with positive diagonal entries. This result is commonly referred to as the Cholesky decomposition of Σ. Roughly speaking, S can be thought of as the "square root" of Σ. Multiplying each side of (9.5) by S^{-1}, the inverse of S, gives

$$S^{-1}\mathbf{Y} = S^{-1}\mathbf{X}\beta + S^{-1}\mathbf{e}$$

Utilizing the result that $\left(S^{-1}\right)' = \left(S'\right)^{-1}$,

$$\text{Var}\left(S^{-1}\mathbf{e}\right) = S^{-1}\,\text{Var}\left(\mathbf{e}\right)\left(S^{-1}\right)' = S^{-1}\sum\left(S^{-1}\right)' = S^{-1}SS'\left(S'\right)^{-1} = \mathbf{I},$$

the identity matrix. Thus, pre-multiplying each term in equation (9.5) by S^{-1}, the inverse of S, produces a linear model with uncorrelated errors. In other words, let

$$\mathbf{Y}^* = S^{-1}\mathbf{Y}, \mathbf{X}^* = S^{-1}\mathbf{X}, \mathbf{e}^* = S^{-1}\mathbf{e}$$

then,

$$\mathbf{Y}^* = \mathbf{X}^*\beta + \mathbf{e}^* \tag{9.6}$$

provides a linear model with uncorrelated errors from which we can obtain the GLS estimate of β using least squares. Let $\hat{\beta}_{LS}^*$ denote the least squares estimate of β for model (9.6), which is a generalization of (9.3) and (9.4). We next show that it equals the GLS estimator of β for model (9.5). Utilizing the result that $\left(AB\right)' = B'A'$

$$\hat{\beta}_{LS}^* = (\mathbf{X}^{*\prime}\mathbf{X}^*)^{-1}\mathbf{X}^{*\prime}\mathbf{Y}^* = \left[\left(S^{-1}\mathbf{X}\right)'\left(S^{-1}\mathbf{X}\right)\right]^{-1}\left(S^{-1}\mathbf{X}\right)'\left(S^{-1}\mathbf{X}\right)$$

$$= \left[\mathbf{X}'\left(S^{-1}\right)'S^{-1}\mathbf{X}\right]^{-1}\mathbf{X}'\left(S^{-1}\right)'S^{-1}\mathbf{Y}$$

$$= \left[\mathbf{X}'\sum{}^{-1}\mathbf{X}\right]^{-1}\mathbf{X}'\sum{}^{-1}\mathbf{Y} = \hat{\beta}_{GLS}$$

noting $\Sigma^{-1} = \left(SS'\right)^{-1} = \left(S'\right)^{-1}S^{-1} = \left(S^{-1}\right)'S^{-1}$, since $\left(A'\right)^{-1} = \left(A^{-1}\right)'$

However, Paige (1979) points out that using (9.6) to calculate the GLS estimates in (9.5) can be numerically unstable and sometimes even fail completely.

Estimating the price elasticity of a food product (cont.)

Given below is the output from R associated with fitting model (9.2) assuming that the errors are AR(1) using least squares based on the transformed versions of the response and predictor variables in (9.6).

---

[1] A lower triangular matrix is a matrix where all the entries above the diagonal are zero.

Output from R

```
Call:lm(formula = ystar ~ xstar - 1)
Coefficients:
                  Estimate  Std. Error  t value  Pr(>|t|)
xstar(Intercept)   4.67566     0.23838   19.614   < 2e-16   ***
xstarlog(Price)   -4.32741     0.56256   -7.692  6.44e-10   ***
xstarPromotion     0.58464     0.16711    3.499   0.00102    **
xstarWeek          0.01252     0.00467    2.681   0.01004    *
```

Comparing the output above with that on a previous page, we see that the estimated
regression coefficients are the same as are the standard errors and t-values.

Figure 9.8 shows plots of the transformed variables from model (9.6). The point
corresponding to Week 1 is highlighted in each plot. It is clearly a very highly influential point in determining the intercept. In view of (9.4) this is to be expected.

We next look at diagnostics based on the least squares residuals from (9.6).
Figure 9.9 shows a plot of the autocorrelation function of the standardized least
squares residuals from model (9.6). None of the autocorrelations in Figure 9.9 are
statistically significant indicating that an AR(1) process provides a valid model for
the errors in model (9.2).

Figure 9.10 contains diagnostic plots of the standardized LS residuals from
model (9.6) plotted against each predictor in its x^* mode. Each of the plots appear
to be random, indicating that model (9.2) with AR(1) errors is a valid model for the
data. However, two outliers (corresponding to weeks 30 and 38) are evident in each
of these plots. These weeks were investigated and the following was found:

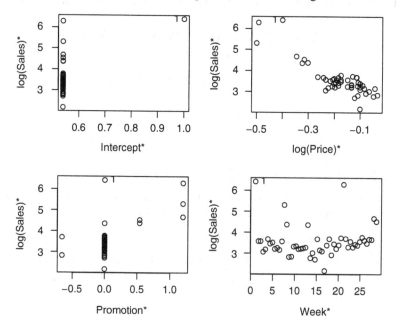

Figure 9.8 Plots of the transformed variables from model (9.6)

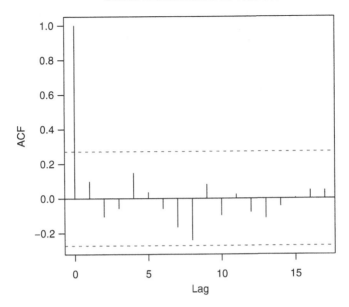

Figure 9.9 Autocorrelation function of the standardized residuals from model (9.6)

- In week 30 another brand ran a promotion along with a price cut and captured a larger than normal share of sales, thus reducing the sales of Brand 1
- In week 38, Brand 1 ran a promotion while none of the brands did, leading to higher sales than expected for Brand 1.

Thus, it seems that the model could be improved by including the prices and promotions of the other brands.

Figure 9.11 contains the diagnostic plots produced by R for the least squares fit to model (9.6). A number of points of high leverage are evident from the bottom right-hand plot in Figure 9.11. Week 38 is a "bad" leverage point and hence it is especially noteworthy. Otherwise the plots in Figure 9.10 provide further support for the assertion that (9.6) is a valid model for the data.

9.3 Case Study

We conclude this topic, by considering a case study using data from Tryfos (1998, p. 162) which demonstrates the hazards associated with ignoring autocorrelation in fitting and when examining model diagnostics. According to Tryfos (1998), the savings and loan associations in the Bay Area of San Francisco had an almost monopolistic position in the market for residential real estate loans during the 1990s. Chartered

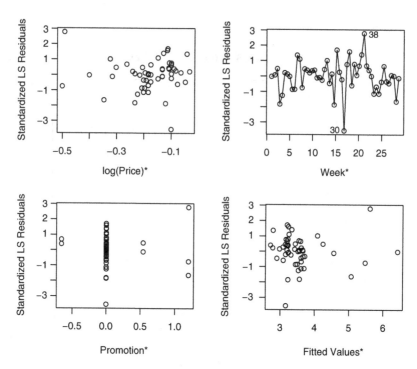

Figure 9.10 Plots of standardized LS residuals from model (9.6)

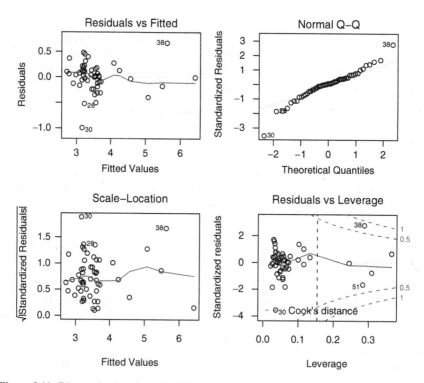

Figure 9.11 Diagnostic plots for model (9.6)

banks had a small portion of the market, and savings and loan associations located outside the region were prevented from making loans in the Bay Area. Interest centers on developing a regression model to predict interest rates (Y) from x_1, the amount of loans closed (in millions of dollars) and x_2, the vacancy index, since both predictors measure different aspects of demand for housing. Data from the Bay Area are available on each of these variables over a consecutive 19-month period in the 1990s. The data can be found on the course web site in the file BayArea.txt.

The scatter plots of the data given in Figure 9.12 reveal a striking nonlinear pattern among the predictors.

Ignoring the autocorrelation effect

In order to demonstrate the effect of ignoring autocorrelation, we shall first fit a model without including it. Thus, we shall consider the model

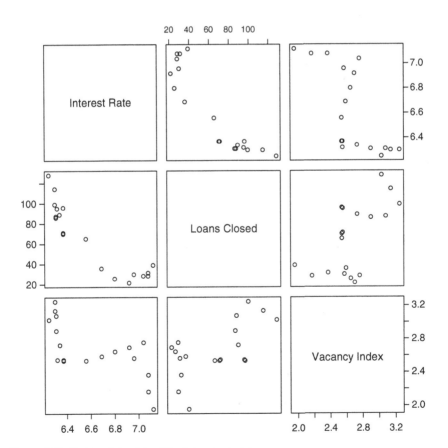

Figure 9.12 Scatter plot matrix of the interest rate data

$$\text{InterestRate}_t = \beta_0 + \beta_1 \text{LoansClosed}_t + \beta_2 \text{VacancyIndex}_t + e \qquad (9.7)$$

We begin somewhat naively by assuming the errors are uncorrelated. Figure 9.13 contains diagnostic plots of the standardized residuals from least squares for model (9.7).

The top left and the bottom left plot in Figure 9.13 are highly nonrandom with an obvious quadratic pattern. The quadratic pattern could be due to the nonlinearity among the predictors and/or the obvious autocorrelation among the standardized residuals.

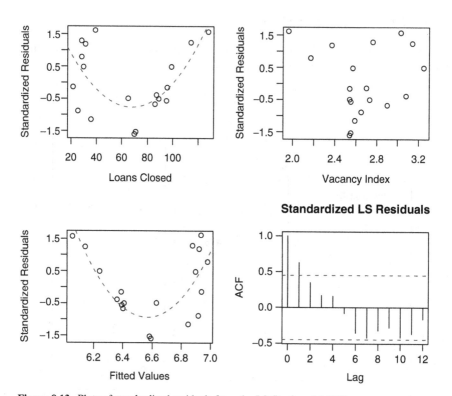

Figure 9.13 Plots of standardized residuals from the LS fit of model (9.7)

Modelling the autocorrelation effect as AR(1)

We next fit model (9.7) assuming the errors are AR(1). Given below is the output from R

Output from R

```
Generalized least squares fit by maximum likelihood

Model: InterestRate ~ LoansClosed + VacancyIndex
Data: BayArea
      AIC          BIC        logLik
-35.30833   -30.58613    22.65416

Correlation Structure: AR(1)
Formula: ~Month
Parameter estimate(s):
       Phi
 0.9572093

Coefficients:
                  Value    Std. Error     t-value     p-value
(Intercept)    7.122990     0.4182065   17.032232      0.0000
LoansClosed   -0.003432     0.0011940   -2.874452      0.0110
VacancyIndex  -0.076340     0.1307842   -0.583710      0.5676

Residual standard error: 0.2377426
Degrees of freedom: 19 total; 16 residual

Approximate 95% confidence intervals

Coefficients:
                     lower              est.              upper
(Intercept)    6.236431638      7.122989795     8.0095479516
LoansClosed   -0.005963412     -0.003432182    -0.0009009516
VacancyIndex  -0.353590009     -0.076339971     0.2009100658

Correlation structure:
         lower          est.          upper
Phi 0.5282504     0.9572093     0.9969078

Residual standard error:
     lower          est.          upper
0.06867346    0.23774259    0.82304773
```

Given below is the output from R associated with fitting model (9.7) assuming that the errors are AR(1) using least squares based on the transformed versions of the response and predictor variables in (9.6). Notice that the results match those in the previous R output.

Output from R

```
Call:
lm(formula = ystar ~ xstar - 1)

Coefficients:
                     Estimate  Std. Error  t value  Pr(>|t|)
xstar(Intercept)     7.122990    0.418207   17.032  1.12e-11   ***
xstarLoansClosed    -0.003432    0.001194   -2.874     0.011   *
xstarVacancyIndex   -0.076340    0.130784   -0.584     0.568
```

Figure 9.14 shows plots of the transformed variables from model (9.7). The point corresponding to Week 1 is highlighted in each plot. It is clearly a very highly influential point, which is to be expected in view of (9.4).

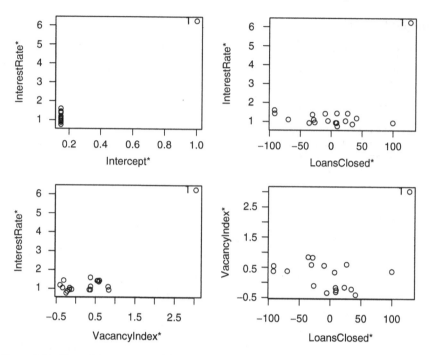

Figure 9.14 Plots of the transformed variables from model (9.7)

Figure 9.15 shows diagnostic plots based on the least squares residuals from (9.6). None of the autocorrelations in the top left plot are statistically significant indicating that an AR(1) process provides a valid model for the errors in model (9.7).

The other plots in Figure 9.15 show standardized LS residuals from model (9.7) plotted against each predictor in its x^* mode. Each of the plots appear to be random, indicating that model (9.7) with AR(1) errors is a valid model for the data. Month 1 again shows up as a highly influential point.

Comparing the top right-hand plot in Figure 9.15 with the top left-hand plot in Figure 9.13 we see that the quadratic pattern has disappeared once we have used generalized least squares to account for the autocorrelated errors.

It is instructive to repeat the analyses shown above after removing the predictor x_2, the vacancy index. The quadratic pattern in the plot of standardized residuals against LoansClosed remains when naively fitting the model which assumes that the errors are independent. This shows that the quadratic pattern is due to the obvious autocorrelation among the standardized residuals and not due to the nonlinearity among the predictors.

This case study clearly shows ignoring autocorrelation can produce misleading model diagnostics. It demonstrates the difficulty inherent in separating the effects of autocorrelation in the errors from misspecification of the conditional mean of Y given the predictors. On the other hand, the case study illustrates the benefit of using least squares diagnostics based on Y^* and X^*.

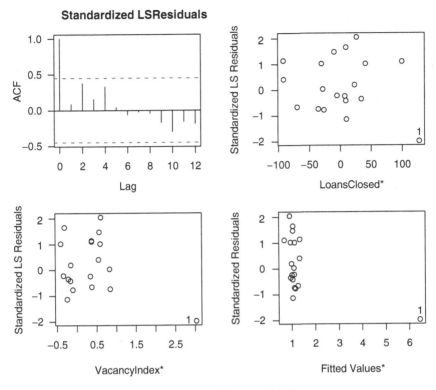

Figure 9.15 Plots of standardized LS residuals from model (9.6)

9.4   Exercises

1. Senior management at the Australian Film Commission (AFC) has sought your
help with the task of developing a model to predict yearly gross box office receipts
from movies screened in Australia. Such data are publicly available for the period
from 1976 to 2007 from the AFC's web site (www.afc.gov.au). The data are given
in Table 9.2 and they can be found on the book web site in the file boxoffice.txt.
Interest centers on predicting gross box office results for 1 year beyond the latest
observation, that is, predicting the 2008 result. In addition, there is interest in
estimating the extent of any trend and autocorrelation in the data. A preliminary
analysis of the data has been undertaken by a staffer at the AFC and these results
appear below. In this analysis the variable Year was replaced by the number of
years since 1975, which we shall denote as YearsS1975 (i.e., YearsS1975 = Year
− 1975).

The first model fit to the data by the staffer was

$$\text{GrossBoxOffice} = \beta_0 + \beta_1 \text{YearsS1975} + e \tag{9.8}$$

Table 9.2 Australian gross box office results

Year	Gross box office ($M)	Year	Gross box office ($M)
1976	95.3	1992	334.3
1977	86.4	1993	388.7
1978	119.4	1994	476.4
1979	124.4	1995	501.4
1980	154.2	1996	536.8
1981	174.3	1997	583.9
1982	210.0	1998	629.3
1983	208.0	1999	704.1
1984	156.0	2000	689.5
1985	160.6	2001	812.4
1986	188.6	2002	844.8
1987	182.1	2003	865.8
1988	223.8	2004	907.2
1989	257.6	2005	817.5
1990	284.6	2006	866.6
1991	325.0	2007	895.4

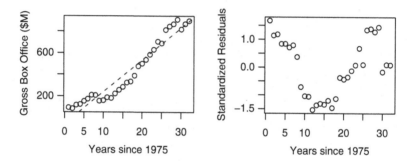

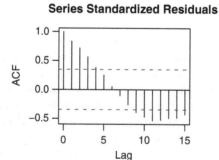

Figure 9.16 Plots associated with the LS fit of model (9.8)

Figure 9.16 shows plots associated with the least squares fit of model (9.8) that
were produced by the staffer. The staffer noted that a number of statistically
significant autocorrelations in the standardized residuals as well as the existence
of an obvious roller coaster pattern in the plot of standardized residuals against

YearsS1975. As such, the staffer decided to fit model (9.8) assuming that the errors are AR(1). Given below is the output from R.

Output from R

```
Generalized least squares fit by maximum likelihood
Model: GrossBoxOffice ~ YearsS1975
Data: boxoffice
    AIC         BIC          logLik
330.3893    336.2522     -161.1947

Correlation Structure: AR(1)
Formula: ~YearsS1975
Parameter estimate(s):
     Phi
0.8782065

Coefficients:
                 Value    Std. Error    t-value    p-value
(Intercept)    4.514082    72.74393    0.062054    0.9509
YearsS1975    27.075395     3.44766    7.853259    0.0000

Correlation:
               (Intr)
YearsS1975     -0.782

Residual standard error: 76.16492
Degrees of freedom: 32 total; 30 residual
```

Given below is the output from R associated with fitting model (9.8) assuming that the errors are AR(1) using least squares based on the transformed versions of the response and predictor variables in (9.6). The staffer was delighted that the results match those in the previous R output.

Output from R

```
Call:
lm(formula = ystar ~ xstar - 1)

Coefficients:
                   Estimate   Std. Error   t value   Pr(>|t|)
xstar(Intercept)     4.514       72.744     0.062       0.95
xstarYearS1975      27.075        3.448     7.853    9.17e-09    ***
```

Figure 9.17 shows diagnostic plots based on the least squares residuals from (9.6). The staffer is relieved that none of the autocorrelations in the right-hand plot are statistically significant indicating that an AR(1) process provides a valid model for the errors in model (9.8). However, the staffer is concerned about the distinct nonrandom pattern in the left-hand plot of Figure 9.17. The dashed line is from a cubic LS fit which is statistically significant (p-value = 0.027). At this stage, the staffer is confused about what to do next and has sought your assistance.

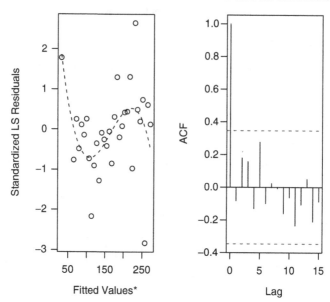

Figure 9.17 Plots of standardized LS residuals from model (9.6)

(a) Comment on the analysis performed by the staffer.
(b) Obtain a final model for predicting GrossBoxOffice from YearsS1975.
     Ensure that you produce diagnostic plots to justify your choice of model.
     Describe any weaknesses in your model.
(c) Use your model from (b) to predict GrossBoxOffice in 2008.
(d) Use your model from (b) to identify any outliers. In particular, decide
     whether the year 2000 is an outlier. There is some controversy about
     the year 2000. In one camp are those that say that fewer people went to the
     movies in Australia in 2000 due to the Olympics being held in Sydney. In
     the other camp are those that point to the fact that a 10% Goods and
     Services Tax (GST) was introduced in July 2000 thus producing an
     increase in box office receipts.

2. This problem is based on an exercise from Abraham and Ledolter (2006,
   pp. 335–337) which focuses on monthly sales from a bookstore in the city
   of Vienna, Austria. The available data consisted of 93 consecutive monthly
   observations on the following variables:

   Sales = Sales (in hundreds of dollars)
   Advert = Advertising spend in the current month
   Lag1Advert = Advertising spend in the previous month
   Time = Time in months
   Month_i = Dummy variable which is 1 for month i and 0 otherwise
   (i = 2, 3, …, 12)

The data can be found on the book website in the file bookstore.txt.

(a) Follow the advice of Abraham and Ledolter (2006, pp. 336–337) and first build a model for Sales ignoring the effects due to Advert and Lag1Advert. Ensure that you produce diagnostic plots to justify your choice of model. Describe any weaknesses in your model.

(b) Add the effects due to Advert and Lag1Advert to the model you have developed in (a). Last month's advertising (Lag1Advert) is thought to have an impact on the current month's sales. Obtain a final model for predicting Sales. Ensure that you produce diagnostic plots to justify your choice of model. Describe any weaknesses in your model.

3. This problem is based on a case involving real data from Tryfos (1998, pp. 467–469). According to Tryfos:

To the sales manager of Carlsen's Brewery, a formal model to explain and predict beer sales seemed worth a try...... Carlsen's Brewery is one of the major breweries in Canada, with sales in all parts of the country, but the study itself was to be confined to one metropolitan area. In discussing this assignment, the manager pointed out that weather conditions obviously are responsible for most of the short-run variation in beer consumption. "When it is hot", the manager said, "people drink more – it's that simple." This was also the reason for confining the study to one area; since weather conditions vary so much across the country, there was no point in developing a single, countrywide model for beer sales. It was the manager's opinion that a number of models should be developed —one for each major selling area.

The available data consisted of 19 consecutive quarterly observations on the following variables:

Sales = Quarterly beer sales (in tons)
Temp = Average quarterly temperature (in degrees F)
Sun = Quarterly total hours of sunlight
Q2 = Dummy variable which is 1 for Quarter 2 and 0 otherwise
Q3 = Dummy variable which is 1 for Quarter 3 and 0 otherwise
Q4 = Dummy variable which is 1 for Quarter 4 and 0 otherwise.
The data can be found on the book web site in the file CarlsenQ.txt.

Develop a model which can be used to predict quarterly beer sales. Describe any weaknesses in your model. Write up the results in the form of a report that is to be given to the manager at Carlsen's brewery.

Chapter 10
Mixed Models

In the previous chapter we looked at regression models for data collected over time. The data sets we studied in Chapter 9 typically involve a single relatively long series of data collected in time order. In this chapter, we shall further consider models for data collected over time. However, here the data typically consist of a number of relatively short series of data collected in time order (such data are commonly referred to as longitudinal data). For example, in the next section we shall consider a real example which involves four measurements in time order collected for each of 27 children (i.e., 16 males and 11 females).

We begin by discussing the concept of fixed and random effects and how random effects induce a certain form of correlation on the overall error term in the corresponding regression model. The term mixed models is used to describe models which have both fixed and random effects. We then show how to fit mixed models with more complex error structures. Finally, we demonstrate the benefits of transforming mixed models into models with uncorrelated errors when it comes to examining model diagnostics.

10.1 Random Effects

Thus far in this book we have looked exclusively at regression models for what are known as fixed effects. The effects are fixed in the sense that the levels of each explanatory variable are themselves of specific interest. For example, in Chapter 1 we were interested in modeling the performance of the 19 NFL field goal kickers who made at least ten field goal attempts in each of the 2002, 2003, 2004, 2005 seasons and at the completion of games on Sunday, November 12 in the 2006 season.

On the other hand, in many studies involving random effects, subjects are selected at random from a large population. The subjects chosen are themselves not of specific interest. For example, if the study or experiment were repeated then different subjects would be used. We shall see in the context of this chapter that what is generally of interest in these situations is a comparison of outcomes within each subject over time as well as comparisons across subjects or groups of subjects. Throughout this section we shall consider the following real example involving random effects.

S.J. Sheather, *A Modern Approach to Regression with R*,
DOI: 10.1007/978-0-387-09608-7_10, © Springer Science+Business Media LLC 2009

Table 10.1 Orthodontic growth data in the form of Distance

Subject	Age = 8	Age = 10	Age = 12	Age = 14
M1	26	25	29	31
M2	21.5	22.5	23	26.5
M3	23	22.5	24	27.5
M4	25.5	27.5	26.5	27
M5	20	23.5	22.5	26
M6	24.5	25.5	27	28.5
M7	22	22	24.5	26.5
M8	24	21.5	24.5	25.5
M9	23	20.5	31	26
M10	27.5	28	31	31.5
M11	23	23	23.5	25
M12	21.5	23.5	24	28
M13	17	24.5	26	29.5
M14	22.5	25.5	25.5	26
M15	23	24.5	26	30
M16	22	21.5	23.5	25
F1	21	20	21.5	23
F2	2 1	21.5	24	25.5
F3	20.5	24	24.5	26
F4	23.5	24.5	25	26.5
F5	21.5	23	22.5	23.5
F6	20	21	21	22.5
F7	21.5	22.5	23	25
F8	23	23	23.5	24
F9	20	21	22	21.5
F10	16.5	19	19	19.5
F11	24.5	25	28	28

Orthodontic growth data

Potthoff and Roy (1964) first reported a data set from a study undertaken at the Department of Orthodontics from the University of North Carolina Dental School. Investigators followed the growth of 27 children (16 males and 11 females). At ages 8, 10, 12 and 14 investigators measured the distance (in millimeters) from the center of the pituitary to the pterygomaxillary fissure, two points that are easily identified on x-ray exposures of the side of the head. Interest centers on developing a model for these distances in terms of age and sex. The data are provided in the R-package, nlme. They can be found in Table 10.1 and on the book web site in the file Orthodont.txt.

Orthodontic growth data: Females

We shall begin by considering the data just for females. Figure 10.1 shows a plot of Distance against Age for each of the 11 females. Notice that the plots have been ordered from bottom left to top right in terms of increasing average value of Distance.

The model we first consider for subject i ($i = 1, 2, ..., 11$) at Age j ($j = 1, 2, 3, 4$) is as follows:

$$\text{Distance}_{ij} = \beta_0 + \beta_1 \text{Age}_j + b_i + e_{ij} \qquad (10.1)$$

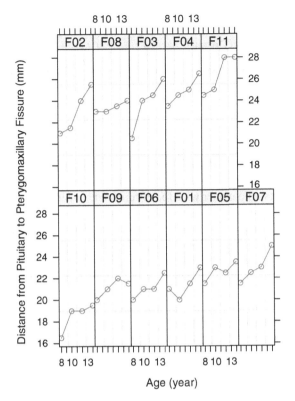

Figure 10.1 Plot of Distance against Age for each female

where the random effect b_i is assumed to follow a normal distribution with mean 0 and variance σ_b^2 (i.e., $b_i \sim N(0, \sigma_b^2)$) independent of the error term e_{ij} which is iid $N(0, \sigma_e^2)$. Model (10.1) assumes that the intercepts differ randomly across the 11 female subjects but that Distance increases linearly with Age at the same fixed rate across the 11 female subjects. Thus, in model (10.1) age is modeled as a fixed effect. Since model (10.1) contains both fixed and random effects, it is said to be a mixed model.

We next calculate the correlation between two distance measurements (at Age j, k such that $j \neq k$) for the same subject (i) based on model (10.1). We shall begin by calculating the relevant covariance and variance terms. Utilizing the independence between the random effect and random error terms assumed in model (10.1) gives

$$
\begin{aligned}
\mathrm{Cov}(\mathrm{Distance}_{ij}, \mathrm{Distance}_{ik}) &= \mathrm{Cov}(b_i + e_{ij}, b_i + e_{ik}) \\
&= \mathrm{Cov}(b_i, b_i) \\
&= \mathrm{Var}(b_i) \\
&= \sigma_b^2
\end{aligned}
$$

and

$$\mathrm{Var(Distance}_{ij}) = \mathrm{Var}(b_i + e_{ij}) = \sigma_b^2 + \sigma_e^2$$

Putting these last two expressions together gives the following expression for the correlation

$$\mathrm{Corr(Distance}_{ij}, \mathrm{Distance}_{ik}) = \frac{\sigma_b^2}{\sigma_b^2 + \sigma_e^2} \qquad (10.2)$$

Thus, the random intercepts model (10.1) is equivalent to assuming that the correlation between two distance measurements (at Age j, k such that $j \neq k$) for the same subject (i) is constant no matter what the difference between j and k. In other words, a random intercepts model is equivalent to assuming a constant correlation within subjects over any chosen time interval. Such a correlation structure is also commonly referred to as compound symmetry.

In order to investigate whether the assumption of constant correlation inherent in (10.1) is reasonable, we calculate the correlations between two distance measurements for the same female subject over each time interval. In what follows, we shall denote the distance measurements for females aged 8, 10, 12 and 14 as DistFAge8, DistFAge10, DistFAge12, DistFAge14, respectively. The output from R below gives the correlations between these four variables. Notice the similarity among the correlations away from the diagonal, which range from 0.830 to 0.948.

Output from R: Correlations between female measurements

	DistFAge8	DistFAge10	DistFAge12	DistFAge14
DistFAge8	1.000	0.830	0.862	0.841
DistFAge10	0.830	1.000	0.895	0.879
DistFAge12	0.862	0.895	1.000	0.948
DistFAge14	0.841	0.879	0.948	1.000

Figure 10.2 shows a scatter plot matrix of the distance measurements for females aged 8, 10, 12 and 14. The linear association in each plot in Figure 10.2 appears to be quite similar. Overall, it therefore seems that the assumption that correlations are constant across Age is a reasonable one for females.

10.1.1 Maximum Likelihood and Restricted Maximum Likelihood

The random effects model in (10.1) can be rewritten as follows:

$$\mathrm{Distance}_{ij} = \beta_0 + \beta_1 \mathrm{Age}_j + \varepsilon_{ij}$$

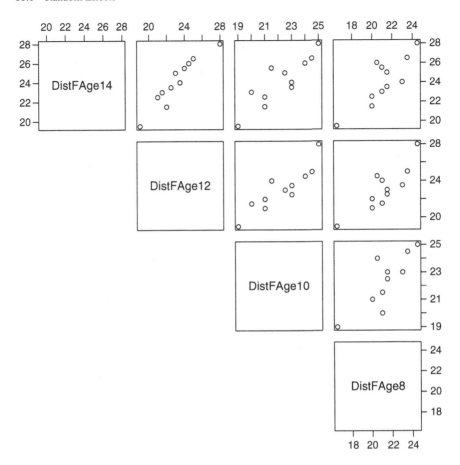

Figure 10.2 Scatter plot matrix of the Distance measurements for female subjects

where $\varepsilon_{ij} = b_i + e_{ij}$. In general matrix notation, this is

$$\mathbf{Y} = \mathbf{X}\beta + \varepsilon \qquad (10.3)$$

where in this example

$$\mathbf{Y} = \begin{pmatrix} y_{1,1} \\ \vdots \\ y_{1,4} \\ \vdots \\ y_{11,1} \\ \vdots \\ y_{11,4} \end{pmatrix} \quad \mathbf{X} = \begin{pmatrix} 1 & x_1 \\ \vdots & \vdots \\ 1 & x_4 \\ \vdots & \vdots \\ 1 & x_1 \\ \vdots & \vdots \\ 1 & x_4 \end{pmatrix}$$

with $Y_{i,j} = \text{Distance}_{ij}, x_j = \text{Age}_j, i = 1,...,11, j = 1,...,4$. We shall assume that

$$\varepsilon \sim N(\mathbf{0}, \Sigma)$$

where in this example Σ is the following symmetric matrix:

$$\Sigma = \begin{pmatrix} \mathbf{D} & 0 & 0 & 0 & \cdots & 0 \\ 0 & \mathbf{D} & 0 & 0 & \cdots & 0 \\ 0 & 0 & \mathbf{D} & 0 & \cdots & 0 \\ 0 & 0 & 0 & \ddots & 0 & 0 \\ \vdots & \vdots & \vdots & \vdots & \mathbf{D} & 0 \\ 0 & 0 & 0 & 0 & 0 & \mathbf{D} \end{pmatrix} \quad \mathbf{D} = \begin{pmatrix} \sigma_b^2 + \sigma_e^2 & \sigma_b^2 & \sigma_b^2 & \sigma_b^2 \\ \sigma_b^2 & \sigma_b^2 + \sigma_e^2 & \sigma_b^2 & \sigma_b^2 \\ \sigma_b^2 & \sigma_b^2 & \sigma_b^2 + \sigma_e^2 & \sigma_b^2 \\ \sigma_b^2 & \sigma_b^2 & \sigma_b^2 & \sigma_b^2 + \sigma_e^2 \end{pmatrix}$$

Estimates of β and Σ can be found using maximum likelihood. However, it is well known that maximum likelihood (ML) estimate of Σ is biased, considerably so in small to moderate sample sizes. Because of this bias, restricted maximum likelihood (REML) is the widely recommended approach for estimating Σ. REML is also referred to as residual maximum likelihood. REML is based on the notion of separating the likelihood used for estimating Σ from that used for estimating β. This can be achieved in a number of ways. One way is to effectively assume a locally uniform prior distribution of the fixed effects β and integrate them out of the likelihood (Pinheiro and Bates, 2000, pp. 75–76). An implication of this separation is that the resulting *REML log-likelihoods for models with different fixed effects are not comparable.*

However, for models with the same fixed effects, the REML log-likelihoods can be used to compare two nested models for Σ. *A likelihood ratio test for two nested covariance models with the same fixed effects is based on comparing twice the difference in the two maximized REML log-likelihoods to a chi-squared distribution with degrees of freedom equal to the difference between the number of variance-covariance parameters in the full and reduced models.*

It can be shown that the log-likelihood function for model (10.3) is given by

$$\log\left(L(\beta, \sigma_b^2, \sigma_e^2 \mid \mathbf{Y})\right)$$

$$= -\frac{n}{2}\log(2\pi) - \frac{1}{2}\log(\det(\Sigma)) - \frac{1}{2}(\mathbf{Y} - \mathbf{X}\beta)'\Sigma^{-1}(\mathbf{Y} - \mathbf{X}\beta)$$

(see e.g., Ruppert, Wand and Carroll, 2003, p. 100). The maximum likelihood (ML) estimates of β and Σ can be obtained by maximizing this function. Alternatively, given the estimated variance–covariance matrix of the error term $\hat{\Sigma}$ obtained from REML, minimizing the third term in the log-likelihood gives $\hat{\beta}_{\text{GLS}}$ the generalized least squares (GLS) estimator of β. It can be shown that

$$\hat{\beta}_{\text{GLS}} = (\mathbf{X}'\hat{\Sigma}^{-1}\mathbf{X})^{-1}\mathbf{X}'\hat{\Sigma}^{-1}\mathbf{Y}$$

For models with the same random effects and hence the same Σ which have been estimated by ML, the ML log-likelihoods can be used to produce a likelihood ratio test *to compare two nested models for fixed effects. This test is based on comparing twice the difference in the two maximized ML log-likelihoods to a chi-squared distribution with degrees of freedom equal to the difference between the number of fixed effects parameters in the full and reduced models. Given that REML log-likelihoods for different fixed effects are not comparable, REML log-likelihoods should not be used to produce a likelihood ratio test to compare two nested models for fixed effects.*

Given below is the output from R associated with fitting model (10.1) to the data on females using REML. The error variance is estimated to be $\hat{\sigma}_e^2 = 0.7800^2 = 0.608$ while the variance due to the random intercept is estimated to be $\hat{\sigma}_b^2 = 2.0685^2 = 4.279$. Utilizing (10.2) we find that the correlation of two measurements within the same female subject is estimated to be

$$\hat{\text{Corr}}(\text{Distance}_{ij}, \text{Distance}_{ik}) = \frac{\hat{\sigma}_b^2}{\hat{\sigma}_b^2 + \hat{\sigma}_e^2} = \frac{4.279}{4.279 + 0.608} = 0.88$$

This result is in line with the sample correlations reported earlier.

Output from R: REML fit of model (10.1) for females

```
Linear mixed-effects model fit by REML
Data: FOrthodont
     AIC       BIC      logLik
149.2183  156.169  -70.60916

Random effects:
Formula: ~1 | Subject
        (Intercept)   Residual
StdDev:    2.06847   0.7800331

Fixed effects: distance ~ age
               Value   Std. Error  DF   t-value   p-value
(Intercept) 17.372727  0.8587419   32  20.230440       0
age          0.479545  0.0525898   32   9.118598       0
Correlation:
     (Intr)
age  -0.674

Number of Observations: 44
Number of Groups:  11
```

Figure 10.3 contains plots of Distance against Age for each female with the straight-line fits from model (10.1) included. Once again these plots have been ordered from bottom left to top right in terms of increasing average value of Distance. We shall see in Table 10.2 that the estimated random intercept is higher than one may initially expect for subject F10 and lower than one may initially expect for subject F11. We shall also see that this is due to so called "shrinkage" associated with random effects.

For comparison purposes, we also fit the following model for subject i ($i = 1, 2,$ ..., 11) at Age j ($j = 1, 2, 3, 4$):

$$\text{Distance}_{ij} = \alpha_i + \beta_1 \text{Age}_j + e_{ij} \tag{10.4}$$

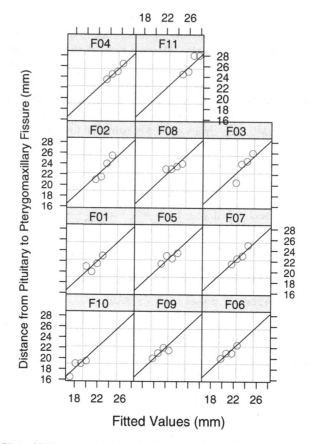

Figure 10.3 Plots of Distance against Age for females with fits from model (10.1)

Table 10.2 Random and fixed intercepts for each female and their difference

Subject	Random intercept	Fixed intercept	Random - fixed
F11	20.972	21.100	–0.128
F04	19.524	19.600	–0.076
F03	18.437	18.475	–0.038
F08	18.075	18.100	–0.025
F07	17.713	17.725	–0.012
F02	17.713	17.725	–0.012
F05	17.351	17.350	0.001
F01	16.144	16.100	0.044
F09	15.902	15.850	0.052
F06	15.902	15.850	0.052
F10	13.367	13.225	0.142

where the fixed effect α_i allows for a different intercept for each subject. Table 10.2 gives the values of the estimates of α_i, that is, estimates of the fixed intercepts in model (10.4) along with the estimated random intercept for each subject from model (10.1), that is estimates of $\beta_0 + b_i$. Also included in Table 10.2 is the difference between the random and fixed intercepts.

Inspection of Table 10.2 reveals that the random intercepts are smaller (larger) than the fixed intercepts when they are associated with subjects with larger (smaller) values of average distance than the overall average value of distance. In other words, there is "shrinkage" in the random intercepts towards the mean. A number of authors refer to this as "borrowing strength" from the mean.

It can be shown that there is more "shrinkage" when n_i, the number of observations on the ith subject is small. This is based on the notion that less weight should be given to the ith individual's average response when it is more variable. In addition, it can be shown that there is more "shrinkage" when σ_b^2 is relatively small and σ_e^2 is relatively large (see for example Frees, 2004, p. 128). This is based on the notion that less weight should be given to the ith individual's average response when there is little variability between subjects but high variability within subjects.

In summary, we have found that the correlation between two distance measurements for female subjects is both relatively constant across different time intervals and high (estimated from model (10.1) to be 0.88). In addition, the fixed effect due to Age in model (10.1) is highly statistically significant.

Orthodontic growth data: Males

We next consider the data just for males. Figure 10.4 shows a plot of Distance against Age for each of the 16 males. Notice that the plots have been ordered from bottom left to top right in terms of increasing average value of Distance.

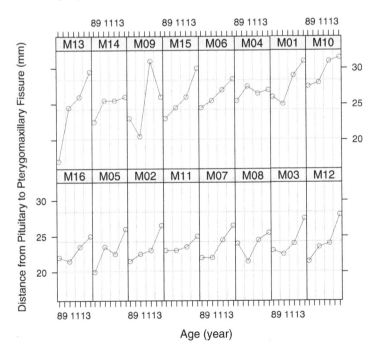

Figure 10.4 Plot of Distance against Age for each male subject

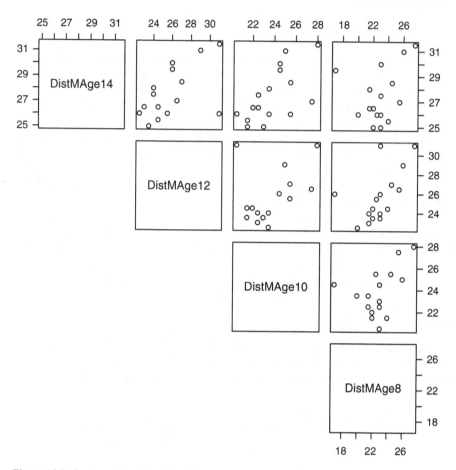

Figure 10.5 Scatter plot matrix of the Distance measurements for male subjects

We again consider model (10.1) this time for subject i ($i = 1, 2, ..., 16$) at Age j ($j = 1, 2, 3, 4$). In order to investigate whether the assumption of constant correlation inherent in (10.1) is reasonable for males, we calculate the correlations between two distance measurements for the same male subject over each time interval. In what follows, we shall denote the distance measurements for males aged 8, 10, 12 and 14 as DistMAge8, DistMAge10, DistMAge12, DistMAge14, respectively. The output from R below gives the correlations between these four variables. Notice the similarity among the correlations away from the diagonal, which range from 0.315 to 0.631.

Output from R: Correlations between male measurements

	DistMAge8	DistMAge10	DistMAge12	DistMAge14
DistMAge8	1.000	0.437	0.558	0.315
DistMAge10	0.437	1.000	0.387	0.631

DistMAge12	0.558	0.387	1.000	0.586
DistMAge14	0.315	0.631	0.586	1.000

Figure 10.5 shows a scatter plot matrix of the distance measurements for males aged 8, 10, 12 and 14. The linear association in each plot in Figure 10.5 appears to be quite similar but much weaker than that in the corresponding plot for females, namely, Figure 10.2. In addition, there are one or two points in some of the plots that are isolated from the bulk of the points. These correspond to subjects M09 and M13 and should in theory be investigated. However, overall, it seems that the assumption that correlations are constant across Age is also a reasonable one for males.

Given below is the output from R associated with fitting model (10.1) to the data on males using REML. The error variance is estimated to be $\hat{\sigma}_e^2 = 1.6782^2 = 2.816$ while the variance due to the random intercept is estimated to be $\hat{\sigma}_b^2 = 1.625^2 = 2.641$. Utilizing (10.2) we find that the correlation of two measurements within the same male subject is estimated to be

$$\hat{\text{Corr}}(\text{Distance}_{ij}, \text{Distance}_{ik}) = \frac{\hat{\sigma}_b^2}{\hat{\sigma}_b^2 + \hat{\sigma}_e^2} = \frac{2.641}{2.641 + 2.816} = 0.48$$

This result is in line with the sample correlations reported earlier.

Output from R: REML fit of model (10.1) for males

```
Linear mixed-effects model fit by REML
Data: MOrthodont
      AIC       BIC      logLik
 281.4480  289.9566  -136.7240
Random effects:
Formula: ~1 | Subject
          (Intercept)   Residual
StdDev:    1.625019     1.67822

Fixed effects: distance ~ age
               Value  Std. Error  DF   t-value  p-value
(Intercept) 16.340625  1.1287202  47  14.477126       0
age          0.784375  0.0938154  47   8.360838       0
Correlation:
     (Intr)
age  -0.914

Number of Observations: 64
Number of Groups: 16
```

Figure 10.6 contains plots of Distance against Age for each male with the straight-line fits from model (10.1) included. Once again these plots have been ordered from bottom left to top right in terms of increasing average value of Distance. Careful inspection of Figure 10.6 reveals that the estimated random intercept is lower than one may initially expect for subject M10, with at least three of the four points lying above the fitted line. This is due to "shrinkage" associated with random effects.

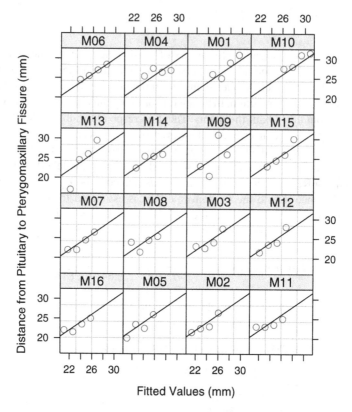

Figure 10.6 Plots of Distance against Age for males with fits from model (10.1)

In summary, we have found that the correlation between two distance measurements for male subjects is both relatively constant across different time intervals and moderate (estimated from model (10.1) to be 0.48). In addition, the fixed effect due to Age in model (10.1) is also highly statistically significant for males.

Table 10.3 gives the estimates of the error standard deviation (σ_e) and the random effect standard deviation (σ_b) for males and females we found earlier in this chapter. Comparing these estimates we see that while $\hat{\sigma}_b$ is similar across males and females, $\hat{\sigma}_e$ is more than twice as big for males as it is for females. Thus, in order to combine the separate models for males and females, we shall allow the error variance to differ with sex, while assuming the random effect variance is constant across sex. The combined model will readily allow us to answer the important question about whether the growth rate differs across sex.

Orthodontic growth data: Males and females

The model we next consider for both male and female subjects i ($i = 1, 2, ..., 27$) at Age j ($j = 1, 2, 3, 4$) is as follows:

Table 10.3 Estimates of the random effect and error standard deviations

	$\hat{\sigma}_b$	$\hat{\sigma}_e$
Males	1.63	1.68
Females	2.07	0.78

$$\text{Distance}_{ij} = \beta_0 + \beta_1\text{Age}_j + \beta_2\text{Sex} + \beta_3\text{Sex} \times \text{Age}_j + b_i + e_{ij}^{\text{Sex}} \quad (10.5)$$

where the random effect b_i is assumed to follow a normal distribution with mean 0 and variance σ_b^2 independent of the error term e_{ij}^{Sex} which is iid $N(0, \sigma_{e\text{Sex}}^2)$, where $\sigma_{e\text{Sex}}^2$ depends on Sex.

Given below is the output from R associated with fitting model (10.5) using REML. The error variances are estimated to be $\hat{\sigma}_{e\text{Male}}^2 = 1.6698^2 = 2.788$ and $\hat{\sigma}_{e\text{Female}}^2 = (0.4679 \times 1.6698)^2 = 0.610$ while the variance due to the random intercept is estimated to be $\hat{\sigma}_b^2 = 1.8476^2 = 3.414$. Utilizing (10.2) we find that the correlation of two measurements within the same male and female subject are estimated to be

$$\hat{\text{Corr}}_{\text{Male}}(\text{Distance}_{ij}, \text{Distance}_{ik}) = \frac{\hat{\sigma}_b^2}{\hat{\sigma}_b^2 + \hat{\sigma}_{e\text{Male}}^2} = \frac{3.414}{3.414 + 2.788} = 0.55$$

and

$$\hat{\text{Corr}}_{\text{Female}}(\text{Distance}_{ij}, \text{Distance}_{ik}) = \frac{\hat{\sigma}_b^2}{\hat{\sigma}_b^2 + \hat{\sigma}_{e\text{Female}}^2} = \frac{3.414}{3.414 + 0.610} = 0.85$$

Thus, allowing for the random effect variance to differ across sex has produced estimated correlations in line with those obtained from the separate models for males and females reported earlier.

Output from R: REML fit of model (10.5) for males and females

```
Linear mixed-effects model fit by REML
 Data: Orthodont
      AIC       BIC    logLik
 429.2205  447.7312  -207.6102

Random effects:
 Formula: ~1 | Subject
         (Intercept)  Residual
StdDev:    1.847570  1.669823

Variance function:
 Structure: Different standard deviations per stratum
 Formula: ~1 | Sex
 Parameter estimates:
      Male      Female
 1.0000000  0.4678944
```

```
Fixed effects: distance ~ age * Sex
                    Value   Std. Error   DF   t-value   p-value
(Intercept)      16.340625   1.1450945   79  14.270111   0.0000
age               0.784375   0.0933459   79   8.402883   0.0000
SexFemale         1.032102   1.4039842   25   0.735124   0.4691
age:SexFemale    -0.304830   0.1071828   79  -2.844016   0.0057

Correlation:
                  (Intr)     age   SexFml
age               -0.897
SexFemale         -0.816    0.731
age:SexFemale      0.781   -0.871  -0.840

Number of Observations: 108
Number of Groups: 27
```

The fixed effect due to the interaction between Sex and Age in model (10.5) is highly statistically significant (p-value = 0.0057). The estimated coefficient of this interaction term is such that the growth rate of females is significantly less than that of males.

We next test whether allowing the error variance to differ across sex is really necessary by comparing the maximized REML likelihoods for model (10.5) and the following model with the same fixed effects but in which the error variance is constant across Sex:

$$\text{Distance}_{ij} = \beta_0 + \beta_1 \text{Age}_j + \beta_2 \text{Sex} + \beta_3 \text{Sex} \times \text{Age}_j + b_i + e_{ij} \qquad (10.6)$$

Given below is the output from R associated with fitting model (10.6) using REML. Notice that the estimates of the fixed effects match those obtained from model (10.5) while the standard errors of these estimates differ a little across the two models.

Output from R: REML fit of model (10.6) for males and females

```
Linear mixed-effects model fit by REML
Data: Orthodont
     AIC        BIC      logLik
 445.7572   461.6236   -216.8786

Random effects:
Formula: ~1 | Subject
        (Intercept)  Residual
StdDev:    1.816214   1.386382

Fixed effects: distance ~ age * Sex
                    Value   Std. Error   DF   t-value   p-value
(Intercept)      16.340625   0.9813122   79  16.651810   0.0000
age               0.784375   0.0775011   79  10.120823   0.0000
SexFemale         1.032102   1.5374208   25   0.671321   0.5082
age:SexFemale    -0.304830   0.1214209   79  -2.510520   0.0141
```

```
Correlation:
                (Intr)      age   SexFml
age             -0.869
SexFemale       -0.638    0.555
age:SexFemale    0.555   -0.638   -0.869


Number of Observations: 108
Number of Groups: 27
```

Given below is the output from R comparing the REML fits of models (10.5) and (10.6). The likelihood ratio test is highly statistically significant indicating that model (10.5) provides a significantly better model for the variance–covariance than does model (10.6).

Output from R: Comparing REML fits of models (10.5) and (10.6)

```
      Model df      AIC       BIC    logLik   Test L.Ratio p-value
m10.6   1   6 445.7572  461.6236 -216.8786
m10.5   2   7 429.2205  447.7312 -207.6102 1 vs 2 18.53677  <.0001
```

10.1.2 Residuals in Mixed Models

In the previous section, we discussed how REML (ML)-based likelihood ratio tests can be used to test nested models for the covariance (fixed effects). However, such tests are of limited value when neither of the models being compared is a valid model for the fixed effects and the covariance. As such, it is clearly desirable to have a set of diagnostics which examine the validity of different aspects of the model under consideration. We begin this discussion by extending the concept of residuals to mixed models.

In models containing random effects, there are at least two types of residuals depending on whether we are considering the data in a conditional or unconditional sense. We shall see that this corresponds to whether we are focusing within subjects or at the population level.

Before we define these two types of residuals, we need to introduce some notation. To keep things as straightforward as possible we shall consider just a single fixed effect and a single random effect. The extension to more than one fixed effect and one random effect will be obvious.

Let Y_{ij} denote the outcome for subject i at fixed effect x_j. Let b_i denote the random effect due to subject i. We shall suppose that the following model is under consideration:

$$Y_{ij} = \beta_0 + \beta_1 x_j + b_i + e_{ij} \qquad\qquad (10.7)$$

where the random effect b_i is assumed to follow a normal distribution with mean 0 and variance σ_b^2 (that is, $b_i \sim N(0, \sigma_b^2)$) independent of the error term e_{ij} which is iid $N(0, \sigma_e^2)$.

The *ij*th *conditional (or within subjects) residual* is the difference between the observed value of Y_{ij} and its predicted value and hence it is given by

$$\hat{e}_{Cij} = Y_{ij} - \hat{\beta}_0 - \hat{\beta}_1 x_j - \hat{b}_i \qquad (10.8)$$

The *ij*th *marginal (or population) residual* is the difference between the observed value of Y_{ij} and its estimated mean and hence it is given by

$$\hat{e}_{Mij} = Y_{ij} - \hat{\beta}_0 - \hat{\beta}_1 x_j \qquad (10.8)$$

In models without random effects, the two sets of residuals are the same since then the predicted value of Y_{ij} equals its estimated mean.

Some authors (e.g., Weiss, 2005, pp. 332–333) define a third residual called *empirical Bayes residuals* which are equal to the estimated random effects and hence given by

$$\hat{e}_{Bi} = \hat{b}_i \qquad (10.9)$$

The standard advice (e.g., Nobre and Singer, 2007, p. 3) for checking the validity of model for the fixed effects is that "plots of the elements of the vector of marginal residuals versus the explanatory variables in X may be employed to check the linearity of *y* with respect to such variables with the same spirit as the usual residuals in standard (normal) linear models. A random behavior around zero is expected when the linear relationship holds".

Furthermore, Nobre and Singer (2007) recommend that the conditional residuals be used to check the usual normality and constant variance assumptions placed on the e_{ij}. In addition, they recommend that the empirical Bayes residuals (i.e., the estimated random effects) be used to identify outlying subjects as well as assessing the normality of the random effects.

However, Weiss (2005, p. 332) and Fitzmaurice, Laird and Ware (2004, p. 238) draw attention to the fact that the marginal residuals are correlated due to the correlation in the model. In particular, Fitzmaurice, Laird and Ware (2004, p. 238) point out that this correlation may produce "an apparent systematic trend in the scatter-plot of the (marginal) residuals against a selected covariate" even when the fixed effects have been modeled correctly. In addition, Weiss (2005, p. 332) also warns readers to be "careful" because the "estimation process introduces correlation into the (conditional) residuals even when none exists in the e_{ij} (using the present notation)."

Orthodontic growth data: Males and females

We shall illustrate the uses and limitations of the three types of residuals (conditional, marginal and empirical Bayes) using (10.5) and (10.6). We begin by considering empirical Bayes residuals.

Figure 10.7 shows a normal Q–Q plot of the empirical Bayes residuals (i.e., the estimated random effects $\hat{b}_j$) for model (10.5). Estimated random effects in the plot

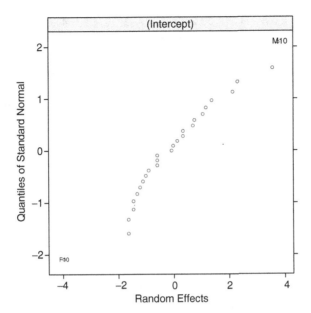

Figure 10.7 Normal Q–Q plot of the estimated random effects from model (10.5)

below the lower or above the upper 2.5% normal critical value are identified as out-liers. In theory, the results for these subjects (namely F10 and M10) should be inves-tigated. Figure 10.7 also shows that there is some skewness in the random effects.

We next demonstrate the fact that the marginal and conditional residuals are cor-related even when we believe we have fitted a valid model (in this case model (10.5)). In what follows we shall denote the marginal residuals for subjects at ages 8, 10, 12 and 14 by MRAge8, MRAge10, MRAge12 and MRAge14, respectively and the corresponding conditional residuals by CRAge8, CRAge10, CRAge12 and CRAge14.

Given below is output from R which gives the correlations of marginal residuals over time. Away from the diagonal the correlations vary from 0.522 to 0.728.

Output from R: Correlations among marginal residuals

	MRAge8	MRAge10	MRAge12	MRAge14
MRAge8	1.000	0.560	0.660	0.522
MRAge10	0.560	1.000	0.560	0.718
MRAge12	0.660	0.560	1.000	0.728
MRAge14	0.522	0.718	0.728	1.000

Figure 10.8 shows a scatter plot matrix of the marginal residuals. The high posi-tive correlations just reported above are clearly apparent in these plots.

Given below is output from R which gives the correlations of conditional residu-als over time. Away from the diagonal the correlations vary from –0.005 to –0.620.

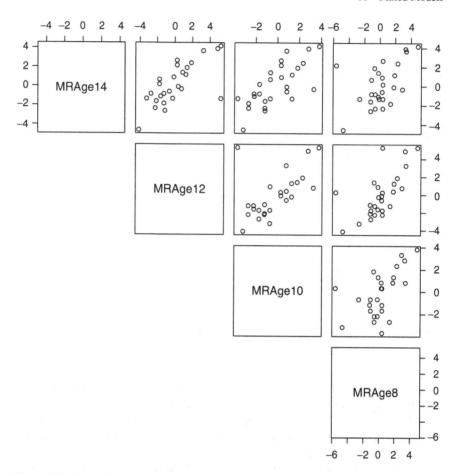

Figure 10.8 Scatter plot matrix of the marginal residuals from model (10.5)

Output from R: Correlations among conditional residuals

	CRAge8	CRAge10	CRAge12	CRAge14
CRAge8	1.000	-0.307	-0.120	-0.620
CRAge10	-0.307	1.000	-0.544	-0.005
CRAge12	-0.120	-0.544	1.000	-0.083
CRAge14	-0.620	-0.005	-0.083	1.000

Figure 10.9 shows a scatter plot matrix of the marginal residuals. The generally negative correlations just reported above are clearly apparent in these plots.

To demonstrate the potential shortcomings of using the conditional residuals to assess the constant variance assumptions placed on the e_{ij}, we shall examine plots of conditional residuals versus fitted values for models (10.6) and (10.5). These plots can be found in Figures 10.10 and 10.11, respectively.

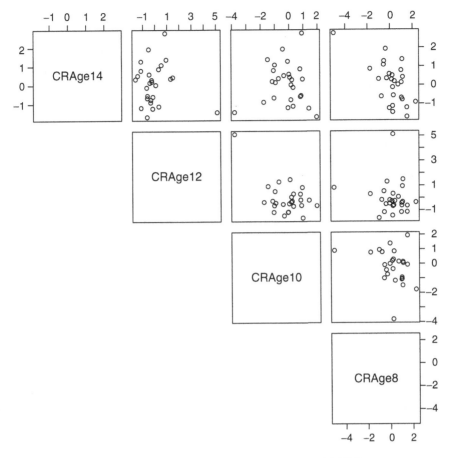

Figure 10.9 Scatter plot matrix of the conditional residuals from model (10.5)

Recall that model (10.6) assumes that the error variability is constant across males and females. Figure 10.10 seems to provide a clear indication that this is not a reasonable assumption since the residuals from this model for males are more variable than those for the females.

On the other hand, model (10.5) allows for the variability of the errors to differ across males and females. Interestingly, Figure 10.11 is remarkably similar to Figure 10.10, with the conditional residuals for males much more variable than those for the females. This example demonstrates that plots of conditional residuals versus fitted values can produce evidence of nonconstant error variance even when a model allowing for differing variances has been fit. In this example, standardizing the conditional residuals by estimates of error variability within sex produces plots which are in line with what we would expect in this case. Thus, this example also illustrates the importance of standardizing conditional residuals when checking assumptions about error variances.

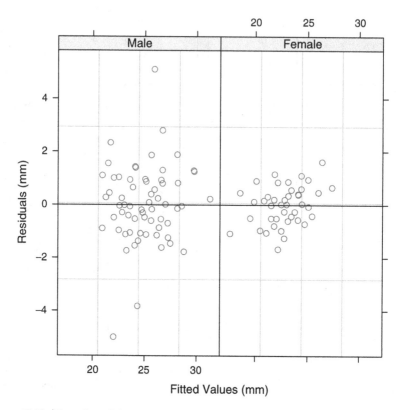

Figure 10.10 Plots of conditional residuals vs fitted values from model (10.6)

Cholesky residuals

In order to overcome the problems associated with the correlation in residuals from mixed models Fitzmaurice, Laird and Ware (2004, p. 238) recommend that the residuals be transformed so that ideally they have zero correlation and constant variance. While there are a number of ways to do this (see Weiss, 2005, pp. 330–331), Fitzmaurice, Laird and Ware (2004, p. 238) consider the transformation based on the Cholesky decomposition of Σ. At this point we need to reintroduce some notation.

Combining the random effects and the error term, we can rewrite (10.7) as

$$Y_{ij} = \beta_0 + \beta_1 x_j + \varepsilon_{ij} \tag{10.10}$$

where $\varepsilon_{ij} = b_i + e_{ij}$ is assumed to follow a normal distribution with mean 0 and variance given by the appropriate element of the variance covariance matrix Σ, which was given below (10.3). In general matrix notation, (10.10) can be written as

$$\mathbf{Y} = \mathbf{X}\beta + \varepsilon \tag{10.11}$$

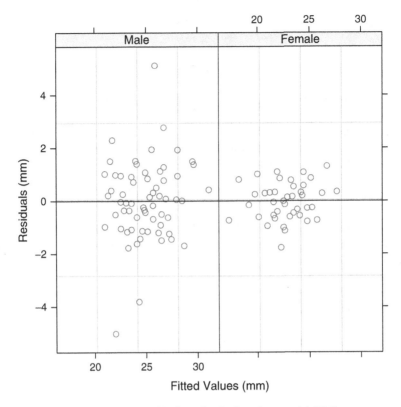

Figure 10.11   Plots of conditional residuals vs fitted values from model (10.5)

Just as we did in Chapter 9, we can express the symmetric positive-definite matrix
Σ as

$$\Sigma = SS'$$

where S is a lower triangular matrix with positive diagonal entries. This result is
commonly referred to as the Cholesky decomposition of Σ. Roughly speaking, S
can be thought of as the "square root" of Σ.

Multiplying each side of (10.11) by S^{-1}, the inverse of S, gives

$$S^{-1}\mathbf{Y} = S^{-1}\mathbf{X}\beta + S^{-1}\varepsilon$$

Notice that

$$\mathrm{Var}\left(S^{-1}\varepsilon\right) = S^{-1}\mathrm{Var}\left(\varepsilon\right)\left(S^{-1}\right)' = S^{-1}\Sigma\left(S^{-1}\right)' = S^{-1}SS'\left(S^{-1}\right)' = \mathbf{I},$$

the identity matrix. Thus, pre-multiplying each term in equation (10.11) by S^{-1}, the
inverse of S, produces a linear model with uncorrelated errors. In other words, let

$$\mathbf{Y}^* = S^{-1}\mathbf{Y}, \mathbf{X}^* = S^{-1}\mathbf{X}, \varepsilon^* = S^{-1}\varepsilon$$

then,

$$\mathbf{Y}^* = \mathbf{X}^*\beta + \varepsilon^* \qquad (10.12)$$

provides a linear model with uncorrelated errors with unit variance.

In practice, we replace the variance–covariance matrix Σ by an estimate $\hat{\Sigma}$ from which we obtain $\hat{S}^{-1}$, the Cholesky decomposition of the inverse of $\hat{\Sigma}$. Instead of fitting (10.12), we could transform $\hat{\varepsilon}$, the residuals from model (10.11) by pre-multiplying them by $\hat{S}^{-1}$ to produce the following set of transformed residuals

$$\hat{\varepsilon}^* = \hat{S}^{-1}\hat{\varepsilon} = \hat{S}^{-1}\left(\mathbf{Y} - \mathbf{X}\hat{\beta}\right)$$

These residuals are also called scaled or *Cholesky residuals*. Their properties were studied by Houseman, Ryan and Coull (2004). In practice, since we use the estimated variance–covariance matrix the Cholesky residuals are not completely uncorrelated.

Figure 10.12 gives separate box plots of the Cholesky residuals from models (10.6) and (10.5) respectively against Sex. It is evident from these box plots that the Cholesky residuals for model (10.6) are more variable for males than females while the same is not true of the Cholesky residuals for model (10.5). This increase in variability in Cholesky residuals from model (10.6) for males is statistically significant (e.g., Levene's

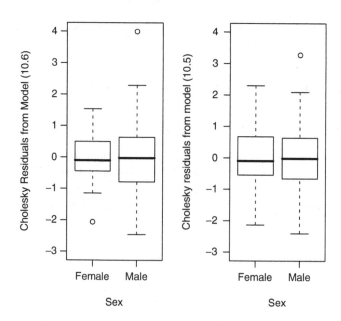

Figure 10.12 Box plots of the Cholesky residuals from models (10.5) and (10.6)

test for equality of variances p-value = 0.012). Thus, the Cholesky residuals readily lead to the correct conclusion that the error variance differs across Sex.

Fitzmaurice, Laird and Ware (2004, p. 238) recommend that the standard set of regression diagnostics based on the Cholesky residuals be applied. In particular, one can check plots of $\hat{\varepsilon}^*$ against $\hat{\mathbf{Y}}^*$ and $\mathbf{X}^*$ which should display no systematic patterns if the model is correctly specified.

10.2 Models with Covariance Structures Which Vary Over Time

Thus far in this chapter we have looked exclusively at regression models with random intercepts. We discovered in Section 10.1 that a random intercepts model is equivalent to assuming a constant correlation within subjects over any chosen time interval. Fitzmaurice, Laird and Ware (2004, p. 78) describe this assumption as "often inappropriate for longitudinal data ... where the correlations are expected to decay with increasing separation over time". Moreover, Weiss (2005, p. 247) points out that the equal correlation assumption "is unlikely for real data measured on human beings over long enough periods of time" with the exception being "measures that are very persistent over the data collection time frame."

In this section we look at models with covariance structures which vary over time. We shall use the following real example to highlight the challenges associated with fitting models to longitudinal data, namely, the choice of model for the fixed effects (i.e., the conditional mean structure) and the choice of the model for the error covariance are inter-reliant.

Pig weight data

Diggle, Heagerty, Liang and Zeger (2002, p. 34) consider a data set provided by Dr. Philip McCloud (when he was a faculty member at Monash University in Melbourne, Australia) on the weights of 48 pigs measured in 9 successive weeks. The data can be found on the book web site in the file pigweight.txt. We seek a model for the fixed effects (i.e., the conditional mean structure of pig weights over time), as well as a model for the error covariance.

Figure 10.13 shows a plot of weight against time (in weeks). Results from the same pig are connected by dotted lines. It is apparent from this figure that there is an increasing trend in both the mean and the variance of weight as time increases.

We next investigate the covariance structure in these data. In what follows, we shall denote the pig weights at weeks 1, 2, ..., 9 by T1, T2, ... T9, respectively. Figure 10.14 shows a scatter plot matrix of the pig weights at weeks 1, 2, ... 9.

The output from R below gives the correlations between T1, T2, ..., T9. Notice how the correlations decay as the time interval between the measurements increase. The decreasing correlation as the time interval between measurements increases is clearly evident in Figure 10.14.

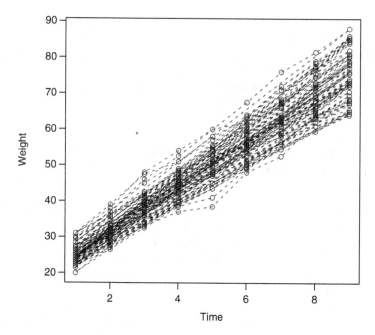

Figure 10.13 Plot of pig weights over time

Output from R: Correlations between measurements

	T1	T2	T3	T4	T5	T6	T7	T8	T9
T1	1.00	0.92	0.80	0.80	0.75	0.71	0.66	0.63	0.56
T2	0.92	1.00	0.91	0.91	0.88	0.84	0.78	0.71	0.66
T3	0.80	0.91	1.00	0.96	0.93	0.91	0.84	0.82	0.77
T4	0.80	0.91	0.96	1.00	0.96	0.93	0.87	0.83	0.79
T5	0.75	0.88	0.93	0.96	1.00	0.92	0.85	0.81	0.79
T6	0.71	0.84	0.91	0.93	0.92	1.00	0.96	0.93	0.89
T7	0.66	0.78	0.84	0.87	0.85	0.96	1.00	0.96	0.92
T8	0.63	0.71	0.82	0.83	0.81	0.93	0.96	1.00	0.97
T9	0.56	0.66	0.77	0.79	0.79	0.89	0.92	0.97	1.00

10.2.1   Modeling the Conditional Mean

When there are relatively few time points and measurements are available for each
subject at all of these time points, it is possible to consider a so-called unstructured
covariance matrix for the error term while concentrating on finding a parsimonious
model for the conditional mean. Having found such a model one can then attempt
to find a parsimonious model for the error covariance. We shall adopt this strategy
for the pig weight data, even though there are nine time points. On the other hand,
when it is not possible to consider an unstructured covariance matrix for the error

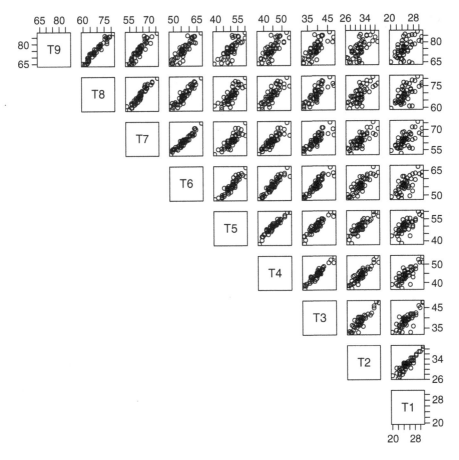

Figure 10.14 Scatter plot matrix of the pig weights at weeks 1 to 9

term the usual practice is to fit a maximal model for the conditional mean and then concentrate on finding a parsimonious model for the error covariance.

A straight line model for the conditional mean

A number of authors including Diggle, Heagerty, Liang and Zeger (2002), Yang and Chen (1995), Ruppert, Wand and Carroll (2003) and Alkhamisi and Shukur (2005) have modeled the conditional mean weight at a given week by a straight line but suggested different error structures. Thus, we shall start with a straight-line model for the conditional mean. Thus we shall consider the following model for the weight Y_{ij} of the ith pig ($i = 1, 2, \ldots 48$) in the jth week x_j ($j = 1, \ldots, 9$):

$$Y_{ij} = \beta_0 + \beta_1 x_j + \varepsilon_{ij}$$

In general matrix notation, this is

$$\mathbf{Y} = \mathbf{X}\beta + \varepsilon \tag{10.13}$$

where in this example

$$Y = \begin{pmatrix} y_{1,1} \\ \vdots \\ y_{1,9} \\ \vdots \\ y_{48,1} \\ \vdots \\ y_{48,9} \end{pmatrix} \qquad X = \begin{pmatrix} 1 & x_1 \\ \vdots & \vdots \\ 1 & x_9 \\ \vdots & \vdots \\ 1 & x_1 \\ \vdots & \vdots \\ 1 & x_9 \end{pmatrix}$$

We shall assume that

$$\varepsilon \sim N(0, \Sigma)$$

where in this example Σ is the following matrix:

$$\Sigma = \begin{pmatrix} D & 0 & 0 & 0 & \cdots & 0 \\ 0 & D & 0 & 0 & \cdots & 0 \\ 0 & 0 & D & 0 & \cdots & 0 \\ 0 & 0 & 0 & \ddots & 0 & 0 \\ \vdots & \vdots & \vdots & 0 & D & 0 \\ 0 & 0 & 0 & 0 & 0 & D \end{pmatrix} \qquad D = \begin{pmatrix} \sigma_1^2 & \sigma_{12} & \sigma_{13} & \cdots & \sigma_{19} \\ \sigma_{12} & \sigma_2^2 & \sigma_{23} & \cdots & \sigma_{29} \\ \sigma_{13} & \sigma_{23} & \sigma_3^2 & \cdots & \sigma_{39} \\ \vdots & \vdots & \vdots & \ddots & \vdots \\ \sigma_{19} & \sigma_{29} & \sigma_{39} & \cdots & \sigma_9^2 \end{pmatrix}$$

with D a positive-definite and symmetric (9×9) matrix. In particular we allow the diagonal entries (i.e., the variances at each time point) to differ. This is in line with the following advice from Fitzmaurice, Laird and Ware (2004, p. 169) that "practical experience based on many longitudinal studies has led to the empirical observation that variances are rarely constant over time."

Given below is the output from R associated with fitting model (10.13) using REML. Notice how in this case the estimated error variance increases over time.

Output from R: REML fit of model (10.13)

```
Generalized least squares fit by REML
Model: weight ~ Time
Data: pigweights
      AIC          BIC         logLik
1635.943    1826.941    -770.9717
Correlation Structure: General
Formula: ~1 | Animal
Parameter estimate(s):
```

```
Correlation:
        1        2        3        4        5        6        7        8
2    0.894
3    0.722    0.895
4    0.767    0.910    0.946
5    0.742    0.879    0.889    0.956
6    0.694    0.836    0.876    0.930    0.923
7    0.650    0.774    0.806    0.863    0.857    0.963
8    0.602    0.720    0.812    0.835    0.810    0.927    0.953
9    0.546    0.669    0.753    0.789    0.788    0.891    0.917    0.968
```

```
Variance function:
Structure: Different standard deviations per stratum
Formula: ~1 | Time
Parameter estimates:
        1            2            3           4           5           6
1.000000     1.133915     1.514700    1.524495    1.825750    1.798021
        7            8            9
2.005047     2.212073     2.561561
```

```
Coefficients:
                 Value    Std. Error    t-value    p-value
(Intercept)   19.072855    0.3292651   57.92552          0
Time           6.174367    0.0791560   78.00249          0
```

```
Residual standard error: 2.476831
Degrees of freedom: 432 total;  430 residual
```

In order to check whether a straight line provides an adequate model for the conditional mean pig weight at a given week we shall examine the Cholesky residuals associated with model (10.13). Figure 10.15 shows a plot of the Cholesky residuals against x^*, the second column of $\mathbf{X}^* = S^{-1}\mathbf{X}$ where S is a lower triangular matrix with positive diagonal entries such that $\hat{\Sigma} = SS'$. Figure 10.15 also includes the loess fit (with $\alpha = \frac{1}{3}$).

If the straight-line model for the fixed effects is valid then there should be no discernible pattern in Figure 10.15. Instead, the loess fit in Figure 10.15 suggests that there is some structure in the Cholesky residuals. In order to check that we are not over interpreting the pattern in Figure 10.15, we fit a fifth-order polynomial fit to the Cholesky residuals in Figure 10.15 as a function of x^*. The resulting overall F-statistic is highly significant (p-value = 0.0002). Thus, there is evidence that model (10.13) is an invalid model for the fixed effects.

In view of this we shall consider an expanded model for the fixed effects. We could include polynomial terms in Time as predictors. In view of the fact that a fifth-order polynomial fit to the Cholesky residuals in Figure 10.15 is highly significant it seems natural to consider a fifth-order polynomial model in Time. However, the resulting regression coefficients of such a high-order polynomial model are difficult to interpret in practice.

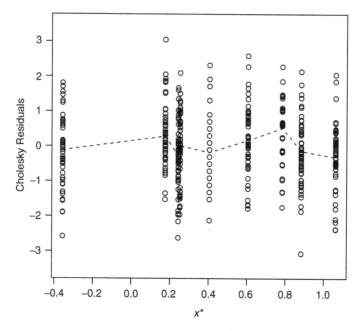

Figure 10.15 Plot of the Cholesky residuals from model (10.13) against x^*

A regression spline model for the conditional mean

Thus, instead, at this point we consider an alternative way of expanding model (10.13) which we shall see is straightforward to interpret, namely, linear regression splines. Put simply, a linear regression spline consists of a series of connected line segments joined together at what are commonly referred to as knots.

In order to proceed we introduce some notation. Define

$$(x-k)_+ = \begin{cases} x-k & \text{if } x>k \\ 0 & \text{if } x \le k \end{cases} \tag{10.14}$$

The left-hand plot in Figure 10.16 provides a graphical depiction of $(x-k)_+$ with k set equal to 5. The inclusion of $(x-k)_+$ as a predictor produces a fitted model which resembles a broken stick, with the break at the knot k. Thus, this predictor allows the slope of the line to change at k. The right-hand plot in Figure 10.16 shows a stylized example of a spline model with a knot at $x=5$ illustrating these points. The exact form of this stylized model is as follows:

$$\mathrm{E}(Y \mid x) = x - 0.75(x-5)_+$$

Utilizing (10.14) we find that in this case

$$\mathrm{E}(Y \mid x) = \begin{cases} x & \text{if } x \le 5 \\ 0.25x & \text{if } x > 5 \end{cases}$$

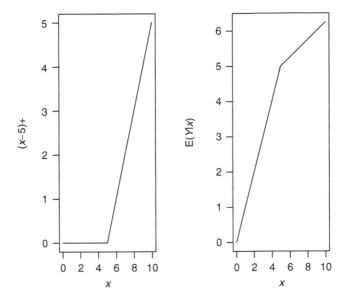

Figure 10.16 Graphical depiction of $(x - 5)_+$ and a stylized linear regression spline

In other words, the model in right-hand plot in Figure 10.16 is made up of connected straight lines with slope equal to 1 if $x \leq 5$ and slope equal to 0.25 if $x > 5$.

We next consider linear regression splines in the context of the pig weight example. In order to make the model for the fixed effects as flexible as possible at this exploratory stage, we shall add knots at all the time points except the first and the last. This will produce a model which consists of a series of connected line segments whose slopes could change from week to week. Thus, we shall add the following predictors to model (10.13) $(x - 2)_+, (x - 3)_+, \ldots (x - 8)_+$ and hence consider the following model

$$\mathbf{Y} = \mathbf{X}\boldsymbol{\beta} + \boldsymbol{\varepsilon} \qquad (10.15)$$

where in this example

$$\mathbf{Y} = \begin{pmatrix} y_{1,1} \\ \vdots \\ y_{1,9} \\ \vdots \\ y_{48,1} \\ \vdots \\ y_{48,9} \end{pmatrix} \qquad \mathbf{X} = \begin{pmatrix} 1 & x_1 & (x_1 - 2)_+ & (x_1 - 3)_+ & \cdots & (x_1 - 8)_+ \\ \vdots & \vdots & \vdots & \vdots & \vdots & \vdots \\ 1 & x_9 & (x_9 - 2)_+ & (x_9 - 3)_+ & \cdots & (x_9 - 8)_+ \\ \vdots & \vdots & \vdots & \vdots & \vdots & \vdots \\ 1 & x_1 & (x_1 - 2)_+ & (x_1 - 3)_+ & \cdots & (x_1 - 8)_+ \\ \vdots & \vdots & \vdots & \vdots & \vdots & \vdots \\ 1 & x_9 & (x_9 - 2)_+ & (x_9 - 3)_+ & \cdots & (x_9 - 8)_+ \end{pmatrix}$$

where Y_{ij} again denotes the weight of the ith pig ($i = 1, 2, ... 48$) in the jth week x_j ($j = 1, ..., 9$). We shall assume that

$$\varepsilon \sim N(\mathbf{0}, \Sigma)$$

where Σ is given below (10.13). Model (10.15) is equivalent to allowing the mean weight to be different for each of the 9 weeks and thus can be thought of as a full model (or a saturated model as it sometimes called).

Given below is the output from R associated with fitting model (10.15) using REML. In the output below $(x-2)_+...(x-8)_+$ are denoted by TimeM2Plus , ..., TimeM8Plus. Looking at the output we see that the coefficients of TimeM3Plus, TimeM7Plus and TimeM8Plus (i.e., of $(x-3)_+,(x-7)_+$ and $(x-8)_+$) are highly statistically significant. The coefficients TimeM3Plus and TimeM8Plus are negative, while the coefficient of TimeM3Plus is positive indicating that the weekly weight gain in pigs both slows and increases rather than remaining constant over the nine week time frame.

Output from R: REML fit of model (10.15)

```
Generalized least squares fit by REML
Model: weight ~ Time + TimeM2Plus + TimeM3Plus + TimeM4Plus +
TimeM5Plus + TimeM6Plus + TimeM7Plus + TimeM8Plus
Data: pigweights
      AIC         BIC      logLik
1613.634    1832.192    -752.817

Correlation Structure: General
Formula: ~1 | Animal
Parameter estimate(s):
Correlation:
        1       2       3       4       5       6       7       8
2 0.916
3 0.802   0.912
4 0.796   0.908   0.958
5 0.749   0.881   0.928   0.962
6 0.705   0.835   0.906   0.933   0.922
7 0.655   0.776   0.843   0.868   0.855   0.963
8 0.625   0.713   0.817   0.829   0.810   0.928 0.959
9 0.558   0.664   0.769   0.786   0.786   0.889 0.917   0.969

Variance function:
Structure: Different standard deviations per stratum
Formula: ~1 | Time
Parameter estimates:
      1         2         3         4         5         6         7         8         9
1.000000 1.130230 1.435539 1.512632 1.836842 1.802349 2.014343 2.197068 2.566113
```

```
Coefficients:
                 Value     Std. Error      t-value     p-value
(Intercept)   18.260417     0.3801327     48.03695      0.0000
Time           6.760417     0.1623937     41.62980      0.0000
TimeM2Plus     0.322917     0.2294228      1.40752      0.1600
TimeM3Plus    -1.552083     0.2925786     -5.30484      0.0000
TimeM4Plus     0.229167     0.2432416      0.94214      0.3467
TimeM5Plus     0.531250     0.3931495      1.35127      0.1773
TimeM6Plus    -0.281250     0.2646021     -1.06292      0.2884
TimeM7Plus     0.833333     0.2974978      2.80114      0.0053
TimeM8Plus    -0.927083     0.2757119     -3.36251      0.0008

Residual standard error: 2.468869
Degrees of freedom: 432 total; 423 residual
```

At this point, we are interested in an overall test which compares models (10.13) and (10.15), that is, we are interested in comparing models with nested fixed effects but the same Σ. As we discussed earlier in Section 10.1.1, the ML log-likelihoods for models with the same Σ can be used to produce a likelihood ratio test to compare two nested models for fixed effects. This test is based on comparing twice the difference in the two maximized ML log-likelihoods to a chi-squared distribution with degrees of freedom equal to the difference between the number of fixed effects parameters in the full and reduced models. We look at such a test next for models (10.13) and (10.15).

Output from R: Comparing ML fits of models (10.13) and (10.15)

```
          Model df     AIC       BIC      logLik    Test   L.Ratio  p-value
m10p13.ML    1 47  1632.303  1823.520  -769.1517
m10p15.ML    2 54  1600.992  1820.687  -746.4958  1 vs 2  45.31183  <.0001
```

The model degrees of freedom reported in the R output include $9 \times (9 + 1)/2 = 45$ associated with estimating each element in Σ. The model degrees of freedom for model (10.13) total 47, due to the two fixed effects in (10.13). The seven extra degrees of freedom in the R output for model (10.15) correspond to the seven additional predictor variables $(x - 2)_+, (x - 3)_+, \dots (x - 8)_+$. The likelihood ratio statistic comparing models (10.13) and (10.15) equals 45.3. Comparing this result to a chi-squared distribution with degrees of freedom equal to 7, results in a p-value < 0.0001. Thus, model (10.15) is to be preferred over model (10.13). This provides further evidence that the straight-line model (10.13) is an invalid model for the data.

However, model (10.15) includes some redundancy due to the fact that a number of the regression spline terms are not statistically significant. Ideally, we would like to remove this redundancy by reducing the number of knots and hence effectively removing some of the $(x - k)_+$ terms from the model.

Knot selection for linear regression splines

In general, deciding which knots (i.e., which of the $(x - k)_+$ terms) to include in a model based on linear regression splines is "mixture of art and science. When it is available, subject-matter knowledge should be brought to bear on the empirical evidence for the most appropriate choice of knot location(s)" (Fitzmaurice, Laird, and Ware, 2004, p. 150). In terms of science, this is a variable selection problem. In practice, stepwise methods based on AIC, AIC_c or BIC such as those discussed in Section 7.2.2 are commonly used for this purpose. This is especially the case when there are many potential locations for knots. If there are a relatively small number of potential knots, then another approach is to remove all the statistically insignificant spline terms and then use a hypothesize test to compare the full and reduced models. Weiss (2005, p. 227) provides an example illustrating the use of the latter approach. We shall consider both approaches. We begin by considering the latter approach first.

Choosing knots by removing statistically insignificant spline terms

We next remove the insignificant spline predictor terms from model (10.15) (i.e., $(x-2)_+, (x-4)_+, (x-5)_+, (x-6)_+$) and thus consider the following model

$$Y = X\beta + \varepsilon \tag{10.16}$$

where in this example

$$Y = \begin{pmatrix} y_{1,1} \\ \vdots \\ y_{1,9} \\ \vdots \\ y_{48,1} \\ \vdots \\ y_{48,9} \end{pmatrix} \qquad X = \begin{pmatrix} 1 & x_1 & (x_1-3)_+ & (x_1-7)_+ & (x_1-8)_+ \\ \vdots & \vdots & \vdots & \vdots & \vdots \\ 1 & x_9 & (x_9-3)_+ & (x_9-7)_+ & (x_9-8)_+ \\ \vdots & \vdots & \vdots & \vdots & \vdots \\ 1 & x_1 & (x_1-3)_+ & (x_1-7)_+ & (x_1-8)_+ \\ \vdots & \vdots & \vdots & \vdots & \vdots \\ 1 & x_9 & (x_9-3)_+ & (x_9-7)_+ & (x_9-8)_+ \end{pmatrix}$$

where Y_{ij} again denotes the weight of the ith pig ($i = 1, 2, \ldots 48$) in the jth week x_j ($j = 1, \ldots, 9$). We shall assume that

$$\varepsilon \sim N(\mathbf{0}, \Sigma)$$

where Σ is given below (10.13).

Recall that the REML log-likelihoods for models with different fixed effects are not comparable. Thus we consider instead the ML log-likelihoods to compare models (10.15) and (10.16). The R output given below shows that the ML likelihood ratio statistic comparing the fixed effects in models (10.16) and (10.15) equals 7.1. Comparing this result to a chi-squared distribution with degrees of freedom equal to 4, results in a p-value $= 0.129$. Thus, based on the result of this hypothesis test, model (10.16) is to be preferred over model (10.15). Notice also that model (10.16) is to be preferred over model (10.15) in terms of having lower values of AIC and BIC.

Output from R: Comparing ML fits of models (10.15) and (10.16)

```
           Model df      AIC       BIC   logLik   Test L.Ratio p-value
m10p16.ML    1 50 1600.125 1803.546-750.0624
m10p15.ML    2 54 1600.992 1820.687-746.4958 1 vs 2 7.133239   0.129
```

Given below is the output from R associated with fitting model (10.16) using REML. Recall that in the output $(x-3)_+$ is denoted by `TimeM3Plus` etc.

Output from R: REML fit of model (10.16)

```
Generalized least squares fit by REML
Model: weight ~ Time + TimeM3Plus + TimeM7Plus + TimeM8Plus
Data: pigweights
    AIC        BIC      logLik
 1608.513   1811.352   -754.2563
Correlation Structure: General
Formula: ~1 | Animal
Parameter estimate(s):
Correlation:
       1       2       3       4       5       6       7       8
2   0.916
3   0.800   0.909
4   0.796   0.909   0.955
5   0.749   0.881   0.924   0.963
6   0.703   0.832   0.906   0.929   0.918
7   0.651   0.771   0.844   0.863   0.849   0.964
8   0.620   0.706   0.816   0.821   0.802   0.927   0.959
9   0.553   0.657   0.769   0.779   0.778   0.889   0.917   0.970

Variance function:
Structure: Different standard deviations per stratum
Formula: ~1 | Time
Parameter estimates:
       1        2        3        4        5        6        7        8        9
1.000001.130610 1.434703 1.512738 1.835018 1.803845 2.019489 2.206996 2.575084

Coefficients:
                 Value     Std. Error    t-value    p-value
(Intercept)   18.256899    0.3550314    51.42334     0.0000
Time           6.820637    0.1377675    49.50832     0.0000
TimeM3Plus    -0.981699    0.1382106    -7.10292     0.0000
TimeM7Plus     0.828698    0.2106252     3.93447     0.0001
TimeM8Plus    -0.767966    0.2497377    -3.07509     0.0022
Residual standard error: 2.46725
Degrees of freedom: 432 total; 427 residual
```

Notice that each of the regression coefficients of x, $(x-3)_+$, $(x-7)_+$ and $(x-8)_+$ are highly statistically significant indicating that there is no redundancy in the fixed effects in model (10.16). Looking further at the output associated with the REML fit of model (10.16) we see the following. The estimated coefficient of x (Time) which measures the overall weekly trend in pig weight is 6.82. The estimated coefficient of $(x-3)_+$ is –0.98, which means that this trend decreases by this amount each week after week 3. The estimated coefficient of $(x-7)_+$ is 0.83, which means that the trend further increases after week 7 by 0.83. Finally, the estimated coefficient of $(x - 8)_+$ is −0.77, which means that the trend further decreases after week 8 by 0.77.

Choosing knots using variable selection

We next consider variable selection methods for choosing the knots. In particular, we consider stepwise methods based on AIC and BIC, since in the current situation the approach based on all subsets would require $2^7 - 1 = 127$ models to be fit. Table 10.4 gives the values of AIC and BIC found from backwards elimination (i.e., start with all knots in the model and at each step eliminate a knot). Note that maximum likelihood was used for each of the stepwise fits, since REML log-likelihoods of models with different fixed effects are not comparable. Highlighted in bold in Table 10.4 are the minimum values of AIC and BIC found from backwards elimination.

We see from Table 10.4 that based on backwards elimination BIC judges the model with knots at weeks 3, 7, 8, that is, model (10.16), to be "best" while AIC has an extra knot at week 5 in its "best" model.

Table 10.5 gives the values of AIC and BIC found from forwards selection (i.e., start with no knots in the model and at each step add a knot). Highlighted in bold are the minimum values of AIC and BIC found from forwards selection.

Table 10.4 Values of AIC and BIC from backwards elimination

Subset size	Knots located at	AIC	BIC
7	2,3,4,5,6,7,8	1600.992	1820.687
6	2,3,5,6,7,8	1599.890	1815.516
5	3,5,6,7,8	1599.524	1811.082
4	3,5,7,8	**1599.149**	1806.639
3	3,7,8	1600.125	**1803.546**
2	3,7	1605.801	1805.154
1	3	1608.973	1804.257

Table 10.5 Values of AIC and BIC from forwards selection

Subset size	Knots located at	AIC	BIC
1	3	1608.973	1804.257
2	3,5	1604.332	**1803.685**
3	3,5,8	1601.273	1804.694
4	3,5,7,8	**1599.149**	1806.639
5	3,5,6,7,8	1599.524	1811.082
6	2,3,5,6,7,8	1599.890	1815.516
7	2,3,4,5,6,7,8	1600.992	1820.687

We see from Table 10.5 that based on forwards selection AIC judges the model with knots at weeks 3, 5, 7 and 8 as "best" while BIC judges the model with knots at weeks 3 and 5 to be "best." Comparing the results in Tables 10.4 and 10.5, we see that while AIC identifies the same model as "best" based on forwards selection and backward elimination, BIC does not. (The lack of complete agreement between the results of forwards and backwards stepwise approaches to choosing knots is not uncommon in practice.) Notice also that the value of BIC for the "best" model from backwards elimination is lower than the corresponding value for forwards selection. Thus, based on BIC, stepwise methods point to the model with knots at weeks 3, 7 and 8, that is, model (10.16), as "best."

We next consider the model judged as "best" from stepwise methods based on AIC and thus consider the following model

$$\mathbf{Y} = \mathbf{X}\beta + \varepsilon \tag{10.17}$$

where in this example

$$\mathbf{Y} = \begin{pmatrix} y_{1,1} \\ \vdots \\ y_{1,9} \\ \vdots \\ y_{48,1} \\ \vdots \\ y_{48,9} \end{pmatrix} \quad \mathbf{X} = \begin{pmatrix} 1 & x_1 & (x_1-3)_+ & (x_1-5)_+ & (x_1-7)_+ & (x_1-8)_+ \\ \vdots & \vdots & \vdots & \vdots & \vdots & \vdots \\ 1 & x_9 & (x_9-3)_+ & (x_9-5)_+ & (x_9-7)_+ & (x_9-8)_+ \\ \vdots & \vdots & \vdots & \vdots & \vdots & \vdots \\ 1 & x_1 & (x_1-3)_+ & (x_1-5)_+ & (x_1-7)_+ & (x_1-8)_+ \\ \vdots & \vdots & \vdots & \vdots & \vdots & \vdots \\ 1 & x_9 & (x_9-3)_+ & (x_9-5)_+ & (x_9-7)_+ & (x_9-8)_+ \end{pmatrix}$$

where Y_{ij} again denotes the weight of the ith pig ($i = 1, 2, \ldots 48$) in the jth week x_j ($j = 1, \ldots, 9$). We shall assume that

$$\varepsilon \sim N(0,\Sigma)$$

where Σ is given below (10.13).

The R output given below shows that the ML likelihood ratio statistic comparing the fixed effects in models (10.17) and (10.16) equals 2.98. Comparing this result to a chi-squared distribution with one degree of freedom, results in a p-value = 0.0845. However, as discussed in Chapter 7, after variable selection the p-values obtained are much smaller than their true values. In view of this, there is little evidence to prefer the more complex model (10.17) over model (10.16). Thus, we will consider model (10.16) as our parsimonious model for the fixed effects.

Output from R: Comparing ML fits of models (10.16) and (10.17)

```
          Model df      AIC      BIC   logLik    Test L.Ratio p-value
m10p17.ML     1 51 1599.149 1806.639 -748.5744
m10p16.ML     2 50 1600.125 1803.546 -750.0624 1 vs 2 2.976025  0.0845
```

The potential problems associated with using stepwise methods to choose knots for regression splines are well-documented in the statistics literature. For example, Zhou and Shen (2001) provide a simple numerical example for iid data which illustrates the tendency of stepwise knot selection methods to not locate the optimal knots and to select more knots than necessary. They conclude that these tendencies are due to the fact that stepwise methods do not take into account the association between knots.

We conclude this discussion of choosing knots for regression splines by reiterating the fact that subject-matter knowledge should be brought to bear and that one can not blindly rely on the output from stepwise selection methods.

Parsimonious models for the error variance–covariance

Given that we have a parsimonious model for the fixed effects (i.e., model (10.16)), the next natural step is to try to replace the unstructured variance–covariance matrix Σ with a more parsimonious one. One of the advantages of imposing structure on Σ is that the precision with which the fixed effects are estimated can be improved. In many situations it is impractical or even impossible to model Σ by an unstructured variance–covariance matrix. Such situations include unbalanced data when there are no more than one data point for an individual at a given time point. Thus, it is often important to choose an appropriate variance–covariance structure from a parameterized family.

Recall that the correlations within individual pigs decay as the time interval between measurements increases (see Figure 10.14). Thus, we shall begin by considering an autoregressive model. The simplest autoregressive model is one of order 1 (denoted by AR(1)). In this situation, the correlation within a subject is given by

$$\mathrm{Corr}(Y_{ij}, Y_{ij+k}) = \rho^k$$

In the pig weight example we shall model Σ by an AR(1) structure with variances which differ across time. Thus, we shall consider model (10.16) with the following error structure

$$\Sigma = \begin{pmatrix} \mathbf{D}_1 & 0 & 0 & 0 & \cdots & 0 \\ 0 & \mathbf{D}_2 & 0 & 0 & \cdots & 0 \\ 0 & 0 & \mathbf{D}_3 & 0 & \cdots & 0 \\ 0 & 0 & 0 & \ddots & 0 & 0 \\ \vdots & \vdots & \vdots & 0 & \mathbf{D}_8 & 0 \\ 0 & 0 & 0 & 0 & 0 & \mathbf{D}_9 \end{pmatrix} \quad \mathbf{D}_i = \sigma_i^2 \begin{pmatrix} 1 & \rho & \rho^2 & \cdots & \rho^8 \\ \rho & 1 & \rho & \cdots & \rho^7 \\ \rho^2 & \rho & 1 & \cdots & \rho^6 \\ \vdots & \vdots & \vdots & \ddots & \vdots \\ \rho^8 & \rho^7 & \rho^6 & \cdots & 1 \end{pmatrix}$$

Given below is the output from R associated with fitting model (10.16) using REML with the AR(1) error structure given above.

Output from R: REML fit of model (10.16) with autoregressive errors

```
Generalized least squares fit by REML
Model: weight ~ Time + TimeM3Plus + TimeM7Plus + TimeM8Plus
Data: pigweights
      AIC        BIC      logLik
  1612.402   1673.254   -791.2011
Correlation Structure: AR(1)
Formula: ~1 | Animal
Parameter estimate(s):
      Phi
  0.9459695
Variance function:
Structure: Different standard deviations per stratum
Formula: ~1 | Time
Parameter estimates:
      1        2        3        4        5        6        7        8        9
1.000000 1.129118 1.370539 1.378911 1.644508 1.563842 1.678001 1.788608 2.080035

Coefficients:
                Value    Std.Error   t-value   p-value
(Intercept)  18.124607   0.3585966  50.54316   0.0000
Time          6.901602   0.1213129  56.89093   0.0000
TimeM3Plus   -1.011721   0.1608429  -6.29012   0.0000
TimeM7Plus    0.953291   0.2509478   3.79876   0.0002
TimeM8Plus   -0.933510   0.3453573  -2.70303   0.0071

Residual standard error: 2.768629
Degrees of freedom: 432 total; 427 residual
```

Comparing the estimates of the fixed effects above with those obtained with the general variance–covariance structure we see that they are quite similar with the biggest difference occurring for the coefficients of $(x-7)_+$ and $(x-8)_+$. With one exception, the standard errors of the estimates of the fixed effects are slightly larger for the model with autoregressive errors. The similarity of the estimates of the fixed effects is to be expected since as we saw in Chapter 9, the estimates of the fixed effects are unbiased even when the error variance-covariance is incorrectly specified. On the other hand, if the error variance–covariance is incorrectly specified the standard errors are not correct leading to the possibility of misleading inferences.

As we discussed earlier in Section 10.1.1, the REML log-likelihoods for models with the same fixed effects can be used to produce a likelihood ratio test to compare two nested covariance models. This test is based on comparing twice the difference in the two maximized REML log-likelihoods to a chi-squared distribution with degrees of freedom equal to the difference between the number of covariance parameters in the full and reduced models. We look at such a test next.

Output from R: Comparing REML fits of (10.16) with different errors

```
            Model df     AIC       BIC      logLik    Test  L.Ratio p-value
m10p16.AR1      1 15  1612.402  1673.254  -791.2011
m10p16          2 50  1608.513  1811.352  -754.2563  1 vs 2 73.88972   1e-04
```

The R output given above shows that the REML likelihood ratio statistic comparing model (10.16) with unstructured and autoregressive errors equals 73.9. Comparing this result to a chi-squared distribution with degrees of freedom equal to 35, results in a p-value = 0.0001. Thus, model (10.16) with the general error structure is to be preferred. We take up the issue of finding a parsimonious model for the error structure for the pig weight data in the exercises.

It is common to want to compare non-nested models for the variance–covariance. For example, imagine a situation in which the competing models are compound symmetry (or equivalently random intercepts) and autoregressive of order 1. In this situation, AIC and BIC can potentially be used, with lower values of AIC and BIC corresponding to better fitting models. However, Fitzmaurice, Laird, and Ware (2004, p. 177) "do not recommend the use of BIC for covariance model selection as it entails a high risk of selecting a model that is too simple or parsimonious for the data at hand." Looking at the R-output that compares model (10.16) with unstructured and autoregressive errors, we see an example of this phenomenon re BIC, namely BIC prefers the autoregressive model, which in this case is statistically significantly worse. On the other hand, AIC prefers the unstructured covariance matrix.

Finally, we briefly mention a promising relatively new graphical technique for identifying a parsimonious model for the variance–covariance structure in a mixed model, known as the regressogram of Σ. A discussion of the details is beyond the scope of this book. Interested readers can find more details on this approach in Pourahmadi (2001, Section 3.5).

10.3   Exercises

1. Consider once again the orthodontic growth data in Section 10.1. In particular, consider model (10.5) which includes random intercepts and an error term whose variance differs across gender. Compare model (10.5) to a model with the same fixed effects but an unstructured covariance matrix, which allows for variances to differ across genders. Test whether the unstructured covariance should be preferred using the maximized REML log-likelihoods.
2. Consider once again the pig weight data in Section 10.2. We shall demonstrate in this exercise that smaller estimated standard errors for the fixed effects in a mixed model does not always correspond to a better model.

   (a) Purely for illustration puposes Ruppert, Wand and Carroll (2003) fit a random intercepts model with constant error variance at each time point to the pig weight data. Fit this model using REML.
   (b) Diggle, Heagerty, Liang and Zeger (2002, p. 77) adopt a model with both random intercepts and random slopes as a "working model." Using their notation, this model can be written as for the weight Y_{ij} of the ith pig in the jth week (x_j):

$$Y_{ij} = \alpha + \beta x_j + U_i + W_i x_j + Z_{ij} \quad j = 1,...,9; \ i = 1,...,48$$

where Y_{ij} is the weight of the ith pig in the jth week (x_j) and where $U_i \sim N(0, \sigma^2)$, $W_i \sim N(0, v^2)$, $Z_{ij} \sim N(0, \tau^2)$ are all mutually independent. Fit this model using REML.

(a) First, compare the models in (a) and (b) in terms AIC and maximized REML likelihoods. Show that the model in (b) is a dramatic improvement over the model in (a).
(b) Next, compare the models in (a) and (b) in terms of the standard errors of the estimates of the fixed effects. Show that the model in (a) produces a much smaller estimate of the standard error of the fixed slope effect.

3. Consider once again the pig weight data in Section 10.2. We showed that model (10.16) with the general error structure is to be preferred over the same model for the fixed effects with an AR(1) error structure. Try to find a parsimonious model for the error structure.

4. Belenky, Wesensten, Thorne, Thomas, Sing, Redmond, Russo and Balkin (2003) examine daytime performance changes of 66 subjects who had spent either 3, 5, 7 or 9 hours daily time in bed for 7 days after having their normal amount of sleep on day 0. We shall just consider the 18 subjects who spent 3 hours in bed. The data consist of the average reaction time on a series of tests given daily to each subject. The data are part of the R-package lme4 and they can be found on the book web site in the file sleepstudy.txt.

(a) Obtain plots of the data and summary statistics such as sample correlations in order to examine the mean structure and the error structure of the data. Identify any unusual data points.

The model we first consider for subject i ($i = 1, 2, \ldots, 18$) at Days j ($j = 0, 1, 2, 3, \ldots 9$) is as follows:

$$\text{Reaction}_{ij} = \beta_0 + \beta_1 \text{Days}_j + \varepsilon_{ij} \qquad (10.17)$$

where ε_{ij} represents a general error term.

(b) Fit model (10.17) with an unstructured covariance matrix, which allows for variances to differ across Days.
(c) Fit model (10.17) with random intercepts and random slopes and an error term whose variance differs across Days.
(d) Fit model (10.17) with random intercepts and an error term whose variance differs across Days.
(e) Compare the models in (b), (c) and (d) in terms of the maximized REML log-likelihoods and the estimated standard errors of the fixed effects.
(f) Expand model (10.17) by adding $(\text{Days} - 1)_+$, $(\text{Days} - 2)_+$, $\ldots$ $(\text{Days} - 8)_+$ as predictors. Fit the expanded model with an unstructured covariance matrix, which allows for variances to differ across Days. Show that it is an improvement on model (10.17). Remove any redundancies in the fixed effects. Finally, attempt to find a parsimonious model for the error structure.

Appendix: Nonparametric Smoothing

In this book we make use of two nonparametric smoothing techniques, namely, kernel density estimation and nonparametric regression for a single predictor. We discuss each of these in turn next.

A.1 Kernel Density Estimation

In this section we provide a brief practical description of density estimation based on kernel methods. We shall follow the approach taken by Sheather (2004).

Let $X_1, X_2, ..., X_n$ denote a sample of size n from a random variable with density function f. The kernel density estimate of f at the point x is given by

$$\hat{f}_h(x) = \frac{1}{nh} \sum_{i=1}^{n} K\left(\frac{x - X_i}{h}\right)$$

where the kernel, K satisfies $\int K(x)\,dx = 1$ and the smoothing parameter, h is known as the bandwidth. In practice, the kernel K is generally chosen to be a unimodal probability density symmetric about zero. In this case, K also satisfies the following condition

$$\int y K(y)\,dy = 0.$$

A popular choice for K which we shall adopt is the Gaussian kernel, namely,

$$K(y) = \frac{1}{\sqrt{2\pi}} \exp\left(-\frac{y^2}{2}\right)$$

Purely for illustration purposes we shall consider a small generated data set. The data consists of a random sample of size $n = 20$ from a normal mixture distribution made up of observations from a 50:50 mixture of $N\left(\mu = -1, \sigma^2 = \frac{1}{9}\right)$ and $N\left(\mu = 1, \sigma^2 = \frac{1}{9}\right)$. The data can be found on the book web site in the file bimodal.txt.

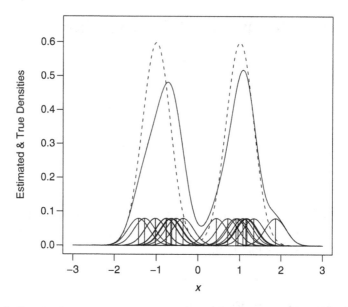

Figure A.1 True density (dashed curve) and estimated density with $h = 0.25$ (solid curve)

Figure A.1 shows a kernel density estimate for these data using the Gaussian kernel with bandwidth $h = 0.25$ (the solid curve) along with the true underlying density (the dashed curve). The 20 data points are marked by vertical lines above the horizontal axis. Centered at each data point is its contribution to the overall density estimate, namely, $\dfrac{1}{nh} K\!\left(\dfrac{x - X_i}{h}\right)$ (i.e., $\dfrac{1}{n}$ times a normal density with mean X_i and standard deviation h). The density estimate (the solid curve) is the sum of these scaled normal densities. Increasing the value of h to 0.6 widens each normal curve producing a density estimate in which the two modes are less apparent (see Figure A.2).

Assuming that the underlying density is sufficiently smooth and that the kernel has finite fourth moment, it can be shown that the leading terms in an asymptotic expansion for the bias and variance of a kernel density estimate are given by

$$\text{Bias}_{\text{asy}}\left\{\hat{f}_h(x)\right\} = \frac{h^2}{2}\mu_2(K)^2 f''(x)$$

$$\text{Var}_{\text{asy}}\left\{\hat{f}_h(x)\right\} = \frac{1}{nh}R(K)f(x)$$

where

$$R(K) = \int K^2(y)\,dy, \quad \mu_2(K) = \int y^2 K(y)\,dy$$

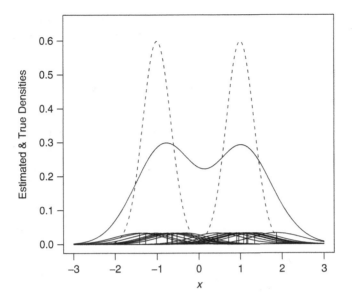

Figure A.2 True density (dashed curve) and estimated density with $h = 0.6$ (solid curve)

(e.g., Wand and Jones, 1995, pp. 20–21). In addition to the visual advantage of being a smooth curve, the kernel estimate has an advantage over the histogram in terms of bias. It can be shown that the bias of a histogram estimator with bandwidth h is of order h, compared to leading bias term for the kernel estimate, which is of order h^2. Centering the kernel at each data point and using a symmetric kernel makes the bias term of order h equal to zero for kernel estimates.

A widely used choice of an overall measure of the discrepancy between $\hat{f}_h$ and f is the mean integrated squared error (MISE), which is given by

$$\text{MISE}(\hat{f}_h) = \text{E}\left\{\int \left(\hat{f}_h(y) - f(y)\right)^2 dy\right\}$$

$$= \int \text{Bias}\left(\hat{f}_h(y)\right)^2 dy + \int Var\left(\hat{f}_h(y)\right) dy$$

Under an integrability assumption on f, the asymptotic mean integrated squared error (AMISE) is given by

$$\text{AMISE}\left\{\hat{f}_h\right\} = \frac{1}{nh}R(K) + \frac{h^4}{4}\mu_2(K)^2 R(f'')$$

The value of the bandwidth that minimizes AMISE is given by

$$h_{\text{AMISE}} = \left[\frac{R(K)}{\mu_2(K)^2 R(f'')}\right]^{1/5} n^{-1/5}.$$

The functional $R(f'')$ is a measure of the underlying curvature. In particular, the larger the value of $R(f'')$ the larger the value of AMISE (i.e., the more difficult it is to estimate f) and the smaller the value of h_{AMISE} (i.e., the smaller the bandwidth needed in order to capture the curvature in f).

There are many competing methods for choosing a global value of the bandwidth h. For a recent overview of these methods see Sheather (2004).

A popular approach commonly called *plug-in methods* is to replace the unknown quantity $R(f'')$ in the expression for h_{AMISE} given above by an estimate. This method is commonly thought to date back to Woodroofe (1970) who proposed it for estimating the density at a given point. Estimating $R(f'')$ by $R(f_g'')$ requires the user to choose the bandwidth g for this estimate. There are many ways this can be done. We next describe the "solve-the-equation" plug-in approach developed by Sheather and Jones (1991), since this method is widely recommended (e.g., Simonoff, 1996, p. 77; Bowman and Azzalini, 1997, p. 34; Venables and Ripley, 2002, p. 129) and it is available in R, SAS and Stata.

Different versions of the plug-in approach depend on the exact form of the estimate of $R(f'')$. The Sheather and Jones (1991) approach is based on writing g, the bandwidth for the estimate $R(\hat{f}'')$, as a function of h, namely,

$$g(h) = C(K)[R(f'') / R(f''')]^{1/7} h^{5/7}$$

and estimating the resulting unknown functionals of f using kernel density estimates with bandwidths based on a normality assumption on f. In this situation, the only unknown in the following equation is h.

$$h = \left[\frac{R(K)}{\mu_2(K)^2 R(\hat{f}''_{g(h)})} \right]^{1/5} n^{-1/5}.$$

The Sheather–Jones plug-in bandwidth, h_{SJ} is the solution to this equation. For hard-to-estimate densities (i.e., ones for which $|f''(x)|$ varies widely due, for example, to the existence of many modes) the Sheather–Jones plug-in bandwidth tends to over-smooth and the method known as least squares cross-validation (Bowman and Azzalini, 1997, p. 32) can be recommended. However, in settings in which parametric regression models are appropriate, the Sheather–Jones plug-in bandwidth appears to perform well.

A.2 Nonparametric Regression for a Single Predictor

In this section we provide a brief practical description of nonparametric regression for a single predictor, which is sometimes called scatter plot smoothing. In this section we are interested in nonparametric estimates of the regression function, $m(.)$ under the assumption of iid errors with constant variance. Thus, in symbols, we assume the following model for $i = 1, \ldots, n$

$$Y_i = m(x_i) + e_i = \mathrm{E}(Y \mid X = x_i) + e_i.$$

We shall consider two classes of estimators, namely, local polynomial kernel estimators and penalized linear regression splines.

A.2.1 Local Polynomial Kernel Methods

Local polynomial kernel methods (Stone, 1977; Cleveland, 1979) are based on the idea of approximating $m(x)$ by a low-order polynomial putting highest weight on the values of y corresponding to x_i's closest to x. According to Cleveland (1979), the idea of local fitting of polynomials to smooth scatter plots of time series, measured at equally spaced time points, dates back to at least the 1930s. The local polynomial estimator $\hat{m}_p(x)$ is the value of b_0 that minimizes

$$\sum_{i=1}^{n} \left\{ y_i - b_0 - b_1(x_i - x) - b_2(x_i - x)^2 - \cdots b_p(x_i - x)^p \right\}^2 \frac{1}{h} K\left(\frac{x_i - x}{h}\right)$$

where once again the kernel, K satisfies $\int K(x)dx = 1$ and the smoothing parameter, h is known as the bandwidth.

The local constant estimator is obtained by setting $p = 0$ in the last equation. Thus, in this case we seek to minimize

$$\sum_{i=1}^{n} \left\{ y_i - b_0 \right\}^2 \frac{1}{h} K\left(\frac{x_i - x}{h}\right)$$

Differentiating with respect to b_0 and setting the result to zero gives

$$-2\sum_{i=1}^{n} \left\{ y_i - b_0 \right\} \frac{1}{h} K\left(\frac{x_i - x}{h}\right) = 0$$

Solving this equation for b_0 gives the local constant estimator $\hat{m}_0(x)$ where

$$\hat{m}_0(x) = \frac{\displaystyle\sum_{i=1}^{n} y_i K\left(\frac{x_i - x}{h}\right)}{\displaystyle\sum_{i=1}^{n} K\left(\frac{x_i - x}{h}\right)}$$

This estimator is also known as the Nadaraya-Watson estimator, as they were the first to propose its use (Nadaraya, 1964; Watson, 1964). It is also possible to derive an explicit regression for the local linear estimator $\hat{m}_1(x)$ (see, e.g., Wand and Jones, 1997, pp. 119, 144).

Choosing a higher degree polynomial leads in principle to a better approximation to the underlying curve and hence less bias. However, it also leads to greater variability in the resulting estimate. Loader (1999, p. 22) provides the following advice:

It often suffices to choose a low degree polynomial and concentrate on choosing the bandwidth to obtain a satisfactory fit. The most common choices are local linear and local quadratic. ... a local constant fit is susceptible to bias and is rarely adequate. A local linear estimate usually performs better, especially at boundaries. A local quadratic estimate reduces bias further, but increased variance can be a problem, especially at boundaries. Fitting local cubic and higher orders rarely produces much benefit.

Based on their experience, Ruppert, Wand and Carroll (2003, p. 85) recommend $p = 1$ if the regression function is monotonically increasing (or decreasing) and $p = 2$ otherwise.

For illustration purposes we shall consider a generated data set. The data consists of $n = 150$ pairs of points (x_i, y_i) where $y_i = m(x_i) + e_i$ with x_i equally spaced from 0 to 1, $e_i \sim N(0, \sigma^2 = 4)$ and

$$m(x_i) = 15\left(1 + x_i \cos(4\pi x_i)\right)$$

The data can be found on the book web site in the file curve.txt.

Figure A.3 shows a local linear regression estimate for these data using the Gaussian kernel with bandwidth $h = 0.026$ (the solid curve) along with the true underlying curve (the dashed curve). The value of the bandwidth was chosen using the plug-in bandwidth selector of Ruppert, Sheather and Wand (2005). Marked as a dashed curve on Figure A.3 is the weight function for each x_i used to estimate the curve at $x = 0.5$, namely, $\dfrac{1}{h}K\left(\dfrac{x_i - 0.5}{h}\right)$ (i.e., a normal density with mean 0.5 and standard deviation h).

Decreasing the value of h fivefold to 0.005, shrinks each normal curve so that each straight line is effectively fit over a very small interval. This produces a curve estimate which is much too wiggly (see the top panel of Figure A.4). On the other hand, increasing the value of h fivefold to 0.132 widens each normal curve so that each straight line is effectively fit over a very large interval. This produces a curve estimate which is clearly over-smoothed, missing the bottom or the top of the peaks in the underlying curve (see the bottom panel of Figure A.4). As the bandwidth h approaches infinity the local linear regression estimate will approach a straight line.

Thus far, in this section we have considered an example based on equally spaced x's. In settings in which parametric regression models are generally appropriate it is common for the x's not to be equally spaced. In particular, outliers, and sparse regions in the x values are common when the distribution of x is skewed. In such situations using a fixed value of the bandwidth h can be problematic, since there may be very few (sometime even no) points in certain regions of the x-axis so that it is not possible to fit a local polynomial for certain values of x. One way of solving this problem is to adjust the bandwidth with the value of x so that the number of points used to estimate $m(x)$ effectively remains the same for all values of x. This is achieved using the concept of the *nearest neighbor bandwidth*.

For $i = 1, 2, ..., n$, let $d_i(x)$ denote the distance x_i is away from x, then

$$d_i(x) = |x - x_i|$$

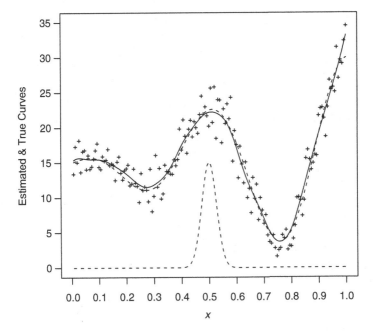

Figure A.3 True curve (dashed) and estimated curve with $h = 0.026$ (solid)

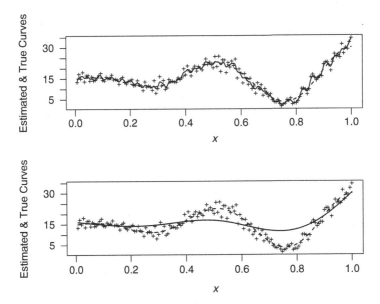

Figure A.4 True curve (dashed) and estimated curves (solid) with $h = 0.005$ (upper panel) and $h = 0.132$ (lower panel)

The *nearest neighbor bandwidth*, $h(x)$ is defined to be the kth smallest $d_i(x)$. In practice, the choice of k is based on what is commonly called the *span* α, namely,

$$k = \lfloor n\alpha \rfloor.$$

Thus, the span plays the role of a smoothing parameter in nearest neighbor bandwidths.

Cleveland (1979) proposed the use of local linear regression estimators based on nearest neighbor bandwidths with the tricube kernel function

$$K(y) = \left(1 - |y|^3\right) I\left(|y| < 1\right).$$

Cleveland (1979) also incorporated a robustness step in which large residuals were down weighted. This estimator is typically referred to as *lowess*. Cleveland and Devlin (1988) studied the properties of local linear regression estimators based on nearest neighbor bandwidths with the tricube kernel without a robustness step. This estimator is typically referred to as *loess*.

Figure A.5 shows the loess estimate based on $p = 2$ (i.e., local quadratic) with span $\alpha = 1/3$ (the solid curve), along with the true underlying curve (the dashed curve). This value of the span was chosen by eye as the value that gave a curve that seemed to best match the data.

The loess estimate with span $\alpha = 1/3$ in Figure A.5 fits the data well. Increasing the span to $\alpha = 2/3$, produces a curve estimate which is slightly over-smoothed, missing

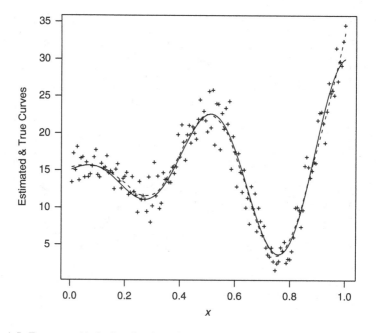

Figure A.5 True curve (dashed) and estimated curve (solid) with span = 1/3

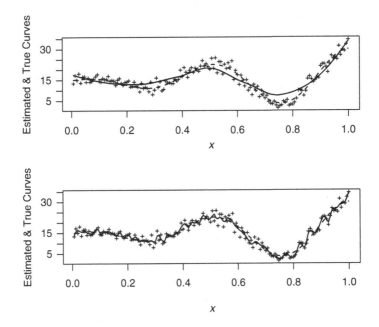

Figure A.6 True curve (dashed) and estimated curves (solid) with span = 2/3 (upper panel) and span = 0.05 (lower panel)

the bottom or the top of the peaks in the underlying curve (see the top panel of Figure A.6). On the other hand, decreasing the value of the span to $\alpha = 0.05$ produces a curve estimate which is much too wiggly (see the bottom panel of Figure A.6).

Nearest neighbor bandwidths do not perform well if the x-space is sparse when the curve is wiggly and/or the x-space is dense when the curve approximates a straight line. Fortunately, this is a highly unusual situation.

The marginal model plot method, proposed by Cook and Weisberg (1997) and described in Chapters 6 and 8, is based on loess fits. This is a natural choice for regression with continuous predictor and outcome variables due to the ability of loess to cope with sparse regions in the x-space. However, its use for binary outcome variables can be questioned, since it seems that no account is taken of the fact that binary data naturally have nonconstant variance. In this situation one could consider a local likelihood estimator, which takes account of the binomial nature of the data (see, e.g., Bowman and Azzalini, 1997, p. 55).

A.2.2 Penalized Linear Regression Splines

Another increasingly popular method for scatter plot smoothing is called penalized linear regression splines, which we discuss in this section. However, we begin by discussing linear regression splines.

Linear regression splines are based on the inclusion of the following term as a predictor

$$(x-c)_+ = \begin{cases} x-c & \text{if } x>c \\ 0 & \text{if } x \leq c \end{cases}$$

The inclusion of $(x-c)_+$ as a predictor produces a fitted model which resembles a broken stick, with the break at c, which is commonly referred to as a knot. Thus, this predictor allows the slope of the line to change at c. (See Figure 10.16 for details.) In order to make the model as flexible as possible, we shall add a large number of knots $c_1, \ldots, c_K$ and hence consider the following model

$$y = \beta_0 + \beta_1 x + \sum_{i=1}^{K} b_{1i}(x-c_i)_+ + e \tag{A.1}$$

We shall see that two approaches are possible for choosing the knots, corresponding to whether the coefficients b_{1i} in (A.1) are treated as fixed or random effects. If the coefficients are treated as fixed effects, then a number of knots can be removed leaving only those necessary to approximate the function. As demonstrated in Chapter 10, this is feasible if there are a relatively small number of potential knots. However, if there are a large number of potential knots, removing unnecessary knots is a "highly computationally intensive" variable selection problem (Ruppert, Wand and Carroll, 2003, p. 64).

We next investigate what happens if the coefficients b_{1i} in (A.1) are treated as random effects. In order to do this we consider the concept of penalized regression splines.

An alternative to removing knots is to add a penalty function which constrains their influence so that the resulting fit is not overfit (i.e., too wiggly). A popular penalty is to ensure that the b_{1i} in (A.1) satisfy $\sum_{i=1}^{K} b_{1i}^2 < C$, for some constant C, which has to be chosen. The resulting estimator is called a *penalized linear regression spline*. As explained by Ruppert, Wand and Carroll (2003, p. 66) adding this penalty is equivalent to choosing $\beta_0 \beta_1, b_{11}, b_{12}, \ldots b_{1K}$ to minimize

$$\frac{1}{n}\sum_{i=1}^{n}\left(y_i - \beta_0 - \beta_1 x_i\right)^2 + \lambda \sum_{i=1}^{K} b_{1i}^2 \tag{A.2}$$

for some number $\lambda \geq 0$, which determines the amount of smoothness of the resulting fit. The second term in (A.2) is known as a roughness penalty because it penalizes fits which are too wiggly (i.e., too rough). Thus, minimizing (A.2) shrinks all the b_{1i} toward zero. Contrast this with treating the b_{1i} as fixed effects and removing unnecessary knots, which reduces some of the b_{1i} to zero.

The concept of random effects and shrinkage is discussed in Section 10.1. In view of the connection between random effects and shrinkage, it is not too surprising that there is a connection between penalized regression splines and mixed models. Put briefly, the connection is that fitting model (A.1) with β_0 and β_1 treated as fixed effects and $b_{11}, b_{12}, \ldots b_{1K}$ treated as random effects is equivalent to minimizing the penalized linear spline criterion (A.2) (see Ruppert, Wand and Carroll, 2003; Section 4.9 for further details).

Speed (1991) explicitly made the connection between smoothing splines and mixed models (although it seems that this was known earlier by a number of the proponents of spline smoothing). Brumback, Ruppert and Wand (1999) made explicit the connection between penalized regression splines and mixed models.

An important advantage of treating (A.1) as a mixed model is that we can then use the likelihood methods described in Sect. 10.1 to obtain a penalized linear regression spline fit.

Finally, one has to choose the initial set of knots. Ruppert, Wand and Carroll (2003, p. 126) recommend that the knots be chosen at values corresponding to quantiles of x_i, while other authors prefer equally spaced knots. Ruppert, Wand and Carroll (2003, p. 126) have found that the following default choice for the total number of knots K "usually works well":

$$K = \min\left(\frac{1}{4} \times \text{number of unique } x_i, 35\right)$$

Figure A.7 shows a penalized linear regression spline fit obtained by fitting (A.1) using restricted maximum likelihood or REML (the solid curve) along with the true underlying curve (the dashed curve). The equally spaced knots, which are 0.02 apart, are marked by vertical lines on the horizontal axis. Notice this is many more knots than is suggested by the rule above and it does not have any adverse effects on the fit. Increasing the spacing of the knots to 0.15 produces a curve estimate which is jagged, missing the bottom or the top of the peaks in the underlying curve and thus illustrating the problems associated with choosing too few knots (see Figure A.8).

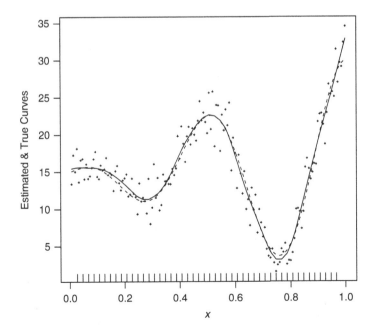

Figure A.7 True curve (dashed) and estimated curve (solid) with knots 0.02 apart

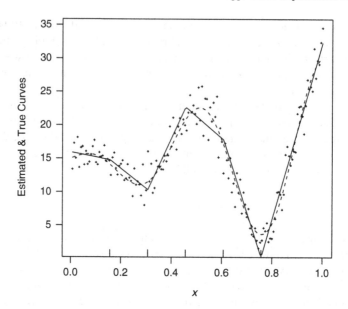

Figure A.8 True curve (dashed) and estimated curve (solid) with knots 0.15 apart

Recently, Krivobokova and Kauermann (2007) studied the properties of penalized splines when the errors are correlated. They found that REML-based fits are more robust to misspecifying the correlation structure than fits based on generalized cross-validation or AIC. They also demonstrated the simplicity of obtaining the REML-based fits using R.

References

Abraham B and Ledolter J (2006) Introduction to regression modeling. Duxbury, MA.

Alkhamisi MA and Shukur G (2005) Bayesian analysis of a linear mixed model with AR(p) errors via MCMC. *Journal of Applied Statistics*, 32, 741–755.

Anscombe F (1973) Graphs in statistical analysis. *The American Statistician*, 27, 17–21.

Anonymous (2005) Michelin guide New York City 2006. Michelin Travel Publications, Greenville, South Carolina.

Atkinson A and Riani M (2000) Robust diagnostic regression analysis. Springer, New York.

Belenky G, Wesensten NJ, Thorne DR, Thomas ML, Sing HC, Redmond DP, Russo MB, and Balkin TJ (2003) Patterns of performance degradation and restoration during sleep restriction and subsequent recovery: a sleep does response study. *Journal of Sleep Research*, 12, 1–12.

Bowman A and Azzalini A (1997) Applied smoothing techniques for data analysis: The Kernel approach with S-plus illustrations. University Press, Oxford.

Box GEP and Cox DR (1964) An analysis of transformations. *Journal of the Royal Statistical Society, Series B*, 26, 211–252.

Bradbury JC (2007) The baseball economist. Dutton, New York.

Brook S (2001) Bordeaux – people, power and politics, pp. 104, 106. Mitchell Beazley, London.

Brumback BA, Ruppert D and Wand MP (1999) Comment on Shively, Kohn and Wood. *Journal of the American Statistical Association*, 94, 794–797.

Bryant PG and Smith MA (1995) Practical data analysis - Cases in business statistics. Irwin, Chicago.

Burnham KP and Anderson DR (2004) Understanding AIC and BIC in model selection. *Sociological Methods & Research*, 33, 261–304.

Carlson WL (1997) Cases in managerial data analysis. Duxbury, Belmont, CA.

Casella G and Berger R (2002) Statistical inference (2nd edn). Duxbury, Pacific Grove, CA.

Chatterjee S and Hadi AS (1988) Sensitivity analysis in linear regression. Wiley, New York.

Cleveland W (1979) Robust locally weighted regression and smoothing scatterplots. *Journal of the American Statistical Association*, 74, 829–836.

Coates C (2004) The wines of Bordeaux – vintages and tasting notes 1952–2003. University of California Press, California.

Cochrane AL, St Leger AS, and Moore F (1978) Health service "input" and mortality "output" in developed countries. *Journal Epidemiol Community Health* 32, 200–205.

Cochrane D and Orcutt GH (1949) Application of least squares regression to relationships containing autocorrelated error terms. *Journal of the American Statistical Association*, 44, 32–61.

Cook RD (1977) Detection of influential observations in linear regression. *Technometrics*, 19, 15–18.

Cook RD and Weisberg S (1994) Transforming a response variable for linearity. *Biometrika*, 81, 731–737.

Cook RD and Weisberg S (1997) Graphic for assessing the adequacy of regression models. *Journal of the American Statistical Association*, 92, 490–499.

Cook RD and Weisberg S (1999a) Graphs in statistical analysis: is the medium the message? *The American Statistician*, 53, 29–37.

Cook RD and Weisberg S (1999b) Applied regression including computing and graphics. Wiley, New York.

Chu S (1996) Diamond ring pricing using linear regression. *Journal of Statistics Education*, 4, http://www.amstat.org/publications/jse/v4n3/datasets.chu.html

Diggle PJ, Heagerty P, Liang K-Y, and Zeger SL (2002) Analysis of Longitudinal Data (2nd edn). Oxford University Press, Oxford.

Efron B, Hastie T, Johnstone I, and Tibshirani R (2004) Least angle regression. *Annals of Statistics*, 32, 407–451.

Fitzmaurice GM, Laird NM, and Ware JH (2004) Applied longitudinal analysis. Wiley, New York.

Frees EW (2004) Longitudinal and panel data. Cambridge University Press, Cambridge.

Flury B and Riedwyl H (1988) Multivariate statistics: A practical approach. Chapman & Hall, London.

Foster DP, Stine RA, and Waterman RP (1997) Basic business statistics. Springer, New York.

Fox J (2002) An R and S-PLUS companion to applied regression. Sage, California.

Furnival GM and Wislon RW (1974) Regression by leaps and bounds. *Technometrics*, 16, 499–511.

Gathje C and Diuguid C (2005) Zagat survey 2006: New York city restaurants. Zagat Survey, New York.

Hald A (1952). Statistical theory with engineering applications. Wiley, New York.

Hastie T, Tibshirani R, and Friedman J (2001) The elements of statistical learning. Springer, New York.

Hesterberg T, Choi NH, Meier L, and Fraley C (2008) Least angle and l1 penalized regression: A review. *Statistics Surveys*, 2, 61–93.

Hill RC, Giffiths WE, and Judge GG (2001) Undergraduate econometrics (2nd edn). Wiley, New York.

Hinds MW (1974) Fewer doctors and infant survival. *New England Journal of Medicine*, 291, 741.

Hoaglin DC and Welsh R (1978) The hat matrix in regression and ANOVA. *The American Statistician*, 32, 17–22.

Houseman EA, Ryan LM, and Coull BA (2004) Cholesky residuals for assessing normal errors in a linear model with correlated errors. *Journal of the American Statistical Association*, 99, 383–394.

Huber P (1981) Robust statistics. Wiley, New York.

Hurvich CM and Tsai C-H (1989) Regression and time series model selection in small samples. *Biometrika*, 76, 297–307.

Jalali-Heravi M and Knouz E (2002) Use of quantitative structure-property relationships in predicting the Krafft point of anionic surfactants. *Electronic Journal of Molecular Design*, 1, 410–417.

Jayachandran J and Jarvis GK (1986) Socioeconomic development, medical care and nutrition as determinants of infant mortality in less developed countries. *Social Biology*, 33, 301–315.

Kay R and Little S (1987) Transformations of the explanatory variables in the logistic regression model for binary data. *Biometrika*, 74, 495–501.

Keri J (2006) Baseball between the numbers. Basic Books, New York.

Krivobokova T and Kauermann G (2007) A note on penalized spline smoothing with correlated errors. *Journal of the American Statistical Association*, 102, 1328–1337.

Kronmal RA (1993) Spurious correlation and the fallacy of the ratio standard revisited. *Journal of the Royal Statistical Society A*, 156, 379–392.

Langewiesche W (2000) The million-dollar nose. *Atlantic Monthly*, 286(6), December, 20.

Leeb H and Potscher BM (2005) Model selection and inference: facts and fiction. *Econometric Theory*, 21, 21–59.

Li KC (1991) Sliced inverse regression (with discussion). *Journal of the American Statistical Association*, 86, 316–342.

Li KC and Duan N (1989) Regression analysis under link violation. *Annals of Statistics*, 17, 1009–1052.

Loader C (1999) Local regression and likelihood. Springer, New York.

Mantel N (1970) Why stepdown procedures in variable selection? *Technometrics*, 12, 621–625.

Maronna RA, Martin RD, and Yohai VJ (2006) Robust statistics: theory and methods. Wiley, New York.

Menard S (2000) Coefficients of determination for multiple logistic regression analysis. *American Statistician*, 54, 17–24.

Montgomery DC, Peck EA, and Vining GG (2001) Introduction to linear regression analysis (3rd edn.). Wiley, New York.

Mosteller F and Tukey JW (1977) Data analysis and regression. Addison-Wesley, Reading, MA.

Nadaraya EA (1964) On estimating regression. *Theory of Probability and its Applications*, 10, 186–190.

Neyman J (1952) Lectures and conferences on mathematical statistics and probability (2nd edn, pp. 143–154). US Department of Agriculture, Washington DC.

Nobre JS and Singer JM (2007) Residual analysis for linear mixed models. *Biometrical Journal*, 49, 1–13.

Paige CC (1979) Computer solution and perturbation analysis of generalized linear least squares problems. *Mathematics of Computation*, 33, 171–183.

Parker RM Jr (2003) Bordeaux – a consumer's guide to the world's finest wines (4th edn). Simon & Schuster, New York.

PearsonK (1897) Mathematical contributions to the theory of evolution: On a form of spurious correlation which may arise when indices are used in the measurement of organs. *Proceedings of the Royal Society London*, 60, 489–498.

Pettiti DB (1998) Hormone replacement therapy and heart disease prevention: Experimentation trumps observation. *Journal of the American Medical Association*, 280, 650–652.

Pinheiro JC and Bates DM (2000) Mixed effects models in S and S-PLUS. Springer, New York.

Potthoff RF and Roy SN (1964) A generalized multivariate analysis of variance model especially useful for growth curve problems. *Biometrika*, 51, 313–326.

Pourahmadi M (2001) Foundations of time series analysis and prediction theory. Wiley, New York.

Prais SJ and Winsten CB (1954) Trend Estimators and Serial Correlation. Cowles Commission Discussion Paper No 383, Chicago.

Pregibon D (1981) Logistic regression diagnostics. *The Annals of Statistics*, 9, 705–724.

Ruppert D, Wand MP, and Carroll RJ (2003) Semiparametric regression. Cambridge University Press, Cambridge.

Sankrithi U, Emanuel I, and Van Belle G (1991) Comparison of linear and exponential multivariate models for explaining national infant and child mortality. *International Journal of Epidemiology*, 2, 565–570.

Schwarz G (1978) Estimating the dimension of a model. *Annals of Statistics*, 6, 461–464.

Sheather SJ (2004) Density estimation. *Statistical Science*, 19, 588–597.

Sheather SJ and Jones MC (1991) A reliable data-based bandwidth selection method for kernel density estimation. *Journal of the Royal Statistical Society, Series B*, 53, 683–669.

Shmueli G, Patel NR, and Bruce PC (2007) Data mining for business intelligence. Wiley, New York.

Siegel A (1997) Practical business statistics (3rd edn). Irwin McGraw-Hill, Boston.

Simonoff JS (1996) Smoothing methods in statistics. Springer, New York.

Simonoff JS (2003) Analyzing categorical data. Springer, New York.

Snee RD (1977) Validation of regression models: methods and examples. *Technometrics*, 19, 415–428.

Speed T (1991) Comment of the paper by Robinson, *Statistical Science*, 6, 42–44.

St Leger S (2001) The anomaly that finally went away? *Journal of Epidemiology and Community Health*, 55, 79.

Stamey T, Kabalin J, McNeal J, Johnstone I, Freiha F, Redwine E, and Yang N (1989) Prostate specific antigen in the diagnosis and treatment of adenocarcinoma of the prostate II, radical prostatectomy treated patients. *Journal of Urology*, 16, 1076–1083.

Stigler S (2005) Correlation and causation: a comment. *Perspectives in Biology and Medicine*, 48(1 Suppl.), S88–S94.

Stone CJ (1977) Consistent nonparametric regression. *Annals of Statistics*, 5, 595–620.

Tibshirani R (1996) Regression shrinkage and selection via the lasso. *Journal of the Royal Statistical Society, Series B*, 67, 385–395.

Tryfos P (1998) Methods for business analysis and forecasting: text & cases. Wiley, New York.

Velilla S (1993). A note on the multivariate Box-Cox transformation to normality. *Statistics and Probability Letters*, 17, 259–263.

Venables WN and Ripley BD (2002). Modern applied statistics with S (4th edn). Springer, New York.

Wand MP and Jones MC (1995) Kernel smoothing. Chapman & Hall, London.

Wasserman L (2004) All of statistics: A concise course in statistical inference. Springer, New York.

Watson GS (1964) Smooth regression analysis. *Sankhya – The India Journal of Statistics Series A*, 26, 101–116.

Weisberg S (2005) Applied linear regression (3rd edn). Wiley, New York.

Weiss RE (2005) Modeling longitudinal data. Springer, New York.

Woodroofe M (1970) On choosing a delta-sequence. *Annals of Mathamatical Statistics*, 41, 1665–1671.

Yang R and Chen M (1995) Bayesian analysis for random coefficient regression models using noninformative priors. *Journal of Multivariate Analysis*, 55, 283–311.

Zhou S and Shen X (2001) Spatially adaptive regression splines and accurate knot selection schemes. *Journal of the American Statistical Association*, 96, 247–259.

Zou H, Hastie T, and Tibshirani R (2007) On the "Degrees of Freedom" of the Lasso. *Annals of Statistics*, 35, 2173–2192.

Index

Statistical Design

George Casella

The goal of this book is to describe the principles that drive good design, paying attention to both the theoretical background and the problems arising from real experimental situations. Designs are motivated through actual experiments, ranging from the timeless agricultural randomized complete block, to microarray experiments, which naturally lead to split plot designs and balanced incomplete blocks.

2008. 307 pp. (Springer Texts in Statistics) Hardcover
ISBN 0-387-75964-7

Statistical Learning from a Regression Perspective

Richard A. Berk

This book considers statistical learning applications when interest centers on the conditional distribution of the response variable, given a set of predictors, and when it is important to characterize how the predictors are related to the response. Real applications are emphasized, especially those with practical implications. The material is written for graduate students in the social and life sciences and for researchers who want to apply statistical learning procedures to scientific and policy problems. Intuitive explanations and visual representations are prominent. All of the analyses included are done in R.

2008, 370 pp. Hardcover
ISBN 978-0-387-77500-5

Introductory Statistics with R
Second Edition

Peter Dalgaard

This book provides an elementary-level introduction to R, targeting both non-statistician scientists in various fields and students of statistics. The main mode of presentation is via code examples with liberal commenting of the code and the output, from the computational as well as the statistical viewpoint. A supplementary R package can be downloaded and contains the data sets. In the second edition, the text and code have been updated to R version 2.6.2. The last two methodological chapters are new, as is a chapter on advanced data handling. The introductory chapter has been extended and reorganized as two chapters. Exercises have been revised and answers are now provided in an Appendix.

2008. Approx. 384 pp. (Statistics and Computing) Softcover
ISBN 978-0-387-79053-4